Cholesterol, Lipoproteins, and Cardiovascular Health

Cholesterol, Lipoproteins, and Cardiovascular Health

Separating the Good (HDL), the Bad (LDL), and the Remnant

Anatol Kontush
Sorbonne University
Paris, France

Library of Congress Cataloging-in-Publication Data:

Names: Kontush, Anatol, author.
Title: Cholesterol, lipoproteins, and cardiovascular health : separating
   the good (HDL), the bad (LDL), and the remnant / Anatol Kontush.
Description: Hoboken, New Jersey : Wiley, [2025] | Includes bibliographical
   references and index.
Identifiers: LCCN 2024022148 (print) | LCCN 2024022149 (ebook) | ISBN
   9781394158379 (hardback) | ISBN 9781394158362 (adobe pdf) | ISBN
   9781394158386 (epub)
Subjects: MESH: Cholesterol | Lipoproteins | Cardiovascular Diseases
Classification: LCC QP752.C5 (print) | LCC QP752.C5 (ebook) | NLM QU 95
   | DDC 572/.5795–dc23/eng/20240625
LC record available at https://lccn.loc.gov/2024022148
LC ebook record available at https://lccn.loc.gov/2024022149

Cover Design: Wiley
Cover Image: Courtesy of Anatol Kontush

Set in 9.5/12.5pt STIXTwoText by Straive, Chennai, India

SKY10083546_090324

Contents

Brief History of Lipoproteins

Cholesterol and LDL

HDL

Remnant Cholesterol

Preface

Why add yet another book about cholesterol?

The answer is easy for someone who has worked on lipoproteins for over 30 years. There are many books about cholesterol but very few about lipoproteins.

You might be wondering what they are – and that's the key point. Many people know about cholesterol, but not many know about lipoproteins. You can ask your doctor to check your good and bad cholesterol, but you might not realize that they are actually just different types of lipoproteins – the particles that carry cholesterol in your blood. Lipoproteins play a big role in heart disease and stroke, which together kill about a third of the world's population. It has made over a billion since 1950, far surpassing all wars combined.

This book offers a comprehensive introduction to blood plasma lipoproteins for the public. These tiny particles play a crucial role in determining our life span and reside in the hidden world of the blood vessels. Plasma lipoproteins are divided into different classes, each with its own distinct universe of particles. Lipoproteins are responsible for transporting fats, also known as lipids, throughout the body. The lipids are central for providing energy and ensuring survival. However, when these particles are altered, they can become harmful, a common occurrence that we should be aware of.

A timeline highlighting significant milestones in our comprehension of cholesterol and lipoproteins.

The book is initially focused on two lipoproteins, commonly known as "bad" and "good" cholesterol. They are not. These complex particles contain many molecules, including cholesterol. They are essential for carrying lipids, and proteins play a major role in this. In this nanoworld we'll meet other members from the lipoprotein family. This includes lipoproteins that are rich in triglycerides, which contain remnant cholesterol. Like other lipids in the body, cholesterol is also essential for life. It is the molecule, however, that damages the arteries, causes stroke and heart disease, and can create other problems.

We will explore the history of lipoproteins and cholesterol by examining the evolution of the science. The history of lipoproteins is filled with unexpected discoveries and unpredicted twists. These enigmatic realms took more than two centuries to unravel. Google Book Viewer says that the word "cholesterol," which first appeared in books published at the end of the 19th century, gained popularity during the first half of the 20th century, and then became really popular after the link between cholesterol and heart diseases was discovered. The clinical role of cholesterol sparked a lot of interest. The development of a technique to isolate the two lipoproteins, low-density (LDL) and high-density (HDL), that transport the majority of cholesterol in the bloodstream, helped increase the popularity of this molecule. It's no surprise that the popularity levels of these three terms are largely correlated.

The use of the terms "cholesterol," "LDL," and "HDL" has significantly risen over time. Source: Google Books Ngram Viewer.

These studies began as basic research, with little or no relevance to medicine. Later, the discovery of the role played by cholesterol and lipoproteins on heart disease shifted the field into clinical science. We will be able to learn about the greatest achievements within their historical context. These studies covered a wide geographical area, from Europe to America and back again, Japan and other parts.

The geographical distribution of the milestones in the field of cholesterol and lipoproteins.

This book provides a comprehensive overview of lipoproteins, including their main classes and the latest biochemistry. It also includes clinical information as well as medical advice. We'll learn why lipoproteins are important, how they function, what they do and why they exist. We'll also see how they stop working and how to fix it with lifestyle and medical treatment. When reading, it is important to bear in mind that the author of this book is a researcher, not a physician.

Despite the challenges, lifestyle changes that correct lipoprotein processing remain the first medical option. The therapies are the last resort, but they are often necessary and can prolong and improve life. The development of new therapies is constant, and they make their way into the clinic. However, it's not always easy to predict how and when they will work. After decades of hardwork, this field of biomedical science at the intersection of biochemistry and biology remains fascinating.

Acknowledgments

I want to express my deepest appreciation to M. John Chapman, who invited me to work on plasma lipoproteins and heart disease at his lab in Paris and supported me over long time. For more than 20 years, John's expertise, guidance, finesse, and friendliness were the foundation of all my scientific studies and writings. I truly had the chance to join the excellent lab of John who introduced me to the world of international research and connected me with top scientists in the field.

I could not have undertaken this journey without Ulrike Beisiegel who led my work on lipoproteins at the University of Hamburg in Germany during the 1990s. Ulrike opened up the world of lipoproteins, atherosclerosis, and heart disease to me and continuously supported me and my family over these years without any doubts, providing me with the unique opportunity to work at her wonderful lab.

I am extremely grateful to all my colleagues from Paris and Hamburg whom I greatly enjoyed working with. My thanks primarily go to Christoph Hübner, Alfried Kohlschütter, Wilfried Weber, Barbara Finckh, Sandrine Chantepie, Marie Lhomme, Isabelle Guillas and Maryam Darabi, but equally to Juliana Bergmann, Nico Donarski, Helga Reschke, Jörg Heeren, Dieter Münch-Harrach, Andreas Niemeier, Alexander Mann, Annette Krapp, Susanne Ahle, Stefanie Koch, Sönke Arlt, Sven Schippling, Dave Evans, Dorte Wendt, Christine Runge, Jan Hinrich Bräsen, Carsten Buhmann, Hans-Jörg Stürenburg, Ulrike Mann, Stephanie Meyer, Barbara Karten, Torsten Spranger, Sirus Djahansouzi, Philippe Giral, Eric Frisdal, Maryse Guerin, Martine Moreau, Roberte Le Galleu, Françoise Berneau, Boris Hansel, Laurent Camont, Alexina Orsoni, Samir Saheb, Carolane Dauteuille, Sora Lecocq, Hala Hussein, Martine Couturier, Murielle Atassi, Paul Robillard, Farid Ichou, Alexandre Cukier, Emilie Tubeuf, Aurelie Canicio, Lucie Poupel, Clement Materne, Alain Carrié, Philippe Couvert, Olivier Bluteau, Thierry Huby, Emmanuel Gautier, Eric Bruckert, Dominique Bonnefont-Rousselot, André Grimaldi, Patrice Therond, Anne Négre-Salvayre, Robert Salvayre, Randa Bittar, Antonio Gallo, Maharajah Ponnaiah, Herve Durand, Sophie Galier and Martine Glorian for their kindness, patience, help, and advice. I am particularly thankful to Philippe Lesnik and Wilfried Le Goff for their continuous support and encouragement of my studies at Sorbonne University and Institut National de la Santé et de la Recherche Médicale (INSERM). I would like to recognize the important support continuously provided by Stéphane Hâtem, Director of the Research Unit 1166 and Head of the IHU ICAN in Paris. I would also like to extend my sincere thanks to all my graduate and doctoral students, postdoctoral researchers, engineers, and technicians whose brightness, enthusiasm, and hard work enriched my vision of the world of research. Many thanks also go out to all my colleagues and friends doing research worldwide in France, Germany, the USA, England, Australia, Canada, Mexico, Italy, the Netherlands, Austria, Sweden, Finland, Switzerland, Croatia, Greece, Brazil, Argentina, Russia, China, Iran, Lebanon, Israel, and other countries whom I relished working with over my career. I would like to acknowledge the generous support from Sorbonne University and INSERM which financed a large part of my research. I would be remiss in not mentioning the University of Odessa in the USSR where I received excellent training.

I am sincerely grateful to Elisa Campos for her outstanding PhD thesis on the history of lipoproteins which greatly aided my research into the past. I also want to thank Børge Nordestgaard for his excellent lectures which inspired the title and cover of this book.

I am truly thankful for the invaluable advice on this manuscript provided by my esteemed colleagues and friends, Philippe Giral and M. John Chapman.

I express my sincere gratitude to all the researchers acknowledged in the manuscript and extend my apologies to those whose contributions could not be cited due to space constraints.

Finally, I want to express my heartfelt thanks to my family. My dear parents, physicists Sergey and Svetlana, first sparked my interest in science. My loving and supportive Varya has always been there for me. My beloved children Michael, Anna and Konstantin have been a constant source of encouragement. My aunt Barbara and grandmother Irina were instrumental in helping me launch my career. My family's unwavering belief in me kept me motivated and upbeat throughout the writing of this book. I also want to thank my childhood friends, who have continued to inspire me over the years.

Abbreviations

Abbreviations are common in every scientific field. While they are helpful for specialists, they can be overwhelming for newcomers. Whether they have a positive or negative impact, it's impossible to understand scientific conversations without them. We will strive to minimize their use and only include essential ones.

Here they are:

4S:	Scandinavian Simvastatin Survival Study
ABC:	ATP binding cassette
ABCA1:	ATP-binding cassette transporter A1
ABCG1:	ATP-binding cassette transporter G1
ACC:	American College of Cardiology
ANGPTL:	angiopoietin-like protein
APO:	apolipoprotein
ATP:	adenosine triphosphate
BET:	bromodomains and extra-terminals
BMI:	body mass index
CETP:	cholesteryl ester transfer protein
GPIHBP1:	glycosylphosphatidylinositol-anchored high-density lipoprotein binding protein 1
HDL:	high-density lipoprotein
HMG-CoA:	hydroxymethylglutaryl coenzyme A
IDL:	intermediate-density lipoprotein
LCAT:	lecithin-cholesterol acyltransferase
ILLUMINATE:	Investigation Of Lipid Level Management To Understand Its Impact In Atherosclerotic Events trial
INTERHEART:	Effect of Potentially Modifiable Risk Factors Associated with Myocardial Infarction study
LDL:	low-density lipoprotein
Lp(a):	lipoprotein (a)
LPL:	lipoprotein lipase
LPR:	LDL receptor-related protein
NADH:	nicotinamide adenine dinucleotide phosphate
mRNA:	messenger RNA
NIH:	National Institute(s) of Health
NMR:	nuclear magnetic resonance

OH:	hydroxyl
PAF-AH:	platelet activating factor-acetylhydrolase
PCSK9:	proprotein convertase subtilisin kexin type 9
PLTP:	phospholipid transport protein
PON:	paraoxonase
PPAR:	peroxisome proliferator-activated receptor
SAA:	serum amyloid A
Sf:	Svedberg flotation unit
SR-BI:	scavenger receptor class B type 1
VLDL:	very low-density lipoprotein

Prologue

The Story of Life and Death

This book isn't just about cholesterol.

It's primarily about the story of life and death, with cholesterol being just one piece of the larger picture.

This story takes place in the invisible realms of the body. This is a network of complex, intricate, invisible worlds. Each has its own complexity. These worlds are made up of visceral organs and functional systems. They also contain interconnected cells and countless molecules. In these worlds, we can find the cardiovascular system and arteries, livers, intestines, hepatocytes, macrophages, endothelial cells, platelets, cholesterol, and lipids. The story of this book is told in these invisible worlds.

This book is centered on cholesterol, which is known for its connection to heart disease. Other molecules, such as proteins and lipids, also play an important role. Hepatocytes, macrophages, endothelial cells, platelets, and other cells are the main players, while the primary affected organs are arteries and heart. These components are all vital to the cardiovascular system.

The book's narrative will be set in these unseen worlds and encompass universal themes such as suffering, healing, mortality, and death. The story will progress as it explores the worlds of heart disease and atherosclerosis, each from a unique perspective. We will investigate the covert world of molecules, the enigmatic world of cells and organs as well as the shadowy world of diseases.

Our story takes place in this setting, where the characters are brought to life.

Like in any story, there are characters who steal the show and those who don't. Lipoproteins are just one of the many residents in these invisible worlds. Two of them have gained international fame, as the "good" and "bad" cholesterol. The rising star of this family, however, is remnant cholesterol.

The three main characters in the story are sometimes referred to by the terms "good", "bad" and "ugly". These labels do not capture their complexity. Each character has both good and bad traits. This makes them more nuanced. The circumstances of each character influence their actions, and the ending of the story is determined by these factors. Each character will do anything to achieve their goals and influence the story in their favor, regardless of their circumstances.

As they navigate through the complex network of organs and cells, the characters interact with one another. They operate in different environments, such as intravascular, intracellular and extracellular.

This book introduces to a wide audience the main characters known as lipoproteins. They are critically important for the development of heart disease and other disorders. The book will show the characters as they move through the different worlds. The book starts by setting the scene, introducing characters, and then diving into the story.

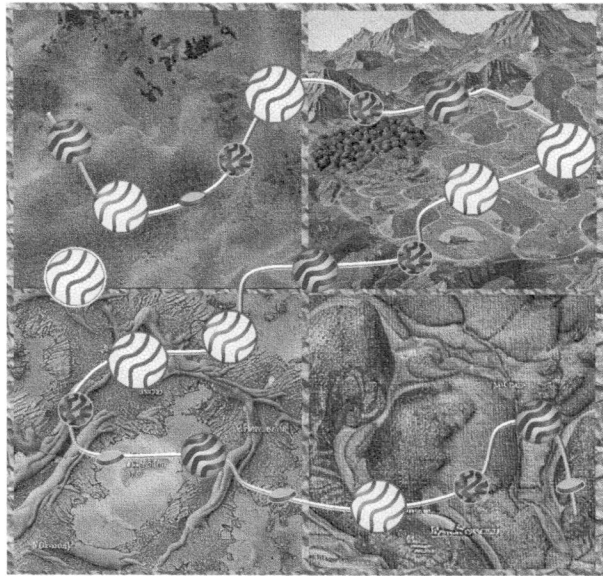

Lipoproteins form the link between the invisible worlds of cholesterol, lipids, cells, organs, and atherosclerosis. They are a key component of this complex web, and weave a long and winding pathway that connects these worlds.

Setting: Realms of Molecules, Cells and Diseases

1

Covert World of Molecules

Cholesterol

Why is cholesterol a problem? Why is this substance so vital? Has this always been the case?

There are both simple and complex answers to these questions. The simple answer is that cholesterol is believed to play a crucial role in determining our lifespan. In other words, cholesterol is thought to be central to the fundamental question of life and death, which is arguably the most important question known to humankind. We believe that cholesterol is responsible for heart disease – an illness that claims more lives than any other cause. This statement may seem paradoxical when we consider that cholesterol is actually an essential substance that the human body requires for good health. How can it be that a fundamental component of the body can become harmful and even life-threatening?

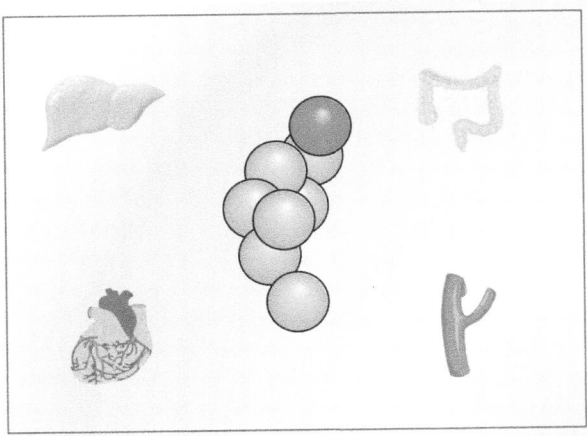

The liver, intestines, arteries, and heart (clockwise from the top left) are the primary organs involved in the processing of cholesterol (center) and its connection to heart disease. The hydrophilic OH group of cholesterol is in dark red.

Cholesterol, a type of fat, is necessary for the proper function of the body. There are biochemical pathways which aim to recycle and spare this compound.

Under normal conditions, it is a solid yellowish substance. Every cell in the body contains cholesterol. It is also an important component of lipoproteins. Animal cells produce cholesterol through a multistep complex process. By contrast, it is almost absent in bacteria.

Cholesterol, Lipoproteins, and Cardiovascular Health: Separating the Good (HDL), the Bad (LDL), and the Remnant,
First Edition. Anatol Kontush.
© 2025 John Wiley & Sons, Inc. Published 2025 by John Wiley & Sons, Inc.

The liver and intestines are the main sites of cholesterol production. The brain, the adrenal glands, and the reproductive organs are also sites where cholesterol is actively synthesized in humans. The brain contains a large amount of cholesterol, which accounts for about one-quarter of the total cholesterol in the human body.

Almost all cells in the body are capable of synthesizing cholesterol. Like other biological molecules, cholesterol also needs to be replaced and destroyed regularly. However, not all cells can do it. Only liver cells are able to degrade cholesterol in large quantities. Transport to the liver is necessary for continual cholesterol turnover within cells.

Cells cannot function properly without cholesterol and will die. In addition, cholesterol is used to make hormones, such as steroid hormones, stress hormones, vitamins, such as vitamin D, and bile acids that help digest food. Finally, cholesterol is important for fighting bacteria and infections.

Cholesterol is essential to separate cells from their surroundings. Cell membranes, thin structures that regulate cellular interactions with their environment, fulfill this function. Membranes are made of a material which cannot easily be destroyed by water. This kind of material is well known to everybody – it is fat. Cell membranes, which are mostly made of fat and protein, include a large amount of cholesterol.

The cholesterol in the body is a fat that is almost insoluble in water. It can also be mixed with other fats. These substances are known as lipophilic, from the Greek words "lipos," meaning "fat" and "philia," meaning "love." Up to 30% of animal cell membranes are composed of cholesterol. The fatty myelin sheath that coats neuronal cells is about one-fifth cholesterol. Communications between neuronal cells critically depend on cholesterol.

Interesting Numbers

The average human body contains 35 grams (0.05%) of cholesterol, which is mostly found in the cell membranes. An average human synthesizes approximately 900 milligrams of cholesterol per day.

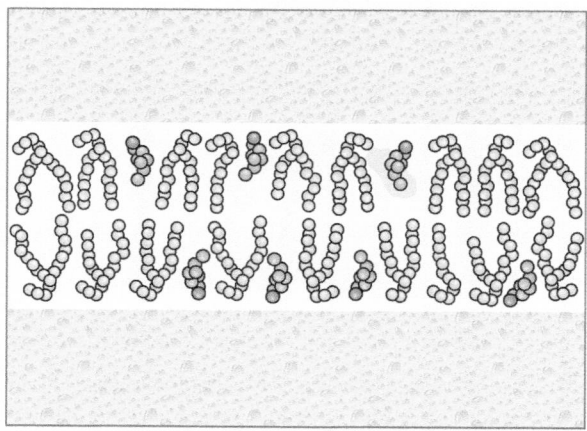

The cell membrane is composed of a bilayer made up of phospholipids and cholesterol. Cholesterol molecules are depicted in yellow, while phospholipids are represented in grey.

We all know that cholesterol can be both good and bad. These terms used by doctors around the world are familiar to us. The terms do not, however, precisely reflect the positive and negative effects of cholesterol on living organisms. As we shall see, they are derived from the measurements of cholesterol levels in blood plasma.

Interesting Numbers

In the United States, a typical daily cholesterol intake is around 300 milligrams.

All animal foods are cholesterol rich, as cholesterol is synthesized in animal cells. The amount of cholesterol in food varies greatly. Cholesterol is a major component of red meat, whole eggs, and egg yolks. In addition to liver, kidneys, giblets, and fish oil, butter also contains significant quantities of cholesterol. Not negligible amounts of cholesterol are also found in human breast milk.

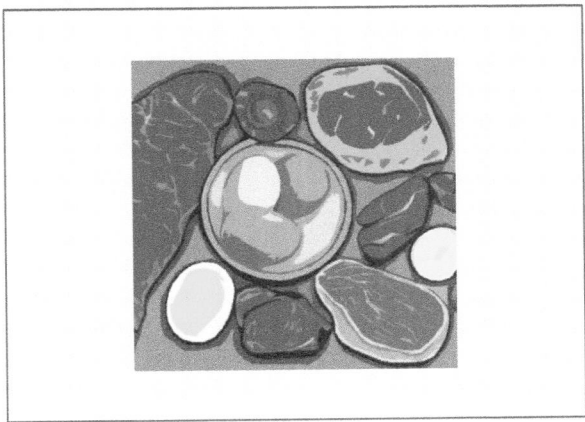

Foods high in cholesterol: red meat, eggs, and butter.

Cholesterol molecule consists of four carbon rings, a hydrophobic side chain and a hydrophilic hydroxyl group OH. The molecule is rigid except for the side chain. The hydroxyl group interacts with water molecules surrounding the membrane, while the side chain is stuck in the cell membrane alongside the hydrophobic chains of phospholipids. This is how cholesterol interacts and works with other lipids to regulate fluidity in cell membranes and to maintain their integrity.

What It Is

Hydrophobic is derived from the Greek words for water (*hydros*) and fear (*phobos*). It means having little or no affinity for water.

Hydrophilic is derived from the Greek for water (*hydros*) and love (*philia*). It means having an affinity for water and being capable of interacting with it.

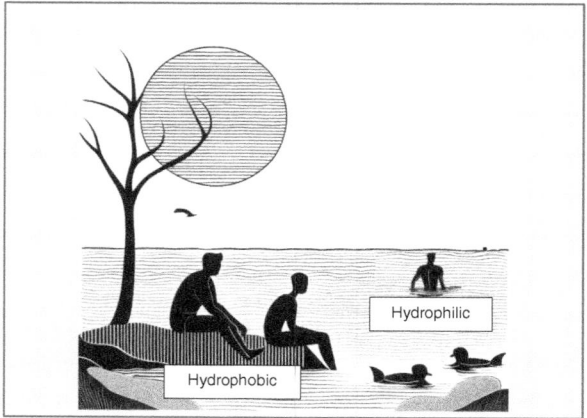

Hydrophilic substances mix easily with water, while hydrophobic substances repel it strongly.

Lipids

Cholesterol is a lipid. The word "lipids" comes from the Greek for fat, "lipos". Oils and fats can be considered lipids, but fats are usually solid at room temperature, while oils tend to be liquid. We attribute cholesterol to fats because it is usually solid.

Lipids make up a vital part of the human body. They are produced by the body or derived from diet. Lipids are essential for energy production, which is widespread, and are present in every cell. They form a main part of every lipoprotein particle. Lipids must be transported to their sites of use and then destroyed to produce energy. Transporting them to the required location is essential.

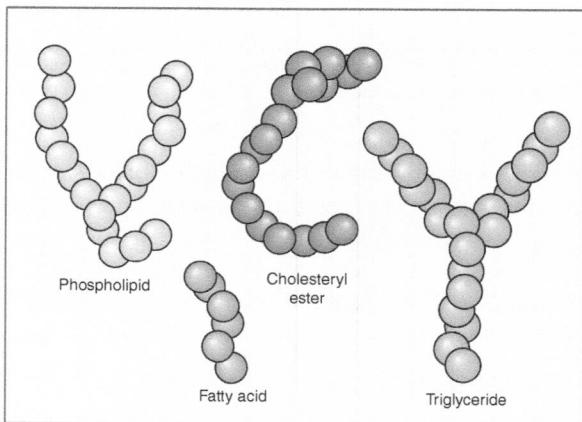

VICY structures of main lipid molecules: V-like phospholipid, I-like fatty acid, C-like cholesteryl ester, and Y-like triglyceride.

Fatty Acids

There are several classes of lipids.

The simplest fatty acids are chains of hydrocarbon units between the terminal hydrocarbon and carbonyl groups. Fatty acids, in the form of triglycerides, account for over 90% of the dietary fat.

The bonds between the groups can be either single or double, which results in different chemical and physical properties. The more double bonds a fatty acid contains, the more fluid it is. Fatty acids that only contain single bonds between carbon atoms are less fluid, and, at physiological temperatures, can even become solid. They are called saturated fatty acids. Stearic acid, which is mostly derived from animal fats, is the most common saturated fatty acid.

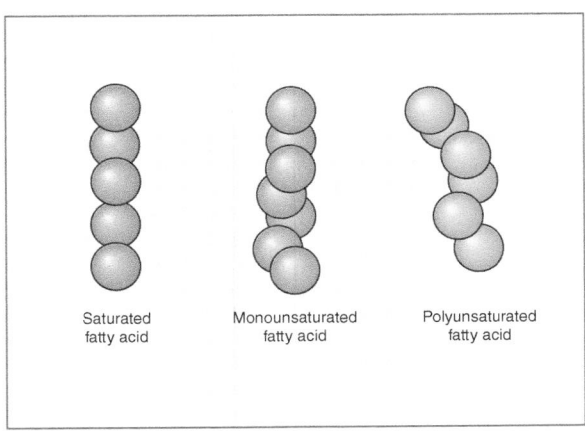

There are three main types of fatty acids: monounsaturated, polyunsaturated, and saturated.

Fatty acids that contain a number of double bonds are fluid and belong to oils. These fatty acids are unsaturated. Unsaturated fats are divided into monounsaturated, which have a single double-bond, and polyunsaturated with two or more. The most common monounsaturated fatty acid is oleic acid which is found in high amounts in olive oil. Polyunsaturated fatty acids are mostly derived from sea animals and vegetables – e.g. linoleic acid from fish oil or sunflower oil.

Triglycerides

Triglycerides, also known as triacylglycerols, are more complex molecules. They are made up of three fatty acids attached to a backbone of glycerol. The molecule of triglyceride is hydrophobic and does not have any hydrophilic components. Triglycerides are a major source of energy that can be either used right away for energy production or stored in the body for meager times.

When the body needs energy, triglycerides derived from the diet or stored in adipocytes (which are fat cells) are broken down into fatty acids and glycerol by enzymes called lipases - proteins that break down lipids. The fatty acids are quickly absorbed by the muscle cells and tissues, where they can be used to generate energy. This is done by producing adenosine triphosphate (ATP), a biological molecule which can be used to power many chemical reactions. The body's process for producing energy from fatty acids is known as beta-oxidation.

What It Is

Adenosine triphosphate (ATP) is a molecule which gives energy to cells in order to perform different functions such as muscle movements, nerve signals, and the production of other biomolecules, like proteins, DNA and RNA. It can be found in all living things and is known as the "molecular currency" that moves energy within cells. When it is used up, it turns into adenosine diphosphate (ADP) or adenosine monophosphate (AMP).

(Continued)

(Continued)

Nicotinamide adenine dinucleotide phosphate (NADP) is a reduced form of a coenzyme which plays a major role in the production of ATP by all living cells. Coenzyme is a molecule that helps enzymes to catalyze biochemical reaction.

Sugars and triglycerides are the primary sources of energy.

The body stores triglycerides as an energy reserve. Excess calories that are not immediately required for energy production can be converted to triglycerides and then stored in fat (also called adipose) tissues for later use. These stored calories can be used during fasting periods or times when energy requirements are higher, like during exercise. Triglycerides are important in providing energy to the body and as a storage for excess calories.

Interesting Numbers

Every day, an adult human uses approximately 50 kilograms of ATP. The majority (about 45 kilograms) is used for the basal metabolism. Approximately 10,000,000 molecules ATP are produced and consumed by each cell per second.

Sugars are another important source of energy. In the process of glycolysis, sugars are broken down into glucose molecules to produce ATP and NADH. This occurs in the cells' cytoplasm and is the initial step of respiration. These products are then used in the subsequent steps of cellular metabolism to produce energy-rich ATP.

Sugars are stored in the body at a much lower rate than lipids. Glycogen is a substance that can be made from sugars in excess and stored by animals and humans. Glycogen is a polymer with a high degree of branching made up of glucose molecules. When the body requires energy, the glycogen is broken into glucose which is then released into the bloodstream and used as fuel by the cells. Glycogen helps regulate blood sugar and can be replenished by carbohydrates.

Phospholipids

Phospholipids, another type of lipid molecule, are a major component of the cell membranes which separate cells. Each phospholipid molecule has hydrophilic and hydrophobic parts which are, respectively, called heads and tails. The cell membrane is made up of two layers of phospholipids with the hydrophilic heads facing the outside and the hydrophobic tails facing the inside. This arrangement creates an obstruction that prevents molecules which are water soluble from passing through the cell membrane. The phospholipid cell membrane contains proteins and other molecules that regulate the movement of substances in and out of the cell. This selective permeability enables cells to maintain an internal environment and carry out vital functions, such as nutrient absorption, waste removal, and communication with other cell types.

Besides their role in the cell membrane, phospholipids are also important in subcellular structures called organelles. Membranes that are similar to the ones that separate cells from their surroundings are used to separate these structures from the rest of a cell. When triglycerides or other hydrophobic fats are stored within a cell, they are surrounded by a monolayer of phospholipids. By providing a barrier to the outside environment, phospholipids play an important role in maintaining cell integrity and function.

Is cholesterol innocent?

Sterols

Sterols are another important class of lipids. Cholesterol is undoubtedly the most known sterol. Sterols play an important role in the structure of cells and are another essential component of cell membranes. Sterols are also required for various physiological processes, such as the synthesis of hormones, vitamin D, and bile acids. Sterols are also found in plants, the most common of which are sitosterol, campesterol, and stigmasterol. These are nutrients that can be absorbed from food, and they are found in the human body at much lower levels than cholesterol.

Sterols are available in two forms: free (unmodified) and esterified. Cholesteryl ester is the most common among esterified sterols. It is composed of cholesterol and a fatty acid. Cholesteryl esters are present in many types of tissues, such as the liver, the adrenal glands, and the intestines. These molecules play an important part in the processing of cholesterol by the body. Cholesteryl esters are often formed when cholesterol is absorbed by the body from food or is produced in the liver.

They can be used for storage and transportation. These molecules, like triglycerides, can be stored within cells until needed for other metabolic processes.

What It Is

Ester is an organic substance that is made by replacing the hydrogen atoms (H) in at least one hydroxyl group (OH) with another chemical group.

Cholesteryl ester has a fatty acid part that takes the place of the hydroxyl group in cholesterol.

Glycolipids cover the surface of cells and lipoproteins in a forest-like fashion.

Glycolipids are a type of lipid molecule containing a sugar (carbohydrate) ring. They play an important role in cell signaling and recognition. Glycolipids are composed of a hydrophilic carbohydrate head and a hydrophobic lipid tail. The tail anchors the molecule onto the membrane while the head is responsible for interactions with the surrounding environment. The carbohydrate head can be of different sizes and complexity. It may consist of a single sugar or a complex chain. Glycolipids play a key role in cell communication. They are markers that appear on the cell surface, which allow cells to interact and recognize one another. Glycolipids play an important role in the signaling pathways of cells. They help to transmit signals between the cell's outside and its inside.

Sphingolipids and glycolipids are two other lipid classes that play an important role in humans. Sphingolipids are important for signaling and communication in cells. They are found in the cell membranes. Sphingolipids are made up of a sphingosine, which is an amino alcohol with a long chain, and a fatty acid chain attached to the amino group. Sphingolipids are involved in many cell processes including cell growth and differentiation, apoptosis, and inflammation. Sphingolipids are also important in the immune system, by regulating both the activation of immune cells and their function.

Sphingomyelin is one of the best-known sphingolipids. It is present in high concentrations within the myelin that surrounds the nerve cells. Sphingomyelin is a sphingolipid that helps insulate nerves and transmit electrical signals along the axons. Ceramide is another important sphingolipid, produced from the breakdown of sphingomyelin. Ceramide is known to regulate cell death, survival, and inflammation.

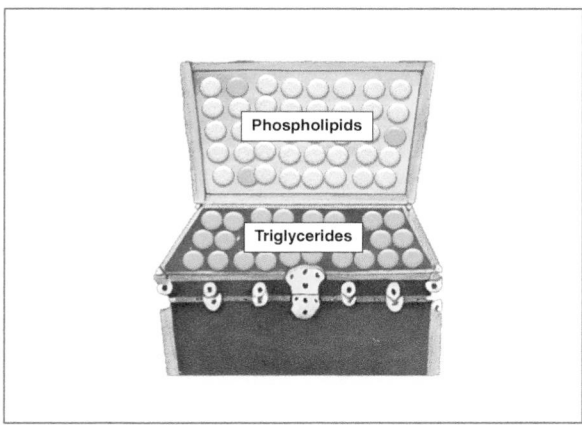

Energy treasure chest: triglycerides, represented in green, are a vital source of energy. They are stored beneath a layer of phospholipids (in grey), with small amounts of cholesterol (in yellow) present within this layer. Within the core of the triglycerides, cholesteryl esters (in pink) can be found.

Interesting Numbers

The average energy content in fats is 9 kcal/g. The average daily intake of fats should not exceed 30% of the total daily calories. This is 70 grams of fat per day based on 2,000 kcal of energy consumed per day.

Although lipids are important energy sources, they differ in terms of their energy-producing capacity. The caloric contents of different lipid classes are quite variable. Triglycerides are the most common type of dietary fat and contain 8–9 kcal/g. The fatty acids that are the building blocks of triglycerides also contain 8–9 kcal/g. The caloric value of fatty acids is affected by their length and saturation. Saturated fatty acid tends to be more calorically dense than unsaturated. In addition, the energy content of fatty acids increases as the chain length increases. Phospholipids have a lower energy density and contain only 4–5 kcal/g. Sterols such as cholesterol do not have any calories because they are not used for energy.

History

Identification of cholesterol

Who	When
Michel Eugène Chevreul	**1815**

Where	What
Paris, France	M. E. Chevreul (1823). "Recherches chimiques sur les corps gras d'origine" Paris: 484p.

Francois Poulletier de la Salle, a French chemist, first identified cholesterol in a solid form in 1769. The chemist named the substance "cholesterol," after the Greek words "stereos" meaning "solid" and "khole" meaning "bile" as its solid form was found in the animal bile. His work was never published and the attribution is only known through his collaborators.

After studying patients suffering from gallbladder stone disease, Michel Eugene Chevreul finally named the compound "cholesterine" in 1815. It was renamed to cholesterol in 1929 due to its chemical composition.

Chevreul was not just any scientist. He believed that man was capable of infinite perfection, both in morals and in physical matters, and that science had unlimited power. He saw the Eiffel Tower as a living synthesis, combining all of the acquired knowledge and experience accumulated over centuries. In his last years, he loved to visit the Champ de Mars and watch the construction of the Universal Exposition of 1889.

He explored all the unknown areas of organic chemistry and biochemistry through his remarkable research into animal fats. He invented stearic candle between 1828 and 1831. This discovery revolutionized domestic lighting and enriched thousands of people without him ever thinking about the profit. His institute only awarded him a 12,000-franc prize.

Chevreul was elected a member of the French Academy of Sciences in 1826. He attended every session and missed none for the next 63 years. Chevreul attributed his remarkable longevity to his love of work, his sobriety, and his culinary principles. On August 31, 1886, the centenary of the "Nestor" of chemistry, who lived under three republics and four kings, was honored with a large ceremony.

Friedrich Reinitzer, a German chemist, published the molecular formula for cholesterine in 1888. It was $C_{27}H_{46}O$. Chevreul, who was alive at the publication of Reinitzer, died at the Jardin des plantes in Paris on Tuesday, 9 April 1889 at 1 o'clock morning after living for 102 years, 7 months, and 8 days. He was the oldest scholar in the world.

It was not until the 20th century that cholesterol was recognized as a component of human blood and tissues, and its role in health and disease was studied.

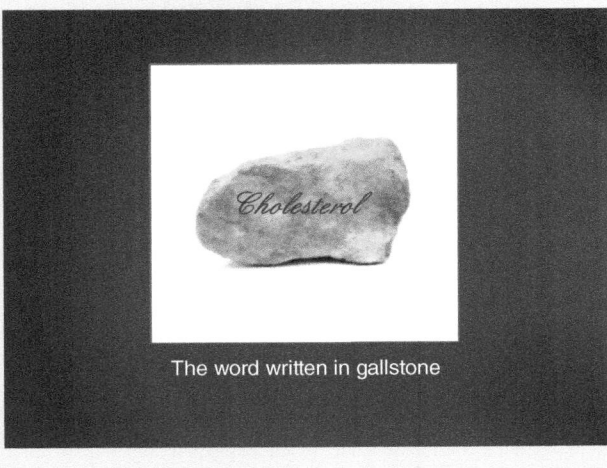

The word written in gallstone

Interesting Numbers

Michel-Eugene Chevreul, one of the pioneers of cholesterol research, is among the 72 scholars who have their names inscribed on the Eiffel Tower's first floor in Paris. On the north-facing side, he is number 14.

Proteins

A deeper understanding of cholesterol requires an appreciation for proteins, another type of biomolecule. Proteins consist of long chains of amino acids linked together and serve various roles within organisms – doing chemical reactions, copying DNA molecules, responding to environmental changes, providing structure and organization in cells and organisms, moving molecules throughout the body as well as transporting essential elements like vitamins. Their sequence is determined by genetic code. Furthermore, polypeptides – long chains of amino acids linked together – make up proteins while some have additional parts attached such as sugars and are, in this case, termed glycoproteins.

Proteins perform various functions. Once produced, however, proteins only last briefly before degrading and recycling by cells. They play an integral part in cell processes with animals needing protein consumption to obtain essential amino acids their bodies cannot produce themselves.

2

Enigmatic World of Cells and Organs

Cells

Lipids are involved in the functioning of every cell. Similarly, certain cells are crucial for the regulation of lipids and lipoproteins. In the context of heart disease, certain cell types are especially important, but we don't fully understand how they work. These include, first of all, hepatocytes, enterocytes, endothelial cells and macrophages.

Hepatocytes

Hepatocytes (also called hepatic cells) are the cells that make up about 80% of the mass of the liver. The name comes from the Greek "hepar," meaning "liver," and "kutos," means "hollow vessel." These cells have many important functions. They make and store proteins, metabolize carbohydrates into other molecules, and produce cholesterol, bile acids, and other lipids. These cells also make bile and help eliminate harmful substances from the body.

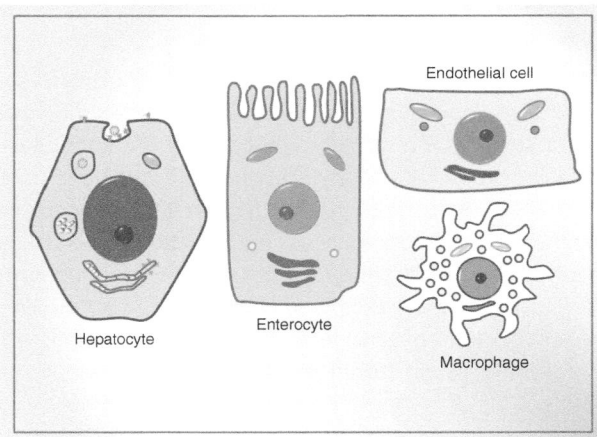

The main cell types involved in cholesterol processing are hepatocytes, enterocytes, endothelial cells, and macrophages.

Hepatocytes are cubical cells with sides that are 20–30 μm wide. Comparatively, they have more smooth endoplasmic membranes (which are involved in producing biomolecules) than other cells. The endoplasmic membrane helps hepatocytes to attach proteins to carbohydrates and lipids.

Cholesterol, Lipoproteins, and Cardiovascular Health: Separating the Good (HDL), the Bad (LDL), and the Remnant, First Edition. Anatol Kontush.
© 2025 John Wiley & Sons, Inc. Published 2025 by John Wiley & Sons, Inc.

Hepatocytes also produce many serum proteins and lipoproteins. The liver secretes hormones known as hepatokines, converts carbohydrates into fatty acids, and makes triglycerides out of fatty acids and glycerol. The liver also makes cholesterol from acetate and produces bile acids. Bile acids can only be synthesized by the liver.

Enterocytes

The most abundant intestinal cells are enterocytes. The name derives from the Greek "enteron," meaning "intestine," and "kutos," meaning "hollow vessel." They are oblong, about 40–50 µm high with an average volume of 1,400 µm^3. The enterocytes are responsible for absorbing nutrients and other substances from food. The surface of enterocytes is coated with digestive enzymes, and they have tiny hairs that protrude into the interior of the small intestine and increase the surface area. These structures are called intestinal villi. Enterocytes can then take up small molecules from the intestines such as proteins, sugars, and lipids. These cells release hormones as well and play several roles. They absorb lipids, take in ions such as sodium and calcium, and absorb water and sugar. They also break down proteins into amino acid chains, reabsorb bile salts, and release immunoglobulins to the intestines. An enzyme called pancreatic lipase breaks down lipids in the intestine, which then move into enterocytes. The smaller lipid molecules are absorbed into the blood vessels of the intestine, while the larger ones are converted into lipoproteins known as chylomicrons. They are then released into the lacteals, which are lymphatic capillaries located in the villi.

> **What It Is**
> **Lacteals** are lymphatic capillaries in the villi of the small intestine.
> **Intestinal villi** are tiny finger-like parts that stick out into the inside of the small intestine.

Endothelial Cells

The endothelial cells found in blood vessels control fluids and substances entering and leaving tissues. The word "endothelium" derives from the Greek words "endon," meaning "inside," and "thele," meaning "teat" or "nipple." These cells, however, are more than just a barrier. They also aid in processes such as blood vessel formation, inflammation, and blood coagulation. Endothelial cells play an important role in maintaining healthy blood vessels and heart. These cells are present in all blood vessels, and they rely on the soluble gas nitric oxide to function.

The endothelium is a single layer of endothelial cells lining the inner surface of blood vessels. It serves as a protective barrier, keeping the blood from coming into direct contact with the rest of the vessel walls. The endothelium has many other functions, including preventing blood clots, regulating blood pressure, and assisting with healing and inflammation.

> **Interesting Numbers**
> The endothelium, although invisible, is a very large part of our body. It is made up of trillions and trillions of cells. On an average, it covers nearly 3 m^2.

Macrophages

Macrophages, which are large white cells in the blood, help to fight infections and injuries. The name comes from the Greek words "makros," meaning "large," and "phagein," meaning "eat." It can be translated as "big eaters." The macrophages work by engulfing and then digesting harmful agents such as cancer cells and bacteria. Macrophages can be found in every tissue and are involved in the body's specific immune response as well as its general defense system. They can reduce inflammation and promote tissue repair. Different macrophages have different functions. These cells have a diameter of about 20 μm and can be distinguished by the proteins they express at their surface.

Platelets

Platelets (or thrombocytes) are white blood cells which help stop bleeding after a blood vessel has been injured. Platelets are white blood cells without a nucleus. They come from bone marrow and lungs, and they travel through the blood. Platelets form a plug when there is an injury or cut to a vessel. Platelets also help prevent blood clots in healthy blood vessels.

Adipocytes

Adipocytes are also known as fat cells. They store energy in lipids. The word comes from the Latin "adeps," which means "fat," or "lard." There are two types of adipocytes: brown and white cells. The white cells are from 20 μm to 0.3 mm wide and contain large droplets of lipid, while the brown cells are smaller. The droplets are primarily composed of triglycerides and also of cholesteryl esters.

Organs

Cells create organs that are involved in the production, transport, and removal of lipids and lipoproteins. The most prominent of these organs are blood arteries. Additionally, three other organs – the liver, heart, and intestine – also play central roles in the regulation of these processes.

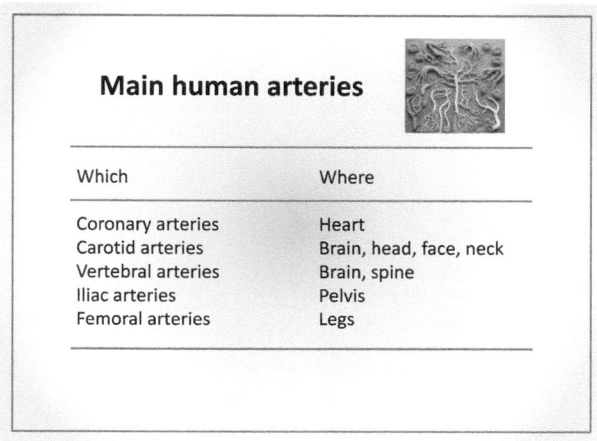

Arteries

The arteries are an important part of cardiovascular systems, as they distribute oxygen-rich blood to all parts of the body. The tube-like blood vessels and muscles within them ensure that organs, tissues, and cells receive the oxygen and nutrients needed to function properly. The heart pumps oxygenated blood into the largest arterial system in the body – the aorta. This vessel then branches into smaller arteries to reach every part of the body. The coronary arteries supply blood to the heart. Other arteries deliver blood to specific organs and areas in the body. The central nervous system sends signals to the arteries that cause them to dilate or constrict, affecting blood pressure. By adjusting the muscle walls, they help regulate blood pressure. Arteries differ from veins which carry deoxygenated blood back to the heart. Veins only have a thin layer of muscle tissue inside and use valves to keep blood flowing.

The arteries have three layers: the inner layer, called tunica intima ("inner coat" in Latin), contains tissue with elastic fibers, the middle layer, called tunica media ("medium coat" in Latin) is mainly smooth muscle, which allows the arteries to be tightened or opened, and finally, there is an outer layer, called tunica adventitia ("additional coat" in Latin), that interacts with other tissues including nerves. The inner layer is composed of one layer of endothelial cells and is supported internally by an elastic lamina. Endothelial cells are directly in contact with blood flow. The inner coat is a transparent, colorless, thin layer that can stretch. It is composed of a layer of flat, oval, or spindle-shaped cells, called pavement endothelium. It also contains a delicate layer of connective tissue, with branched, flat cells between them, and an elastic layer.

Arteries can be arranged in arterial trees and classified according to their size, function, and location. Elastic arteries are the largest arteries of the body. The aorta is one of these arteries, as it is the main blood vessel that carries the blood from the heart to other parts of the body. Elastic arteries are characterized by a high concentration of elastin, a protein that allows them to contract and expand with every heartbeat. This allows for a constant flow of blood to be maintained throughout the body.

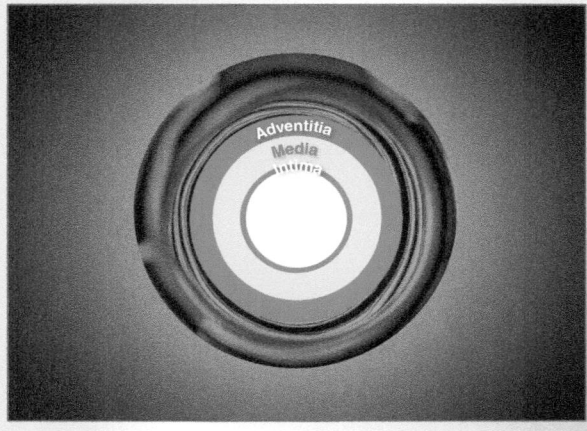

The arterial wall is composed of three layers: the intima, media, and adventitia. The intima is the layer just beneath the endothelial cells, followed by the media and adventitia.

Muscular arteries are smaller than elastic arteries but larger than arterioles. Their walls have more smooth muscles, which allows them to contract or dilate in response to changes in blood pressure. Muscular arteries deliver blood to specific organs or tissues. Arterioles are small branches from muscular arteries leading to capillaries. They are characterized by a thin layer of smooth muscle in their walls, which regulates blood flow to capillary beds. Capillaries are tiny vessels that connect arterioles and venules. They have thin walls which allow the exchange of oxygen and nutrients between tissues and blood.

Heart

The heart is a muscular organ that pumps blood throughout the body. The heart brings oxygen and nutrients to the tissues and removes waste products like carbon dioxide. The heart in humans is the size of a closed fist. The heart beats because of an electric current generated by pacemaker cells.

The heart is a double pump. It has four chambers which work together to pump the blood around the body. The right side of the heart pumps oxygen-poor blood to the lungs where it is oxygenated. This oxygen-rich blood is then pumped back to the body by the left side of the heart via a network of arterial vessels. These arteries are divided into capillaries that supply oxygen and nutrients directly to the cells. The heart collects the deoxygenated vein blood and sends it to the lungs.

> **Interesting Numbers**
>
> Heart pumps five liters of blood per minute. The diameter of the coronary arteries is between 2 and 5 mm. They pump from 0.25 to 1.0 liters of blood per minute.

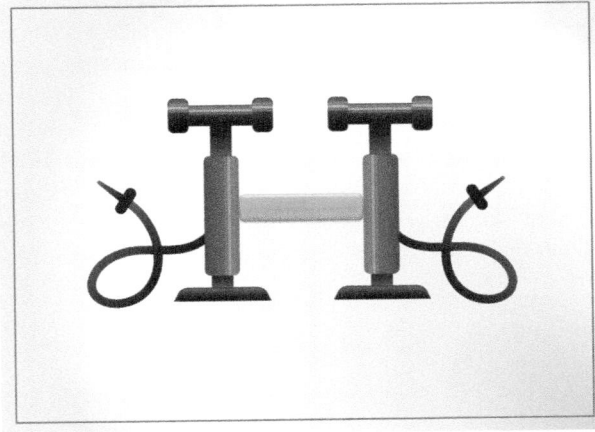

The heart can be visualized as a double H-pump, where two pumps operate in perfect coordination.

The heart regulates metabolism by responding to the changes in energy demand. At rest, the heart normally beats 72 times per minute. However, exercise can increase that rate in order to provide more oxygen and nutrients for working muscles. During sleep or at rest, the heart rate decreases to conserve energy. The heart is also crucial in maintaining fluid balance. It regulates blood pressure by changing its pumping rate and strength in response to changes within blood vessels.

Liver

The liver is the heaviest of all the organs in the body and also the largest gland. Hepatocytes make up the majority of the liver's volume (70–85%). The liver plays a principal role in how the body uses lipids. The liver helps take up, produce, and secrete various types of lipids within the body. This is done by four processes: burning fat as energy, producing different types of lipoproteins, making cholesterol and phospholipids, and converting carbohydrates into triglycerides to store.

The liver makes and secretes different types of lipoproteins. It also removes lipoproteins from the blood using special protein receptors. Bile, a yellow–green liquid that serves to break down fats, is produced by the liver to aid digestion. Some of the bile is sent straight to the small intestinal tract, while some remains in the gallbladder.

What It Is

Bile, also called gall, is a yellowish-green liquid constantly produced by the liver to help break down fats in the intestine. The gallbladder stores and concentrates it and then releases it into the small intestine.

Gallbladder is a small organ which holds bile and then releases it into the small intestine for fat digestion. It is pear-shaped and it sits below the liver. The gallbladder receives the bile from the liver and then sends it into the small intestine. Gallstones may form in the gallbladder, causing pain.

Intestine

The small intestine, which is part of the digestive tract, helps absorb nutrients from food. The small intestine is 6 m long and located between the stomach and the large intestine. It has three sections – the duodenum, jejunum, and ileum. The duodenum prepares nutrients for absorption, the jejunum absorbs nutrients, and the ileum takes up bile salts and leftover digestion products. Pancreatic lipase breaks down fats in the intestine into fatty acids and glycerol. Bile is produced by the liver, and it helps the lipase to do its work. It attaches to fats to make them easier to break down. The lipase, however, is hydrophilic while the fats, on the other hand, are not. Bile therefore helps to mix the two together so that the fats may be broken down and absorbed.

Villi at the surface of the intestine help absorb lipids. Each villus measures between 0.5 and 1.6 mm in length, with many smaller projections known as microvilli. The microvilli are what create the striated or brush border on the surface. This increases the amount of surface that can absorb nutrients.

An artistic view of the surface of a hepatocyte, an endothelial cell, an enterocyte, and a macrophage covered with proteoglycans (clockwise from the top left).

3

Shadowy World of Diseases

Atherosclerosis

The reason why we are interested in hepatocytes, enterocytes, macrophages, and other cells and organs dealing with cholesterol, lipids, and lipoproteins is that their malfunction can cause atherosclerosis. This word is derived from the Greek "athero" meaning "paste" or "gruel", and "sclerosis," which means "hardness." The Greek suffix "osis," which describes a diseased state, is used here, too. Atherosclerosis, then, is a disease that involves the hardening of arterial walls due to the accumulation of cholesterol. Parallel to this, other substances deposit in the arterial walls. Over a long time, atherosclerosis can lead to heart attack and stroke.

Atherosclerotic Lesions

Lesions and plaques are the terms used to describe sites affected by atherosclerosis. These lesions are found in the inner linings of arteries that have been damaged. Atherosclerotic lesion slowly develops over many years to become life-threatening. Different factors can accelerate this pathological progression, including genetic deficiencies or an unhealthy lifestyle.

Atherosclerosis involves a buildup of lesions inside the arteries. This causes the latter to narrow and become harder. The lesion is composed of cholesterol and other lipids and substances which accumulate over time on the inner wall of the arteries. The lesions develop in the arterial intima and are covered by a cup made of proteins. The more lipid is present in the lesion, the more likely it is to rupture. The rupture can block blood flow to organs like the heart, brain, and kidneys as it continues to accumulate. This can cause serious health issues such as heart attack, stroke, and kidney failure.

Atherosclerotic lesions are microscopic injuries of the arterial walls caused by lipids. The lipids act like foreign bodies and cause irritation to the arterial tissue, resulting in a type of chronic inflammation. We have all experienced acute inflammation. It can cause achy backs, abscesses in the mouth, or skin rashes. It is visible, uncomfortable, and often painful. Redness is due to an influx of blood into the area. Swelling is the result of immune cells that are working to heal and prevent infection or injury. These symptoms are part of the natural healing process, and they stop once the process is complete. Chronic inflammation, on the other hand, does not cease for a very long time. It continues when healing is impossible. Atherosclerotic lesions are injuries to the arterial walls where chronic inflammation occurs.

Atherosclerotic lesions develop over a lifetime, starting early. Prenatal atherosclerosis may also affect the development and growth of the arteries within the fetus before birth. This condition may have serious consequences for the developing fetus. Prenatal atherosclerosis may cause a variety

Cholesterol, Lipoproteins, and Cardiovascular Health: Separating the Good (HDL), the Bad (LDL), and the Remnant, First Edition. Anatol Kontush.

of complications including fetal development restriction, preterm delivery, and stillbirth. Prenatal atherosclerosis is diagnosed using ultrasound imaging, which measures blood flow in the fetal arteries. The discovery of prenatal atherosclerosis followed that of atherosclerotic lesions in young adults made in the 1950s in young victims of the Korean War.

Children can be affected by atherosclerotic lesions. However, the lesions more often develop in later life. This causes a large number of adults to have this disease, which is usually silent. In fact, the noninvasive imaging of the Progression of Early Subclinical Atherosclerosis (PESA) study showed that 71% of men in their middle age and 43% of women had signs of early atherosclerosis.

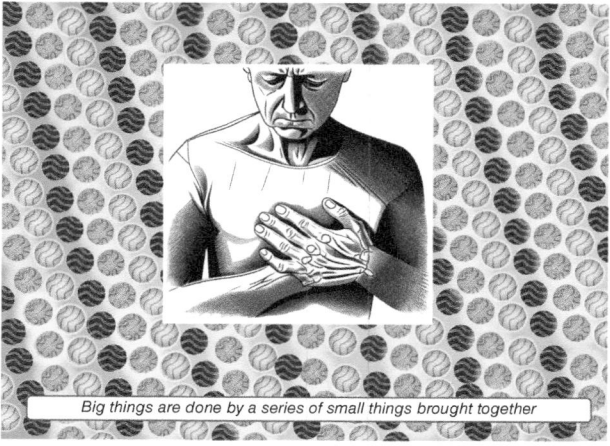

Big things are done by a series of small things brought together

It takes decades for myriads of lipoprotein particles to have a negative effect on health.

In a clinic, atherosclerosis can be quantified using several methods. These include measuring the thickness of the arterial walls or the presence of calcium that accumulates during lesion buildup.

Practical Aspects: How to Measure Atherosclerosis

Intima-media thickness (IMT) measures how thick the two inner layers of an arterial wall are. It is usually done using ultrasound from outside the body. IMT is typically measured in carotid arteries that supply the head and neck with oxygenated blood. The carotid IMT is used to find out if someone has atherosclerosis and to see if it is getting better or worse. Although carotid IMT can predict future heart problems, some studies have found that changes in carotid IMT over time may not be a good way to predict them. Some guidelines recommend using carotid IMT for high-risk patients or those with intermediate risk if other methods are not enough.

Intravascular ultrasound (IVUS) is another ultrasound imaging method that uses a special catheter with a small ultrasound probe at the end. The other end of the catheter is connected to a computerized ultrasound machine. This allows doctors to use ultrasound technology to see inside blood vessels and look at the inner walls. IVUS is most commonly used to look at the arteries of the heart to see how much plaque has built up in them. IVUS can be helpful when other imaging methods do not work well, such as in areas where there are overlapping artery segments.

Coronary calcium scan is a specialized computer tomography of the heart that looks for calcium deposits in the coronary arteries. This type of scan can detect the presence of coronary artery disease before symptoms appear, and the results can help determine the risk of heart attacks or strokes. Additionally, these results may be used to plan or adjust treatment for coronary artery disease.

How to evaluate atherosclerosis

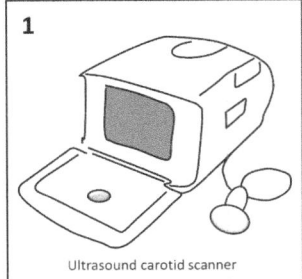

Ultrasound carotid scanner

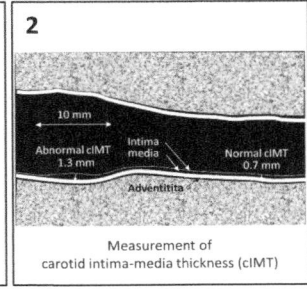

Measurement of
carotid intima-media thickness (cIMT)

Retention of Lipids

Atherosclerotic lesions are formed by a complex process that occurs over time. Atherosclerosis begins with the accumulation of lipoproteins rich in cholesterol in the arterial walls. This occurs due to lipoprotein retention. Lipoprotein particles become trapped in the arterial walls where they can undergo unwanted modifications. These undesirable modifications promote inflammation and oxidative stress, which leads to the formation of fatty streaks, and ultimately atherosclerotic lesions.

Lipoproteins are retained in the arterial intima in a number of ways. Lipoproteins first must penetrate the endothelial layer that covers the inner surface of blood vessels. It can happen through cell gaps or active transport mechanisms. Once inside the arterial intima, lipoproteins bind to different components, such as proteoglycans and extracellular matrix proteins.

What It Is

Proteoglycans are a class of high-molecular-weight proteins that are found in the extracellular space of connective tissue, which provides support to the body's structure. These highly glycosylated proteins make up a significant portion of the extracellular matrix and fill the spaces between the cells.

The size and the concentration of lipoproteins play a major role in their retention. Elevated concentrations of lipoproteins in the blood cause their accelerated penetration in the arterial walls and this facilitates their retention. Lipoprotein concentrations are directly related to retention: the higher they are, the stronger the retention is. Lipoprotein size, on the other hand, is negatively correlated with retention. In fact, larger particles have a harder time penetrating the endothelial surface of the arterial wall, while smaller particles do so more easily. In the arterial wall, lipoproteins are retained by specific interactions with local components.

Oxidative Stress

Oxidative stress is another factor that can contribute to lipoprotein retention. Lipoprotein particles exposed to reactive oxygen species can undergo modifications, such as oxidation and glycation. This makes them more likely for the retention in the arterial walls. Oxidation can be compared to the deterioration of metals or apple slices. Glycation is the process of attaching glucose to biomolecules and proteins, which can affect their function in a similar way to oxidation.

Reactive oxygen species can be generated by various sources such as smoking, hypertension, diabetes, and high blood cholesterol. Oxidized lipoproteins are more apt to bind to the proteoglycans of arterial walls, resulting in increased retention. Inflammatory processes in the arterial wall may also promote lipoprotein retention by altering the expression of proteoglycans and increasing oxidative stress. It occurs when reactive oxygen species are more powerful than antioxidants, which protect lipoproteins against oxidation. There are many antioxidants in the bloodstream that protect lipoproteins. However, the antioxidants can be consumed by the reactive oxygen species produced locally by the cells at certain sites within the arterial wall. Here is where lipoprotein oxidation occurs. Finally, changes in the blood flow and the shear stress placed on endothelial cell linings of the arterial walls can affect lipoprotein accumulation by altering the expression of proteoglycans and promoting inflammation.

> **What It Is**
>
> **Oxidative stress** is caused by an imbalance in the production of reactive oxygen species in tissues and cells, on the one hand, and in the ability of these tissues and cells to eliminate them, on the other hand. In simpler terms, it happens when there are too many oxygen-derived molecules and not enough capacity to remove them.

Nitric oxide is the main factor that determines the way cells in the blood vessels function. This small molecule is able to relax blood vessels quickly and reduce blood pressure. Nitric oxide is a highly reactive free radical that has an unpaired electron. However, it does not cause harm to the body. It is a member of a group known as reactive oxygen species. Most of these free radicals are harmful to health and can destroy the nitric oxide when they react with it.

> **What It Is**
>
> **Reactive oxygen species** are chemical species with at least one oxygen atom. These include free radicals such as superoxide anion and hydroxyl radical which have unpaired electrons. Oxygen, water, and hydrogen peroxide are typically combined to form reactive oxygen species which are highly reactive. These species play an important role in many biological processes as they can be used to turn on or off various processes.

These chemicals can reduce the effectiveness of nitric oxide when there are too many reactive oxygen species present in blood vessel walls. This can occur in conditions such as high blood cholesterol or diabetes. The relaxation of the endothelial cell layer on the inner surface of the arteries can be disrupted. This leads to a condition known as endothelial dysfunction, which increases the risk for heart disease. This dysfunction can be caused by a variety of factors, such as inflammation in blood vessel cells or the death of specific cell types.

The composition of lipoproteins also affects their retention in arterial walls. Lipoprotein composition can influence their ability to bind to arterial wall proteoglycans and thus become retained in the arterial walls. The structure of the arterial walls, such as the presence of extracellular matrix and proteoglycans, can also influence lipoprotein retention.

The next step involves the recruitment of immune cell types such as macrophages and monocytes to the injury site in a protective response. These cells absorb and digest oxidized lipid particles, resulting in the formation of so-called foam cells. Atherosclerotic lesions are characterized by a constant inflammatory response within the walls of the arteries. As the process continues, immune cells continue to be recruited and release cytokines – inflammatory molecules which further intensify the inflammation. The chronic inflammation in the arterial wall is caused by a continuous cycle of lipid buildup, immune cell recruitment, and cytokine production. This chronic inflammation can lead to plaques forming within the arterial walls, which may rupture over time and cause a heart attack or stroke. This is followed by the deposit of blood cells and protein on the damaged inner surface. The protective process is similar to that of wound healing and is designed to limit blood flow but can block the artery, causing arterial occlusion. Elevated blood pressure causes the rupture of tissues around it, potentially leading to death.

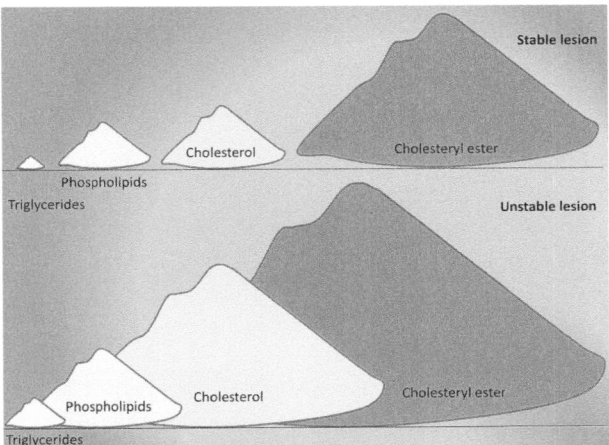

Lipids that build up atherosclerotic lesion. The accumulation of all lipid classes occurs in parallel to the progression of lesions from stable form to unstable. Cholesterol in its free and esterified forms dominates the lipids. It accounts for 80% of lipid weight.

Foam cells release cytokines, chemokines, and other substances that draw more immune cells toward the injury site. Foam cells contribute to the growth of plaques and are a characteristic feature of atherosclerotic lesion. The cells can slowly lose their ability to digest harmful particles and sometimes die, causing lipids to build up outside of them. Eventually, these deposits can turn into cholesterol crystals. Smooth muscle cells proliferate as the lesion advances. They migrate from the media to the intima layers. They produce extracellular matrix protein such as collagen or elastin which forms a fibrous layer over the lipid-rich core. This fibrous cap prevents rupture of the lesion. If the inflammation persists, however, the lesion can become unstable and more prone to rupture.

The fibrous cap may also become weakened by calcium deposits. Plaque can rupture or erode as it grows. This exposes its contents to blood components like platelets and clotting factors. Platelets clump together at the site and release clotting factors that encourage fibrin deposition. This leads to a series of events which result in thrombosis, or the formation of blood clots. These can cause further tissue damage.

Platelets are involved in the rupture and development of artery blockages. Blood clot formation in the arteries can be affected by substances like tissue factors that promote clotting and substances like tissue factor pathway inhibitor and tissue plasminogen activator that prevent clotting. When a blockage breaks, platelets form a clot of fibrin and platelets.

Atherosclerotic lesions can become more unstable during infection. Atherosclerosis and infection are linked by inflammation, blood clotting, and problems with endothelial cells that line blood vessels. The molecule lipopolysaccharide, which is present in some bacteria and makes them dangerous, can be released when bacteria multiply or split apart, causing an inflammatory response.

Atherosclerotic lesions are complex structures involving multiple cell types, molecular pathways, and other factors. The main components of the lesions include lipids and immune cells. They also contain extracellular matrix proteins and calcium deposits. Lipids of atherosclerotic lesions are mainly cholesterol and phospholipids.

Cholesterol, a major component of atherosclerotic lesions, is derived primarily from lipoproteins. Phospholipids play a vital role in the structural integrity of cell membranes. In atherosclerotic lesion, reactive oxygen species produced by immune cells like macrophages or neutrophils can oxidize phospholipids. Oxidized phospholipids can activate inflammation pathways and promote further accumulation of lipids. Atherosclerotic plaques also contain minor lipid species, such as sphingomyelin and ceramides. These lipids may also contribute to inflammation and plaque progression.

Site Specificity

Atherosclerosis affects thick-walled, elastic arteries. The thick walls allow them to withstand high blood pressure from the heart. Atherosclerosis affects not all arteries in the same way and is location specific. The disease alters certain arteries more often than others. Atherosclerosis can cause chest pain, also known as angina, or a heart attack when coronary arteries are damaged. Atherosclerosis may also affect other arteries, including those in the legs and the carotid artery that leads to the brain. The disease can cause different symptoms and complications depending on the arteries affected.

Atherosclerosis affects specific arteries.

Shear Stress

Atherosclerosis is influenced by shear stresses within the arteries. Shear stress is frictional force acting on the walls of blood vessels caused by the blood flow. The friction between the blood and endothelial lining of the arterial walls causes this force to be perpendicular in relation to the direction that blood flows.

Shear stress plays a vital role in maintaining the health of coronary arteries. It helps maintain the integrity and function of the endothelial layer, regulates blood flow, promotes vessel relaxation, helps prevent plaque accumulation, and encourages endothelial growth and repair. However, shear stress that is abnormal can lead to atherosclerosis and cardiovascular diseases.

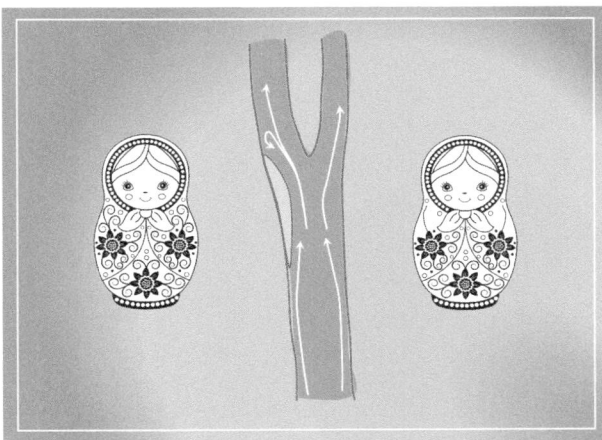

The point where a common carotid artery splits into the internal and external carotid arteries has a slow and turbulent movement of blood. This split also has a low shear stress that causes lipoproteins to penetrate the arterial wall. As a result, atherosclerotic lesions form in the arterial wall on the side opposite to where the artery splits. The turbulent flow of blood contributes to the development of atherosclerosis. The low shear stress also causes a temporary change in blood flow during each heartbeat.

Shear stress is affected by the shape of the arteries. Straight arteries have a higher shear stress than curved arteries. This force can be reduced in coronary arteries because they are highly curved.

Shear stress is primarily a problem for endothelial cells lining the inner vessel surface. Endothelial cells play an important role in maintaining the health and function of coronary vessels. Shear stress can lead to endothelial cell deformation and stretching, resulting in inflammation and oxidative stress. This can cause damage to the arterial wall and an accumulation of cholesterol and other substances. These processes promote inflammation and increase the likelihood of plaque formation.

Open Question

Can shear stress at specific sites be modified to stop atherosclerosis?

Atherosclerotic lesions are more likely to form in areas with low or disturbed shear stress than high shear stress. Low shear stress can cause endothelial cell stagnation and debris accumulation. This can lead to plaque formation and inflammation. Maintaining an optimal level of shear stress ensures proper coronary artery function.

History

Description of Cholesterol-Containing Arterial Lesions

Who	When
Rudolf Virchow	**1857**

Where	What
Berlin, Germany	Virchow, R. (1858) Die Cellularpathologie in ihrer Begrundung auf physiologische und pathologische Gewebelehre. Berlin: Verlag von August Hirschwald.

Albrecht von Haller, a Swedish anatomist, used the term "atheroma" to describe vascular lesions in 1755. The term is derived from the Greek word "atheroma," which means "tumor full of matter that resemble gruel," from Greek "athēra" for "gruel." Jean-Friedrich Lobstein was a French pathologist who proposed the term "arteriosclerosis," to describe these lesions, in 1833. Felix Marchand, the German pathologist, proposed changing this term to "atherosclerosis." It took over 50 years before the World Health Organization adopted it in 1957.

Rudolf Virchow was a German pathologist who demonstrated in 1850 that the atherosclerotic vascular lesion contained cholesterol. He conducted a thorough biological study on human atherosclerosis and distinguished the major phases of the disease. In the middle of the 19th century, his pathological research led him to develop the inflammatory theory for atherosclerosis. Virchow demonstrated that the lesion was located in the inner layer of the arterial walls. He said that the lesion begins by absorbing blood components and then grows through proliferation. This process was similar to an inflammatory response, which led Virchow into thinking that the atheroma is a localized inflammation of the inner arterial wall.

This concept was in opposition to the thrombogenic atherosclerosis theory, which Karl von Rokitansky advanced, an Austrian pathologist who described the deposition of blood proteins on the surface of the arterial endothelium. Both thrombogenic and inflammatory pathways are now widely accepted as playing key roles in the atherogenesis. Both theories have become classics.

The Virchow theory has been helpful in understanding how atherosclerosis occurs, but it was a long while before the pathological mechanisms were fully understood. Virchow's ideas were largely forgotten for many years because scientists thought that atherosclerosis is caused only by cholesterol accumulation. Russell Ross and John Glomset, in the 1970s, proposed their response-to-injury hypothesis which put inflammation back on the stage.

Cardiovascular Disease

Atherosclerosis is not a topic that most people are familiar with, but everyone knows about myocardial infarction and stroke. Atherosclerosis can be clinically silent, but it manifests as cardiovascular diseases, most commonly a heart attack. When atherosclerosis is not treated properly, it can lead to a heart attack and stroke.

A group of diseases that affect the heart and the blood vessels is called cardiovascular disease. There are several types of them. Coronary artery diseases are the most common cardiovascular disease. They occur when the arteries supplying blood to the heart narrow or block, causing chest pain or heart attack. Myocardial infarction is usually the cause of this condition, which causes localized damage to the heart muscle as a result of a ruptured coronary blood vessel. The stroke, or cerebrovascular disease, is caused by a rupture or blockage of a brain blood vessel, causing damage or death to the surrounding brain cells.

Heart failure, arrhythmia, aortic aneurism, peripheral artery disease, arteritis, rheumatic heart disease, congenital heart conditions and deep vein thrombosis are also common cardiovascular diseases. Heart failure is caused when the heart cannot pump enough blood for the body. This can lead to fatigue, shortness in breath, and swelling of the legs. An arrhythmia is an abnormal heartbeat that can be fast, slow, or irregular. Aortic aneurysms are bulges in the aorta wall that can rupture, causing life-threatening bleeding. A buildup of atherosclerotic plaque called peripheral artery disease can cause pain and numbness to occur in the arms and legs. Arteritis, a medical condition that is marked by considerable inflammation of the blood vessels, can affect any artery in the body but most commonly the large and medium arteries found in the neck, arms, and head. The rheumatic heart disease, which is caused by streptococcal bacteria, can cause damage to heart muscles and valves. Birth defects can affect normal heart development and function. Blood clots that can move from the leg veins to the heart or lungs cause deep vein thrombosis.

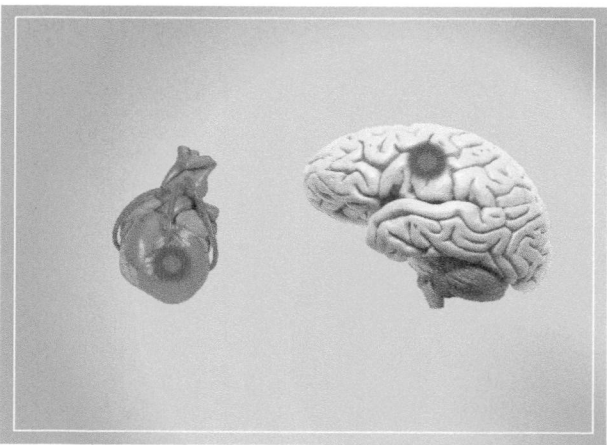

Myocardial infarction (left) and stroke (right) are both caused by an occlusion or rupture of a blood vessel. In myocardial infarction, one of the coronary arteries that supplies blood to the heart is damaged. In stroke, damage occurs to the cerebral arteries.

Atherosclerosis is intimately related to most cardiovascular diseases. The formation of atherosclerotic lesion leads to the narrowing of arteries, which can lead to arterial occlusion. Untreated ischemia can lead to tissue damage or even death. It can affect any part of the human body including the brain, heart, limbs, and organs. This condition is commonly associated with peripheral artery disease, stroke, and coronary artery disorders.

What It Is

Ischemia occurs when blood flow is reduced to a certain part of the body. This is usually caused by a narrowing or blocking of an artery. Under ischemia, organs and cells do not get enough oxygen and nutrients, which can cause them to not function properly.

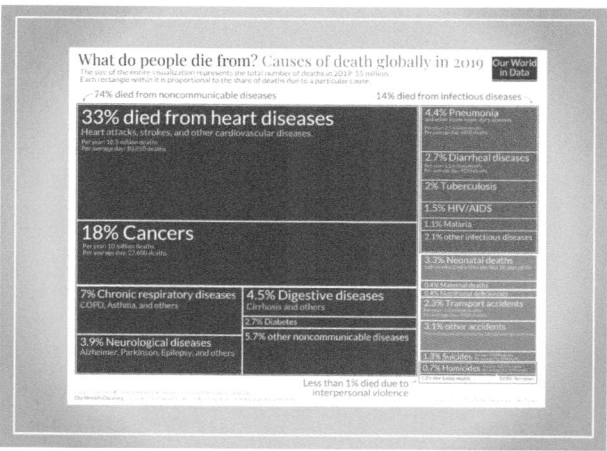

Cardiovascular disease is a leading cause of death around the world, accounting for about one-third of all deaths in 2019. Source: Our World in Data/CC-BY 4.0.

Cardiovascular Mortality

Why are we so concerned about cardiovascular diseases? The answer is straightforward: they are the biggest killers in recent human history. In the 30 years between 1990 and 2019, almost half a billion people have died from cardiovascular causes. This number is greater than one billion if we look at a longer time frame from 1950 to 2020. In comparison, World War II saw up to 60 million casualties, which is more than 10 times lower. This extraordinary number of deaths can only be beaten by death from infectious diseases which killed a majority of humans over the overwhelming part of history until the invention of antibiotics. Nothing however compares in this regard to cardiovascular causes in the second half of the 20th and early 21th century.

> **Interesting Numbers**
>
> In the 30 years between 1990 and 2019, 448 million people worldwide died of cardiovascular causes.

The three leading causes of death in the world are cardiovascular diseases, oncological diseases, and respiratory diseases.

At the moment, cardiovascular diseases are the leading cause of death in the world, with approximately 18.5 million deaths per year. The high mortality rate (which is calculated as the number of deaths per a certain number of people) of cardiovascular diseases remains a major concern for public health worldwide. According to the Institute for Health Metrics and Evaluation (IHME), cardiovascular diseases are responsible for almost 33% of all deaths in the world. This share is impressive, as it exceeds the combined shares of oncological, respiratory and digestive diseases, which are the next three on the list.

> **Interesting Numbers**
>
> In the United States, someone dies from cardiovascular causes every 34 seconds. Globally, this happens less than every two seconds, meaning that five people have died from cardiovascular causes while you read this text.

This means that in the race to reduce world population, cardiovascular diseases are followed by oncological and respiratory disorders.

Future fairy tales.

In 2017, there were an estimated 8.9 million fatal heart attacks and 6.2 million fatal strokes. By 2020, it was estimated that 523 million people had some form of cardiovascular disease. With the world population estimated at 7.82 billion in 2020, approximately 6.7% of the world was living with cardiovascular disease. This places significant strain on national health systems.

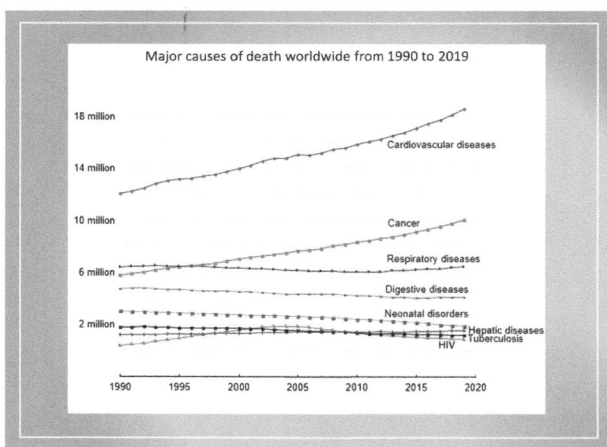

The mortality rate due to cardiovascular disease and cancer has increased rapidly over the past 30 years, reaching respectively 18 and 10 million deaths per year. Source: Our World in Data/CC-BY 4.0.

As the population grows, so does the number of deaths. The cardiovascular mortality rate is steadily increasing, and the same trend can be observed with major cancer types. The increasing cardiovascular mortality is primarily due to the growing number of elderly people around the world. Cardiovascular diseases affect all age groups but are more prevalent in older adults. It is no surprise that older adults accounted for over half of all cardiovascular deaths in both 1990 and 2019. Calculations show, however, that the age-standardized cardiovascular mortality worldwide (which is a mortality of a specified age group) decreased by 14.5% from 2006 to 2016. The data also highlight the role played by the ageing population on the rise in total mortality.

The increase in mortality for respiratory and liver disorders was also seen, but the increases were less dramatic. In contrast, mortality from digestive disorders, neonatal illnesses, and tuberculosis has decreased in the past decades.

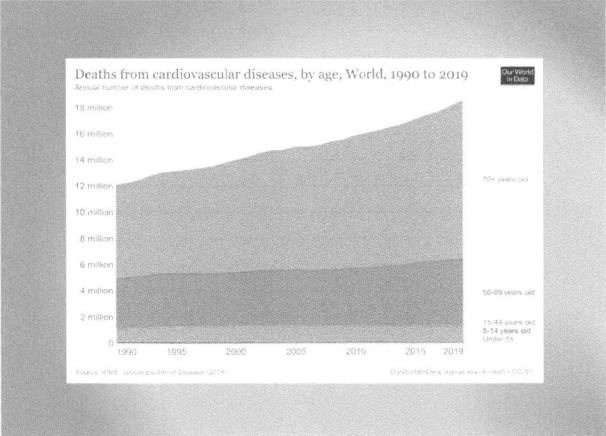

Cardiovascular disease is a major concern for people older than 70 years. Source: Our World in Data/CC-BY 4.0.

Many cardiovascular diseases are responsible for death. Mortality associated with different types of cardiovascular disease is very different. The myocardial infarction, which accounts for a little more than half of all cardiovascular death worldwide, is the most deadly type of this disease. Stroke is ranked second in this grim list, accounting for more than a third of all cardiovascular deaths. The two most dangerous cardiovascular disorders together are responsible for over four-fifths of all cardiovascular mortality. The other types, which include hypertensive and cardiomyopathic heart diseases, atrial fibrillation, rheumatic disease, and cardiomyopathy, are much less dangerous, in terms of their contribution to mortality.

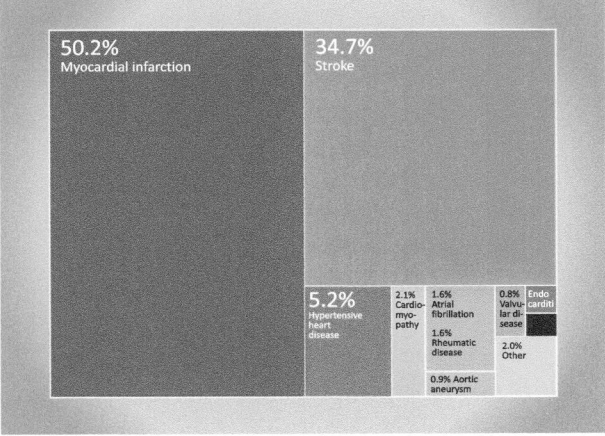

A number of diseases can cause cardiovascular mortality. In 2017, the combined deaths from myocardial infarction (8,930,400 cases) and stroke (6,167,300 cases) accounted for over 80% of all 17,790,900 cases worldwide. Source: Adapter from Our World in Data.

Economic Costs

According to the World Heart Federation, the cost of cardiovascular diseases in the world was 863 billion US dollars in 2010. This is expected to increase to over 1 trillion US dollars by 2030.

Heart disease and stroke are also costly in the United States. The healthcare system spends 216 billion US dollars a year on heart diseases and strokes, and 147 billion US dollars is lost in productivity. Costs related to cardiovascular disease in the European Union increased from 106 billion euro in 2009 to over 210 billion euro in 2017. The burden of cardiovascular diseases is 53% attributed to healthcare costs directly, 26% attributed to productivity losses at work, and 21% ascribed to informal care.

Cardiovascular disease is disproportionately more prevalent in low- and middle-income countries than high-income ones. In fact, over 80% of all cardiovascular deaths occur in countries with low and middle incomes. In sub-Saharan Africa, for example, half of all cardiovascular deaths are in the age range 30–69, 10 years before high-income countries. Total economic losses due to cardiovascular disease in low- and middle-income countries were estimated at $3.7 trillion from 2011 to 2015. This staggering number is approximately half of the total economic burden caused by noncommunicable diseases and represents 2% in gross domestic product for low- and medium-income countries.

Risk Factors

The vast majority of those who suffer from cardiovascular disease are predisposed to it by a variety of factors. These people are at elevated cardiovascular risk, meaning they have a higher chance of getting heart disease or having a stroke or heart attack. Cardiovascular risk factors are conditions that increase the likelihood of cardiovascular disease. High blood cholesterol, smoking, obesity, and diabetes are all risk factors. These factors can increase inflammation and oxidative stress in the arterial walls, resulting in an accelerated growth of atherosclerotic plaques.

Our arteries are at the mercy of cardiovascular risk factors. These risk factors were discovered through large-scale epidemiological research conducted over the past 70 years. Framingham Heart Study, Seven Countries Study, and the Effect of Potentially Modifiable Risk Factors Associated with Myocardial Infarction (INTERHEART) Study are the most influential and important studies of this type.

Practical Aspects: How to Perform a Large-Scale Epidemiological Study

An epidemiological study looks at how health issues are spread and what causes them in specific groups of people. This helps us figure out how to control health problems. There are three main types of epidemiological studies: cross-sectional, case–control, and prospective cohort studies. Cross-sectional studies look at both exposures and outcomes in a population at one time. Case–control studies look at the exposure of interest after the outcome has already happened. Prospective cohort studies follow a group of people over time to see what happens to them. A large-scale prospective study recruits a large study population and evaluates them at the beginning to determine if they have a risk factor for a specific disease. The population is then followed over a long period of time to track cases of the disease and determine if they occur more frequently in subjects with or without the risk factor at the start of the study. In the example shown above, having the risk factor at the baseline doubled the incidence of the disease from 25% to 50%.

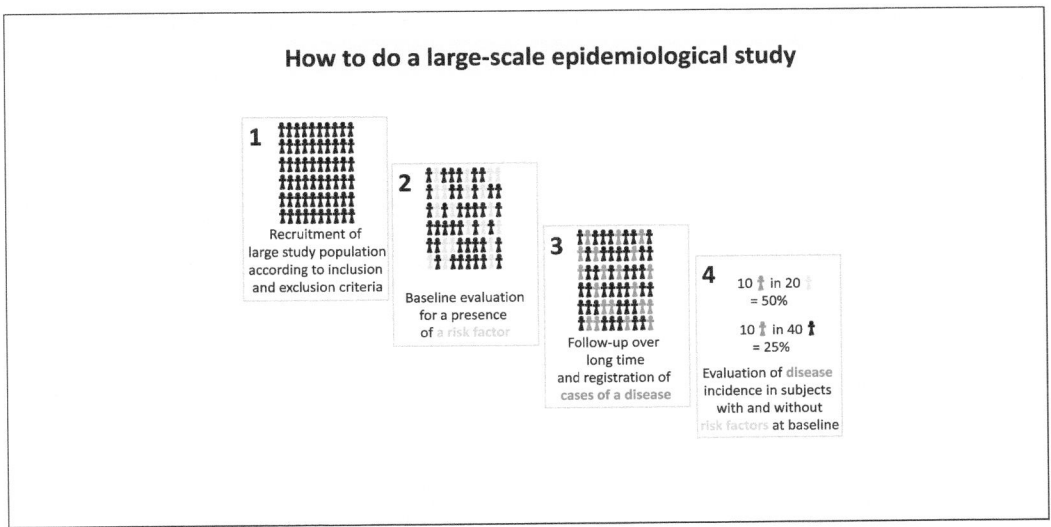

Cardiovascular diseases are caused by a number of risk factors. Some of these factors can be changed, but others are not. Lifestyle choices like smoking, inactivity, poor diet, and excessive alcohol are all modifiable factors. The most important modifiable cardiovascular disease risk factor is smoking. Smoking damages the linings of blood vessels and increases heart rate and blood pressure. It also reduces the oxygen supply to the body. Inactivity and a poor diet can lead to obesity, high blood cholesterol, high blood pressure, and hypertension. All of these factors increase the risk for cardiovascular disease. Alcohol consumption can have a negative impact on plasma lipids.

Age, gender, ethnicity, and family history are all non-modifiable cardiovascular risk factors. The risk of cardiovascular disease increases as people age. Men are more susceptible than women to develop cardiovascular diseases. The risk is also increased by a family history of cardiovascular diseases. Some ethnic groups, such as African Americans or Hispanics, have a greater incidence of cardiovascular diseases than others. The minimal cardiovascular risk is therefore only associated with age and gender.

Understanding both modifiable and non-modifiable risks is important for the prevention of heart disease. Modifiable risk factors can be controlled or changed through lifestyle changes, such as quitting tobacco, exercising regularly, eating healthy, maintaining a healthy body weight, and reducing alcohol intake. These factors, when properly addressed, can prevent heart attacks and strokes and delay or even stop the progression of atherosclerosis.

Open Question

How do prevention measures differ in their capacity to reduce cardiovascular disease?

It is important to note that non-modifiable risk factors (such as genetics and family history) can be not only detrimental but also beneficial. In fact, genetic inheritance can protect some people from heart disease.

Risk factors for cardiovascular disease are often referred to as silent killers. The main risk factors include age, smoking, hypertension, diabetes, dyslipidemia, and male sex.

Age

The main factor that affects cardiovascular risk is age. The cardiovascular risk of older people increases for several reasons. Atherosclerotic lesions are the main cause. As we age, our arteries become more susceptible to damage. It can cause plaques to accumulate in the arteries. This can reduce blood flow and narrow the arteries.

Blood pressure also increases with age. The structure and function of blood vessels change with age, which causes the blood pressure to rise. High blood pressure damages the arteries which become less elastic, increasing the risk of heart disease.

Age is also associated with changes in cholesterol levels. As people age, their cholesterol levels tend to be upward. High cholesterol levels can lead to atherosclerotic plaques in the arteries. Physical activity decreases as we age, which can cause weight gain, high blood pressure, and other cardiovascular risk factors.

The age is also linked to an increased risk for other health conditions, including diabetes, obesity, and kidney disease that can increase cardiovascular risk.

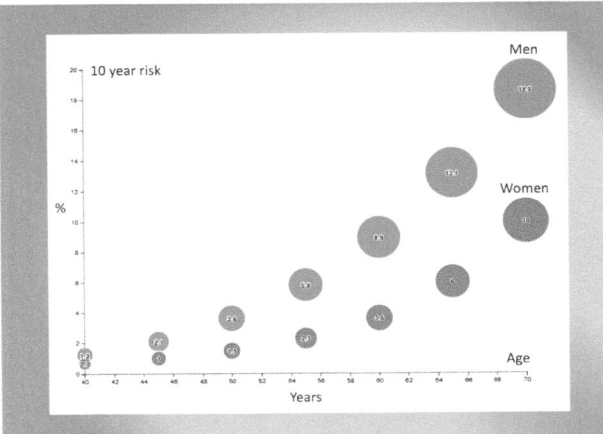

With age, the risk of cardiovascular disease increases. Risk is strongly influenced by sex, and men are at a higher risk than women. After 65, the difference between men and women's risk becomes smaller. According to the American College of Cardiology's (ACC) Risk Calculator, the lifetime risk of developing this disease goes from 39% for a 40-year-old woman to 46% for a 40-year-old man without risk factors. The size of the circle is proportional to risk at any given age. Source: ASCVD Risk Calculator of the American College of Cardiology at https://tools.acc.org/ascvd-risk-estimator-plus/#!/calculate/estimate/.

Gender

After menopause, women's risk of cardiovascular disease increases. This is due to hormonal differences between women and men, with women having higher levels of estrogen. Estrogen has a protective effect on the cardiovascular system, keeping blood vessels flexible and helping to prevent the buildup of plaques.

In contrast, testosterone, which is found in greater amounts in men, increases the risk of cardiovascular disease by promoting the formation of atherosclerotic plaques. In addition, men tend to have higher cholesterol levels than women. Smoking, poor nutrition, lack of physical activity, and stress are all modifiable lifestyle factors that increase the risk of heart disease in both men and women. These factors can cause high blood pressure and high cholesterol, as well as obesity, diabetes, or other conditions that increase heart disease risk.

Hypertension 10,800,000	Dyslipidemia 3,810,000	Air pollution 3,130,000	Smoking 2,370,000
	Diabetes 2,300,000	Household air pollution 1,610,000	Lead exposure 1,570,000
Dietary habits 6,580,000	Obesity 1,950,000	Cold 1,020,000	Alcohol 407,000
	Kidney dysfunction 1,870,000	Passive smoking 743,000	Low physical activity 397,000 / Heat

The most common modifiable cardiovascular risk factor is hypertension, followed by unhealthy eating habits, dyslipidemia, air pollution and smoking. The numbers are the number of deaths due to a risk factor around the world in 2021.

Smoking

Smoking is one of the major causes of cardiovascular disease. Tobacco smoke contains harmful chemicals that damage the linings of the arteries and cause them to narrow and become stiff. It reduces the blood flow to organs and the heart, which increases the risk of stroke, heart attack, and peripheral arterial disease. Smoking increases carbon monoxide levels in the blood which decreases oxygen transport by red blood cells. Smoking can put additional pressure on the heart, resulting in angina or heart failure.

Smoking also raises blood pressure, increasing the risk of blood clots in the arteries. These clots can block blood flow to vital organs like the heart and the brain. The good news is that quitting smoking lowers the risk of cardiovascular disease. After quitting, blood circulation improves and blood pressure drops, reducing the risk of stroke and heart attack. Quitting smoking is especially beneficial for those with cardiovascular disease.

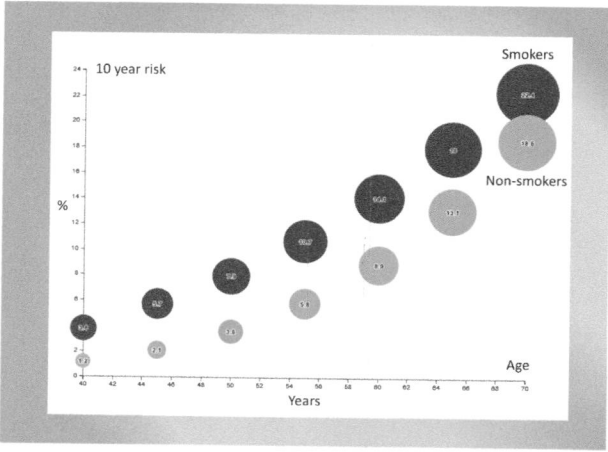

Smokers are at a higher risk for cardiovascular disease than nonsmokers. This difference is more pronounced in younger age, when age is not yet a significant cardiovascular risk factor. According to the ACC Risk Calculator, the lifetime risk of developing this disease goes from 46% for a male 40-year-old non-smoker up to 50% for a male 40-year-old smoker without other risk factors. The size of the circle is proportional to risk at any given age.

Hypertension

High blood pressure or hypertension is a significant risk factor for cardiovascular diseases. Around one-third of adults in the United States suffer from hypertension. This number is on the rise worldwide. Consistently high blood pressure can damage arterial walls and cause atherosclerosis. This can increase the risk of heart attacks, strokes, and other cardiovascular conditions. Hypertension can cause the heart pumping blood to the rest of the body to work harder. This can cause an enlarged and less-functional heart over time. It increases the risk for heart failure. Hypertension has direct effects on the cardiovascular system, but it is also linked to other cardiovascular risk factors such as diabetes, obesity, and high cholesterol.

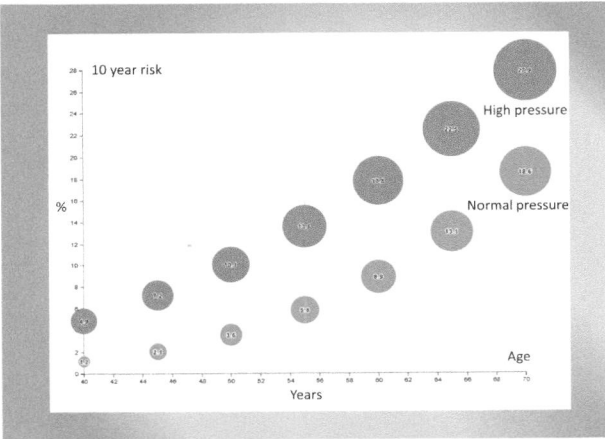

High blood pressure increases the risk of cardiovascular disease. The risk increase is higher in younger age. According to the ACC Risk Calculator, the lifetime risk of developing this disease increases from 46% for a 40-year-old man without hypertension up to 69% for a 40-year-old man with systolic blood pressure of at least 150 mmHg and no other risk factors. The size of the circle is proportional to risk at any given age.

Diabetes

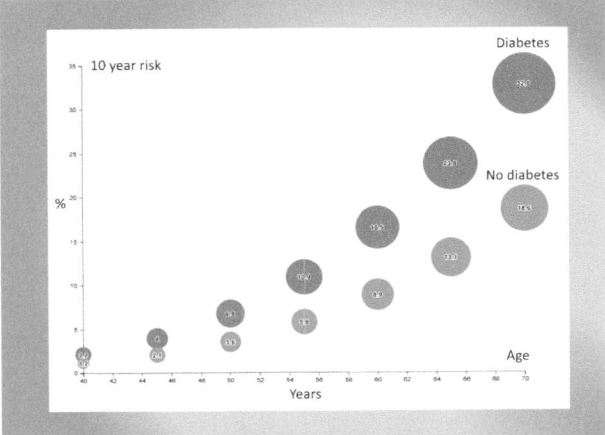

Diabetes increases the risk of cardiovascular disease. According to the ACC Risk Calculator, the lifetime risk of developing cardiovascular disease goes from 46% for a 40-year-old man without diabetes to 50% for a 40-year-old man with diabetes and without any other risk factors. The size of the circle is proportional to risk at any given age.

Diabetes is a chronic metabolic disease that affects the body's ability of producing or using insulin, the hormone that regulates blood sugar level. Diabetes is a leading risk factor for cardiovascular diseases. Diabetes increases cardiovascular risk by damaging blood vessel walls and increasing inflammation. Insulin resistance is the condition that precedes and causes diabetes. It occurs when the cells of the muscles, the adipose tissues, and the liver are not responsive to insulin. They cannot readily absorb glucose from the blood in response to insulin, which increases blood glucose. Insulin resistance is often associated with abnormal pancreatic function, where the beta-cells do not secrete enough insulin.

High glucose levels in the blood may cause damage to blood vessel linings, increasing their susceptibility to plaque accumulation and narrowing. Diabetes can also worsen hypertension. Hypertension increases the risk of stroke and heart attack by putting extra pressure on the heart and blood vessels. Diabetic patients are more likely to experience abnormal cholesterol levels. This can lead to the buildup of plaque in the arteries, increasing the risk of heart disease.

Dyslipidemia

Dyslipidemia, or abnormal levels of lipids in the blood, is a medical condition. This condition is one of the major risk factors for cardiovascular diseases such as coronary artery disease, stroke, and peripheral artery disease. A high level of blood lipids is responsible for at least half of all heart-related deaths. Dyslipidemia plays a multifaceted role in cardiovascular disease. High cholesterol levels can cause plaques to form in the walls of the arteries. This can reduce blood flow and narrow the arteries. These plaques may also rupture and lead to blood clots, which can completely block blood flow and cause a stroke or heart attack.

Triglycerides can also contribute to cardiovascular diseases. High levels of triglycerides can increase the risk of metabolic syndrome. This is a group of conditions including high blood pressure, overweight, and insulin resistance. The metabolic syndrome is one of the major risk factors for cardiovascular disease.

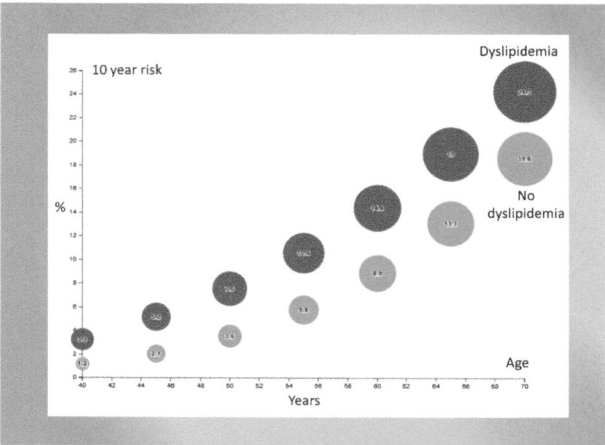

The risk of cardiovascular disease is higher in people with high cholesterol levels in the blood. According to the ACC Risk Calculator, the lifetime risk of developing cardiovascular disease increases from 46% for a 40-year-old man with LDL-cholesterol of 120 mg/dl up to 50% for a 40-year-old man with LDL-cholesterol of 190 mg/dl. The size of the circle is proportional to risk at any given age.

Dyslipidemia is detrimental to heart health. This is illustrated by the relationship between the concentration of cholesterol in the blood and the severity of atherosclerotic lesions. The more cholesterol there is, the worse the condition is. This relationship is interestingly curvilinear. Individuals with moderately high cholesterol levels may not necessarily have more severe lesions. For severe atherosclerosis and coronary events to occur, high cholesterol levels are needed. This observation is explained by an elevated risk that occurs only at a critical lesion stage.

Nutrition

The role of nutrition in the development of cardiovascular disease is crucial. Diets high in sodium, sugars, and animal fats (saturated or trans), as well as processed foods, can increase the risk of heart disease. These foods can increase cholesterol levels and blood pressure, while also promoting inflammation. A healthy diet, on the other hand, can reduce the risk of heart disease. It does this by lowering the blood pressure, cholesterol, and inflammation levels in the body. Diets rich in fruits, vegetables, and whole grains as well as lean protein, omega-3 fatty acids, and healthy fats can improve heart health. These foods contain antioxidants, fiber, vitamins, and minerals that help protect the heart and blood vessels. By contrast, omega-6 fatty acids, which are very similar to omega-3, can aggravate inflammation and heart disease. It is therefore important to maintain a balance between omega-3 fatty acids and omega-6 fatty acids.

What It Is

Omega-3 and omega-6 are types of polyunsaturated fatty acids – those that contain more than one double bond. The main difference between omega-3 and omega-6 is where the first double bond happens in the molecule. In omega-3 fatty acids, the first double bond is on the third carbon atom, while in omega-6, it is on the sixth carbon atom from the end.

Foods that are ultra-processed have been through several manufacturing stages. These foods are high in sugar and salt, and they may also contain preservatives and additives. These foods lack fibers and nutrients found in fruits, vegetables, plain yogurt, and homemade bread. In recent years, the global consumption of highly processed foods such as cereals and protein bars, fizzy beverages, ready meals, and fast food has increased. The average American and British diet is now made up of more than half ultra-processed foods.

Interesting Numbers

Some groups of people, particularly those who are young, poorer, or come from disadvantaged areas, have a diet that contains up to 80% ultra-processed foods.

New research is adding to the growing evidence of harmful effects caused by ultra-processed foods. Recent studies should be a wake-up call that ultra-processed foods are associated with a higher risk of heart disease, stroke, high blood pressure, and heart attack. In one study, an increase of 10% in daily calorie consumption of ultra-processed foods was linked to a 6% higher risk of heart disease. According to research conducted in China, those who consume ultra-processed foods at a rate of less than 15% are least likely to suffer from heart disease.

A second study, which tracked 10,000 women over 15 years, found that those who consumed the most ultra-processed foods were 39% more susceptible to high blood pressure. In addition, the gold-standard meta-analysis of over 325,000 men, women, and children showed that those who consumed ultra-processed foods were 24% more at risk for cardiovascular events such as heart attacks, strokes, and angina.

In striking contrast, nutritional habits such as the Mediterranean diet typical for Italy and Greece are linked to a reduced risk of heart disease and mortality from all causes. This diet is popular in the islands of the Mediterranean Sea and in Japan. This diet is known for its high amounts of fruit, vegetables, and fish. It also includes moderate alcohol consumption and lots of olive oil.

Interesting Numbers

Okinawa, Nicoya, and Sardinia have very low rates of heart disease. These islands, located in Japan, Costa Rica, and Italy, respectively, are part of the Blue Zones. This is a group of areas where life expectancy is above average. In Okinawa, a woman 60 years old has a 9.3% non-negligible chance of becoming a centenarian. Nicoya has a probability of 4.3% and Sardinia 1.4%. Comparatively, in Sweden, the chance of a woman aged 60 living to 100 years is just 1.1%. Men's chances are lower, but still better on the islands – 2.1% for Okinawa, 4.8% for Nicoya, and 0.7% for Sardinia. In Sweden, they only have 0.2%.

History

Experimental Induction of Atherosclerosis

Who	When
Nikolai Anitschkow	**1913**

Where	What
St Petersburg, Russia	Anitschkow N. N. (1913) Ueberdie Verenderungen der Kaninchenaorta bei experimentteller Cholesterinesteatose. " Beitr Pathol Anal 56: 379-404.

In 1910, German scientist Adolf Windaus conducted the first chemical analysis of atherosclerotic lesions. His studies revealed that the lesions had much higher cholesterol levels than a normal arterial wall. He found that the lesions contained 6 times as much free cholesterol and 20 times more esterified cholesterol than the normal arterial walls. Most likely, the cholesterol came from blood circulation.

Felix Boudet discovered cholesterol in the serum in 1833. The French chemist isolated a substance by treating an alcoholic serum extract with cold alcohol. It was a substance similar to cholesterol that appeared at the bottom in the form of small crystalline plates. So, scientists began trying to make the disease happen by giving the suspected agent, following Robert Koch's famous postulates. One of the rules said that the agent should cause disease when put into a healthy organism. The first tests like this were done with eggs and rabbits in 1900 by Karl Aschoff in Germany.

In the past few years, animal studies had shown that eating too much meat, milk, or eggs could lead to atherosclerosis. Researchers had also found that atherosclerosis patients had high cholesterol levels in their blood. Nikolai Anitschkow, a Russian scientist, later came up with an idea that too much cholesterol can lead to atherosclerosis. In 1913, he was a pathologist working at the Military Medical Academy of Saint Petersburg. He reported the experimental induction of atherosclerosis in rabbits. The historical context of these experiments was controversial. In the early 20th century, cholesterol was believed to be beneficial for health. Animal extract liquors containing cholesterol were prescribed as a treatment against gastroenteritis and tuberculosis.

In 1908, a Russian physician Alexander Ignatowski, working at the same institution, reported that feeding rabbits a diet consisting of eggs, meat, and full-fat dairy products caused experimental atherosclerosis. Anitschkow, a researcher who followed up on this work, was able to induce atherosclerosis by feeding rabbits cholesterol, or more specifically, oil and fat. This regimen caused hypercholesterolemia, which preceded appearance of atherosclerotic lesions. Foam cells containing cholesterol and lipids accumulated in large quantities in the arterial wall. The more cholesterol the animals consumed and the longer their exposure to it, they became worse.

The work, performed in collaboration with Semen Khalatov, a student at the Military Medical Academy, established the causative link between cholesterol and atherosclerosis using an animal model. It was questioned whether cholesterol is beneficial to human health. This discovery was published in the German journal *Zentralblatt für allgemeine Pathologie und pathologische Anatomie* just before World War I, in 1913.

These studies have shown that there are distinct differences in the atherosclerotic lesions between rabbits and humans. Moreover, similar experiments performed on rats and dogs did

(Continued)

(Continued)

not yield the same results. This led some to doubt that cholesterol is the cause of atherosclerosis. Later, it was discovered that rats and dogs are less affected than other animals by a high-cholesterol diet. It is now known that this reflects distinct lipid metabolic processes in these biological species. It is possible that the susceptibility to dietary cholesterol of rabbits was due to their herbivore diet, which contains very little cholesterol. Carnivorous animals, on the other hand, consume much higher amounts of dietary cholesterol, which is metabolized in a different way.

Anitschkow presented the results of his research at international conferences in the 1920s. At the time, this work was not well known in America or Europe. Before 1950, it appears that the only English-language reference to his atherosclerosis theory was in a 1933 chapter he wrote for the first edition of *Arteriosclerosis* by Edmund Cowdry. After several decades, other scientists began to realize the importance of his research. Anitschkow's early experiments received global attention in 1951 when John Gofman's team published their results on lipoprotein fractionation in *Science*. They highlighted their discovery that giving cholesterol to rabbits caused atherosclerosis. Gofman's lab confirmed Anitschkow was correct by using his method. William Dock, a Stanford University Medical School professor, recognized in 1958 the importance of Anitschkow's theory of atherosclerosis. Dock wrote in *Annals of Internal Medicine* that Anitschkow's work was comparable to Harvey's on circulation and Lavoisier's on the exchange of oxygen and carbon dioxide in the lungs.

The European Atherosclerosis Society (or EAS) established the Anitschkow Award in 2007 to honor Nikolai Anitschkow's contribution to medicine. This prize is awarded to outstanding scientists working in the field of atherosclerosis, metabolic disorders, and related diseases.

Obesity

Obesity can be a significant risk factor in cardiovascular disease. A body mass index (BMI) of at least 30 is considered obese. The risk of cardiovascular disease is increased by obesity in multiple ways. It can cause other risk factors, such as high blood cholesterol, high blood pressure, and diabetes. These conditions can damage blood vessels, increasing the risk of stroke and heart attack. Second, obesity can change the structure and functionality of the heart. When a person becomes obese, the heart must work harder to pump blood throughout the body. This can cause an enlarged heart (also known as cardiomegaly) and an increased risk for heart failure. Third, being overweight can lead to inflammation. Chronic inflammation damages blood vessels, increasing the risk of atherosclerosis. Obesity can also influence the way blood clots are formed. Obese people are more likely than nonobese individuals to have blood which clots easily. It increases the risk of blood clots causing heart attacks and strokes.

Practical Aspects; How to Evaluate Obesity

Obesity can be evaluated using body mass index (BMI). This is a way to see if someone is obese or overweight. It is not perfect, but it can help find people who might get sick because of their weight. BMI is calculated by dividing weight (in kilograms) by height (in meters) squared. If BMI is less than 18.5, this is underweight. A BMI of 18.5 to <25.0 is considered a healthy weight range. A BMI of 25.0 to <30.0 is considered overweight, and a BMI of 30.0 or higher is considered obese. Waist circumference can also be measured to evaluate obesity.

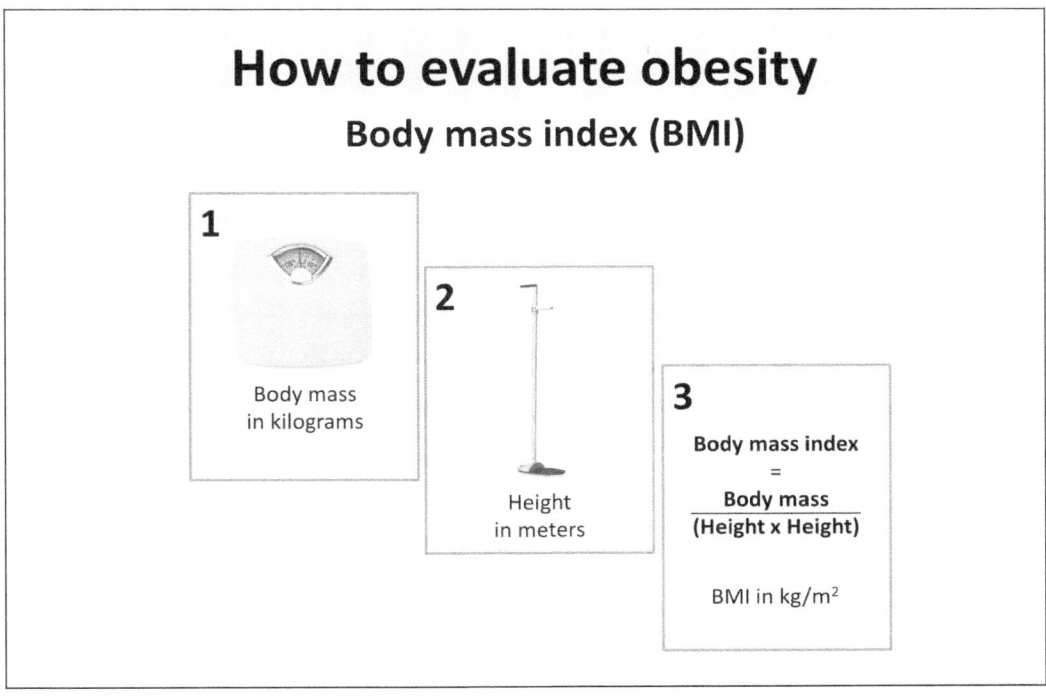

Inactivity

Inactivity is a major factor in the development of cardiovascular disease. Regular exercise improves heart health because it strengthens the heart muscle and reduces blood pressure. It also increases blood flow. Exercise also reduces the appearance of cardiovascular risk factors, such as diabetes, obesity, and high cholesterol. Even moderate physical activity has been shown to have significant cardiovascular benefits. A daily 30-minute jog or walk can lower blood pressure by as much as 30% and reduce heart disease risk. Regular exercise can reduce stress, improve mood, and boost self-esteem. It is striking that the absence of moderate physical activity increases risk in a similar way to smoking.

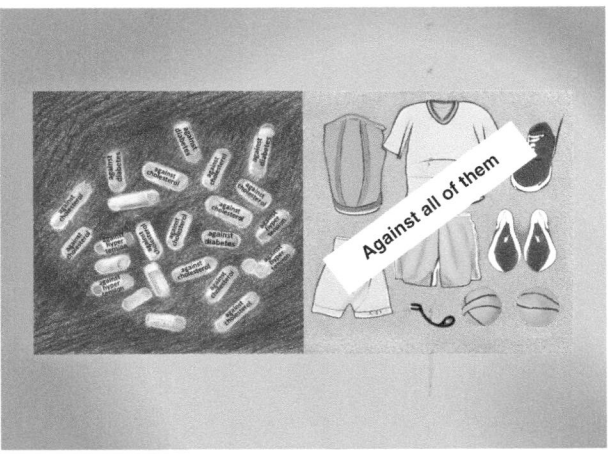

Physical activity is an excellent way to protect against heart disease and stroke.

Social Factors

Psychosocial stress is a major risk factor in cardiovascular disease. Stressful life experiences, including job loss, financial problems, divorce, and bereavement have been linked to an increased risk of cardiovascular disease. Stress can have a variety of effects on the cardiovascular system. Stress can cause an increase in heart rate and blood pressure, which over time can damage the heart and blood vessels. Chronic stress can cause inflammation in the body. This is one of the key factors in cardiovascular disease. Depression and anxiety are also linked to a higher cardiovascular risk. These conditions can lead to unhealthy behaviors like smoking, poor eating habits, and lack of exercise, which increase the risk for heart attacks.

The prevention and management of cardiovascular disease is largely dependent on social development. Social development is the process through which people acquire social skills, values, and knowledge that allow them to effectively interact with others within their community. It includes access to education, healthcare, and social networks. Social development is associated with a lower risk of cardiovascular disease. It is because these individuals are more likely than others to adopt healthy habits such as regular physical activity, healthy eating, and refraining from excessive alcohol and smoking. Social support networks are also a great way to help people manage their stress, which is a risk factor for heart disease. Social support may come from friends, family, or local groups. Access to healthcare is also an important part of social development. Regular health checkups can identify cardiovascular risk factors early and offer appropriate interventions.

A variety of factors have been implicated in the increased risk of heart disease associated with urbanization. People who move to urban areas are exposed to different lifestyles that can negatively impact their health. Air pollution is a major cause of cardiovascular disease. Air pollution is a common problem in cities and can cause inflammation and oxidative damage to the body. The risk of cardiovascular disease can be increased by this. Urban areas are also characterized by sedentary living. In cities, many people spend a lot of time sitting at their desks and in public transport or cars. This lack of exercise can lead to obesity, high blood cholesterol, and other cardiovascular risk factors. Urbanization can also lead to changes in eating habits. When people move to the city from rural areas, they may have less access than before to fresh fruit and vegetables, but more access to processed food that is high in sugar, salt, and unhealthy fats. These changes in diet can increase cardiovascular risk.

Additive Effects

All major risk factors are additive and increase the risk. Insulin resistance can worsen hypertension. Some types of dyslipidemia can cause insulin resistance. Stress can also increase cholesterol levels, and so on. According to the American College of Cardiology's Risk Calculator, a 40-year-old male smoker with hypertension and dyslipidemia, as well as diabetes, has a 45.6% chance of developing cardiovascular disease in the next 10 years. This is a very unfavorable figure compared to the 1.2% risk that a man of the same age would have without any risk factors. In this example, the risk of developing a disease in your lifetime has also increased, but less dramatically. It is now 69%, up from 46%. The age of the person is a major risk factor. Therefore, the risk increases to an incredible 82.2% if the high-risk man reaches the age of 70. Risk factors can act in synergy, enhancing each other.

Open Question

Can specific prevention methods differentially reduce stroke versus heart disease?

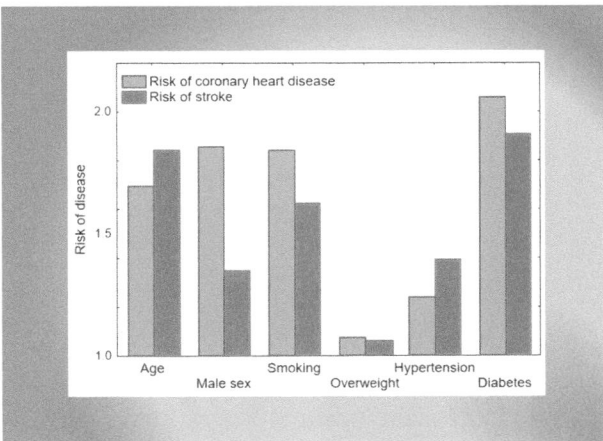

Different risk factors for cardiovascular disease have varying impacts on the likelihood of experiencing a heart attack or stroke. For example, being male and smoking pose higher risks for heart attacks, whereas hypertension is more commonly associated with strokes.

Risk factors differ markedly depending on the type of cardiovascular disease. They are associated in a different way with myocardial infarction and stroke. The risk of heart disease is higher in men, smokers, and diabetics, while hypertension and age are more significant factors for stroke.

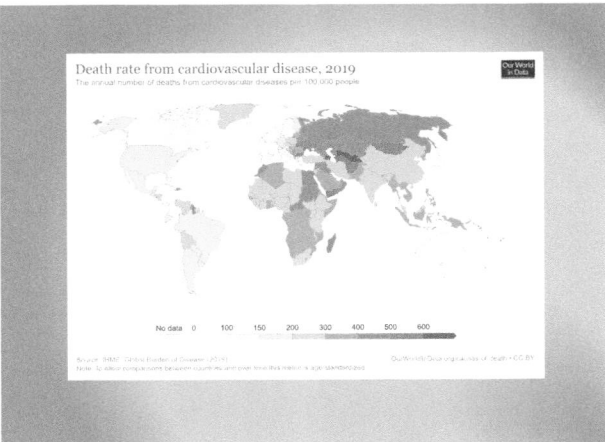

The global distribution of cardiovascular mortality is markedly different. Low- and middle-income nations have a higher rate than high-income ones. Source: Our World in Data/CC-BY 4.0.

Economic Status

In countries with low and middle income, cardiovascular diseases are more common. According to the World Health Organization, over three-quarters of all cardiovascular deaths occur in these countries. Cardiovascular disease is less common in Western countries but has increased in recent years. It occurs more frequently in Eastern Europe, the Middle East, Northern Africa, and Oceania. This negative development is due to the increased incidence of alcohol consumption, smoking, and sedentary lifestyle. In contrast, cardiovascular disease is not as common in Japan, Korea, and France. It it remarkable that main risk factors for heart disease do not however change across different countries. This was revealed by the landmark INTERHEART study which identified the same nine major risk factors for the heart attack across 32 countries.

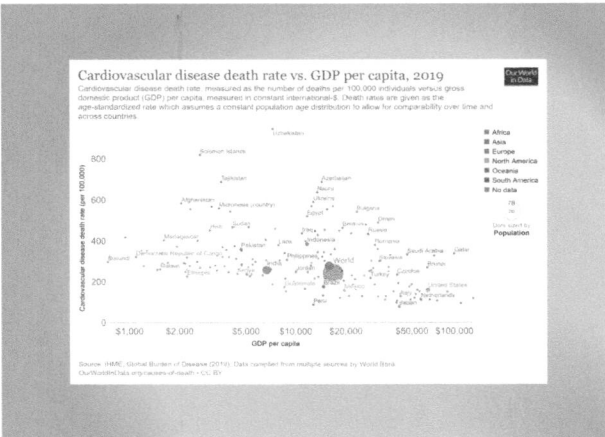

The number of deaths from cardiovascular diseases tends to decline with an increase in gross domestic product per capita. This means that more wealthy countries have fewer people dying of cardiovascular causes. Source: Our World in Data/CC-BY 4.0.

Cardiovascular diseases are frequently associated with poverty in low- and middle-income countries. In these countries, cardiovascular disease risk factors such as smoking, poor diet, inactivity, and obesity are more prevalent. These countries may also have limited resources to prevent and treat cardiovascular disease. High-income countries, on the other hand, tend to have a lower rate of cardiovascular disease due to their better access to healthcare resources. In general, there is an obvious relationship between income and cardiovascular disease. However, some high-income countries face difficulties in reducing the prevalence of cardiovascular disease due to lifestyle factors, such as unhealthy eating habits and sedentary living. Even though the USA's public health crisis improved, the number of deaths due to heart disease increased again in 2020 and remained high in 2022.

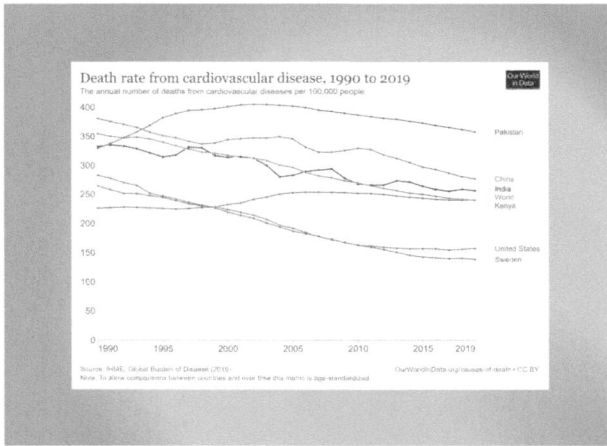

The death rate from cardiovascular diseases has generally decreased over the past three decades in many countries. Source: Our World in Data/CC-BY 4.0.

> **Interesting Numbers**
>
> In the United States, heart disease-related deaths decreased by 8.9% between 2010 and 2019, but increased by 9.3% in 2022. It means that nearly 10 years' progress was lost in reducing the number of deaths. There were also 228,000 deaths more than expected.

People with a high social status have a better chance of achieving good health. This includes fewer cardiovascular diseases. This could be due to many factors including better access to healthcare, higher income and education levels, and more social support. The relationship between social standing and cardiovascular health may be explained by the fact that people with higher status tend to adopt healthier behaviors, such as exercising regularly and eating a balanced diet. Also, they may have better access to preventative health services like regular screenings and checkups. A high social standing can also provide a buffer to the negative effects that stress has on the body. By contrast, people with a lower social standing may be under chronic stress because of financial insecurity and discrimination. This can lead to cardiovascular disease. High social status can also provide more opportunities for social connection and support, which may have positive effects on mental well-being. This can then lead to improved cardiovascular health.

However, fortune and wealth are not a guarantee against cardiovascular disease. The list of famous and wealthy people who have died from cardiovascular causes is long. Winston Churchill is on the list, as are Federico Fellini and Mao Zedong. Dwight Eisenhower is also on it, along with Elvis Presley, Nikita Khrushchev, Stanley Kubrick, Mark Twain, Guglielmo Marconi, Christian Dior, Louis Armstrong and many others.

Invalidity

Cardiovascular diseases, in addition to death, are the leading cause of disability worldwide. They can significantly impact an individual's life quality. Physical limitations can be caused by cardiovascular diseases, such as difficulties with daily activities including walking and climbing stairs. People may also experience emotional distress if they are afraid of having another cardiac episode or need to continue medical treatment. Cardiovascular diseases have a significant impact on the economy, in addition to their personal impact. These conditions can put a burden on the healthcare system and family members.

Noncommunicable diseases like heart attack, stroke, and cancer account for over 60% of disability-adjusted life years globally, 70% of all deaths, and more than 80% of years of disability. Cardiovascular diseases account for 24% of the disability-adjusted years of noncommunicable diseases. Ischemic heart disease and cerebrovascular disorders are two of the major causes of global disability.

> **What It Is**
>
> **Disability-adjusted life years (DALYs)** measure the total impact a disease has on a population. This is done by adding the years lost to premature death and years with a disability.

Global trends in disability-adjusted years of life and years lost have increased dramatically from 1990 to 2019. Years lived with disability also doubled, from 17.7 to 34.4 million. Since 1990, the number of disability-adjusted life years caused by heart disease has steadily increased. In 2019, there were 9.14 million deaths and 197 million prevalent cases of heart diseases. Since 1990, the number of disability-adjusted life years due to stroke has also increased. This total now stands at 143 million disability-adjusted life years. In 2019, there will be 6.55 million deaths and a staggering 101 million cases of stroke.

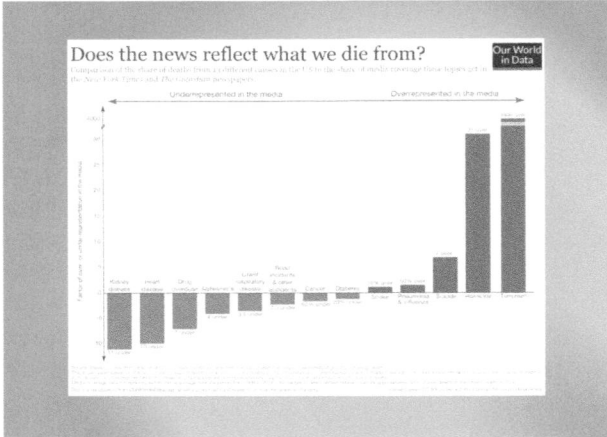

News reports do not accurately reflect the main causes of death. Source: Our World in Data/CC-BY 4.0.

Public Awareness

The public is not fully aware of the dangers of cardiovascular disease despite their critical importance. News does not accurately reflect the causes of death. In fact, the public's attention is largely focused on dramatic causes of deaths such as terrorism and homicide. Unfortunately, news outlets tend to ignore heart disease.

Many people do not realize the dangers these diseases pose to their health, despite the fact that they are prevalent. It is important to raise public awareness about cardiovascular diseases in order to prevent their occurrence and reduce their impact on society and individuals. People can learn how to reduce the risk of these diseases by focusing on factors like smoking, high blood pressure, high cholesterol, diabetes, obesity, and physical inactivity. Community events like health fairs and screenings can raise public awareness in addition to education campaigns. These events offer people the chance to find out about their health and get information on how they can improve it. The healthcare professionals play an important role in raising awareness of cardiovascular diseases. They can give patients information about how to prevent these diseases and manage risk factors.

It is surprising that people are unaware of cardiovascular disease and atherosclerosis, which have been known for centuries. Egyptian mummies have revealed the presence of atherosclerotic lesions. Ancient Egyptian records mention that heart disease was treated using herbal remedies. The Ebers Papyrus (c. 1550 BCE) contains descriptions of treatment for chest pains and other symptoms related to heart disease. Hippocrates, a famous physician in ancient Greece wrote about how diet and exercise can prevent heart disease. He also acknowledged that some people were more susceptible to it.

Indian Ayurvedic traditions also recognize the importance of diet, lifestyle, and exercise in preventing cardiovascular disease. Ayurvedic texts suggest a low-fat diet high in fruits, vegetables, and grains as well as regular physical activity. These ancient civilizations may not have understood atherosclerosis or cardiovascular disease as well as we do, but their knowledge of herbal treatments, diet, and lifestyle factors could have prevented or treated these conditions.

The paintings of the Renaissance period also show signs of abnormal lipid metabolic processes. In fact, the famous Mona Lisa del Giocondo portrait contains lipoma as well as xanthelasma. Lipoproteins are the primary carriers of cholesterol to xanthoma. The appearance of the xanthoma can be viewed as a symptom of high blood cholesterol and dyslipidemia.

What It Is

Lipoma is a benign tumor composed of lipids, primarily triglycerides.
Xanthoma is a benign tumor made primarily of cholesterol. This is a small, yellowish deposit that appears under the skin when blood cholesterol levels are high.

Here is arguably the most famous case of xanthelasma and lipoma in the world as reported by researchers from the University Hospitals in Leuven, Belgium and the University of Palermo in Italy. Source: Dequeker J et al Isr Med Assoc J 2004;6:505, pubmed.ncbi.nlm.nih.gov/15326839/.

Further Reading

Bergmark, B. A., N. Mathenge, P. A. Merlini, M. B. Lawrence-Wright and R. P. Giugliano (2022). "Acute coronary syndromes." Lancet 399(10332): 1347–1358.

Bergström, G., M. Persson, M. Adiels, et al. (2021). "Prevalence of subclinical coronary artery atherosclerosis in the general population." Circulation 144(12): 916–929.

Björkegren, J. L. M. and A. J. Lusis (2022). "Atherosclerosis: Recent developments." Cell 185(10): 1630–1645.

Bos, D., B. Arshi, Q. J. A. van den Bouwhuijsen, M. K. Ikram, M. Selwaness, M. W. Vernooij, M. Kavousi and A. van der Lugt (2021). "Atherosclerotic carotid plaque composition and incident stroke and coronary events." Journal of the American College of Cardiology 77(11): 1426–1435.

Delgado-Lista, J., J. F. Alcala-Diaz, J. D. Torres-Peña, G. M. Quintana-Navarro, F. Fuentes, A. Garcia-Rios, A. M. Ortiz-Morales, A. I. Gonzalez-Requero, A. I. Perez-Caballero, E. M. Yubero-Serrano, O. A. Rangel-Zuñiga, A. Camargo, F. Rodriguez-Cantalejo, F. Lopez-Segura, L. Badimon, J. M. Ordovas, F. Perez-Jimenez, P. Perez-Martinez, J. Lopez-Miranda and C. Investigators (2022). "Long-term secondary prevention of cardiovascular disease with a Mediterranean diet and a low-fat diet (CORDIOPREV) a randomised controlled trial." Lancet 399(10338): 1876–1885.

Finicelli, M., A. Di Salle, U. Galderisi and G. Peluso (2022). "The Mediterranean diet: An update of the clinical trials." Nutrients 14(14): 21.

Forman, H. J. and H. Q. Zhang (2021). "Targeting oxidative stress in disease: Promise and limitations of antioxidant therapy." Nature Reviews Drug Discovery 20(9): 689–709.

Fuchs, F. D. and P. K. Whelton (2020). "High blood pressure and cardiovascular disease." Hypertension 75(2): 285–292.

Golaszewski, N. M., A. Z. LaCroix, J. G. Godino, M. A. Allison, J. E. Manson, J. J. King, J. C. Weitlauf, J. W. Bea, L. Garcia, C. H. Kroenke, N. Saquib, B. Cannell, S. Nguyen and J. Bellettiere (2022). "Evaluation of social isolation, loneliness, and cardiovascular disease among older women in the US." Jama Network Open 5(2): 12.

Hu, E. A., L. M. Steffen, J. Coresh, L. J. Appel and C. M. Rebholz (2020). "Adherence to the healthy eating index-2015 and other dietary patterns may reduce risk of cardiovascular disease, cardiovascular mortality, and all-cause mortality." Journal of Nutrition 150(2): 312–321.

Khan, S. U., A. N. Lone, M. S. Khan, S. S. Virani, R. S. Blumenthal, K. Nasir, M. Miller, E. D. Michos, C. M. Ballantyne, W. E. Boden and D. L. Bhatt (2021). "Effect of omega-3 fatty acids on cardiovascular outcomes: A systematic review and meta-analysis." EClinicalMedicine 38: 100997.

Ko, C. W., J. Qu, D. D. Black and P. Tso (2020). "Regulation of intestinal lipid metabolism: Current concepts and relevance to disease." Nature Reviews Gastroenterology & Hepatology 17(3): 169–183.

Libby, P. (2021). "The changing landscape of atherosclerosis." Nature 592(7855): 524–533.

Libby, P. (2021). "Inflammation during the life cycle of the atherosclerotic plaque." Cardiovascular Research 117(13): 2525–2536.

Lopez-Jimenez, F., W. Almahmeed, H. Bays, A. Cuevas, E. Di Angelantonio, C. W. le Roux, N. Sattar, M. C. Sun, G. Wittert, F. J. Pinto and J. P. H. Wilding (2022). "Obesity and cardiovascular disease: Mechanistic insights and management strategies. A joint position paper by the World Heart Federation and World Obesity Federation." European Journal of Preventive Cardiology 29(17): 2218–2237.

Luo, J., H. Y. Yang and B. L. Song (2020). "Mechanisms and regulation of cholesterol homeostasis." Nature Reviews Molecular Cell Biology 21(4): 225–245.

Malik, V. S. and F. B. Hu (2022). "The role of sugar-sweetened beverages in the global epidemics of obesity and chronic diseases." Nature Reviews Endocrinology 18(4): 205–218.

Mason, R. P., P. Libby and D. L. Bhatt (2020). "Emerging mechanisms of cardiovascular protection for the omega-3 fatty acid eicosapentaenoic acid." Arteriosclerosis Thrombosis and Vascular Biology 40(5): 1135–1147.

Pirillo, A., M. Casula, E. Olmastroni, G. D. Norata and A. L. Catapano (2021). "Global epidemiology of dyslipidaemias." Nature Reviews Cardiology 18(10): 689–700.

Powell-Wiley, T. M., Y. Baumer, F. O. Baah, A. S. Baez, N. Farmer, C. T. Mahlobo, M. A. Pita, K. A. Potharaju, K. Tamura and G. R. Wallen (2022). "Social determinants of cardiovascular disease." Circulation Research 130(5): 782–799.

Reitsma, M. B., M. B. Reitsma, P. J. Kendrick, et al. (2021). "Spatial, temporal, and demographic patterns in prevalence of smoking tobacco use and attributable disease burden in 204 countries and territories, 1990–2019: A systematic analysis from the Global Burden of Disease Study 2019." Lancet 397(10292): 2337–2360.

Ridker, P. M., D. L. Bhatt, A. Pradhan, R. J. Glynn, J. G. MacFadyen, S. E. Nissen and Prominent Reduce IT and Strength Investigators (2023). "Inflammation and cholesterol as predictors of cardiovascular events among patients receiving statin therapy: A collaborative analysis of three randomised trials." Lancet 401(10384): 1293–1301.

Shaya, G. E., T. M. Leucker, S. R. Jones, S. S. Martin and P. P. Toth (2022). "Coronary heart disease risk: Low-density lipoprotein and beyond." Trends in Cardiovascular Medicine 32(4): 181–194.

Sheena, B. S., L. Hiebert, H. Han, et al. (2022). "Global, regional, and national burden of hepatitis B, 1990–2019: A systematic analysis for the Global Burden of Disease Study 2019." Lancet Gastroenterology & Hepatology 7(9): 796–829.

Tall, A. R., D. G. Thomas, A. G. Gonzalez-Cabodevilla and I. J. Goldberg (2022). "Addressing dyslipidemic risk beyond LDL-cholesterol." Journal of Clinical Investigation 132(1): 11.

Tsao, C. W., A. W. Aday, Z. I. Almarzooq, et al. (2022). "Heart disease and stroke statistics-2022 update: A report from the American Heart Association." Circulation 145(8): E153–E639.

Tsao, C. W., A. W. Aday, Z. I. Almarzooq, et al. (2023). "Heart disease and stroke statistics-2023 update: A report from the American Heart Association." Circulation 147(8): E93–E621.

Witkowski, M., T. L. Weeks and S. L. Hazen (2020). "Gut microbiota and cardiovascular disease." Circulation Research 127(4): 553–570.

Characters: Lipoproteins of Human Blood

In all human cells, cholesterol plays a number of important roles. In the body, cholesterol originates from both external and internal sources. The first is a result of the digestion of food in the intestine, and the second is a result of cholesterol production in the liver. The molecules from the liver and intestine must travel through the bloodstream into the cells. Although cholesterol is not soluble in water, it must still be transported through living organisms that are largely composed of water. Special transport proteins bind the cholesterol and carry it throughout body fluids. These structures that contain both proteins and lipids are called lipoproteins.

Lipoproteins are the main characters of this book. They are special particles that transport lipids in water. Lipoproteins are usually spherical, containing multiple molecules of proteins and lipids and are rarely shaped like discs. They are mainly composed of lipids and proteins and contain small amounts of carbohydrates. The particles are made up of all the major classes of lipids including triglycerides, phospholipids, and sterols. In human lipoproteins, cholesterol is much more abundant than other sterols.

Lipoproteins are primarily responsible for transporting triglycerides and cholesterol. Triglycerides are vital for energy production and survival. The transport of cholesterol is also important due to its ubiquitous role within cell membranes and other biological functions. Lipoproteins contain cholesterol and other sterols in both their esterified and free forms.

The transformation of free cholesterol (also called nonesterified cholesterol) into cholesteryl esters hides cholesterol in the core of lipoprotein particles and facilitates its transport. Both triglycerides and cholesteryl ester are hydrophobic and therefore are hidden from water contact, forming a nonpolar core for lipoproteins. At the surface of lipoproteins, a monolayer of more polar phospholipids and nonesterified cholesterol is formed. The phospholipids and cholesterol form a layer that separates hydrophobic lipids and the aqueous milieu from each other. This monolayer is similar to the one-molecule-wide cover on either side of a cell membrane.

Apolipoproteins are a main subset of the proteins found in lipoproteins. Apolipoproteins bind lipids. This is a family of proteins that have a structure dictated by minimizing the exposure of hydrophobic surfaces to water. Apolipoproteins have specialized roles in the lipid processing. Those which are structural ensure that lipoproteins have a unique organization. Apolipoproteins which are regulatory control lipid processing in the blood. Apolipoproteins are able to bind lipids using their beta-sheets and alpha-helixes, which contain both hydrophilic and hydrophobic parts.

Cholesterol, Lipoproteins, and Cardiovascular Health: Separating the Good (HDL), the Bad (LDL), and the Remnant,
First Edition. Anatol Kontush.
© 2025 John Wiley & Sons, Inc. Published 2025 by John Wiley & Sons, Inc.

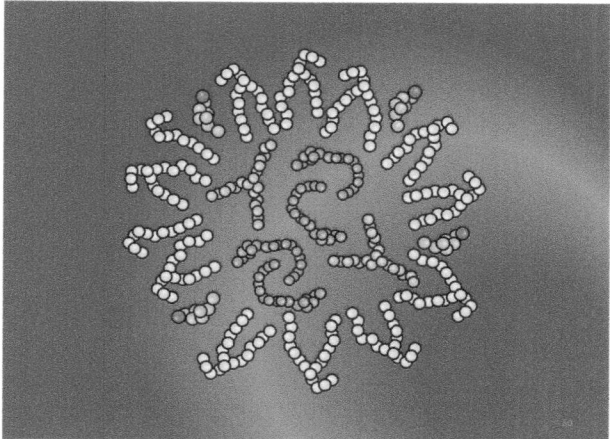

The arrangement of lipids in a lipoprotein involves the formation of a nonpolar core consisting of hydrophobic molecules of triglycerides (green) and cholesteryl esters (pink). On the surface, there is a monolayer arrangement of more hydrophilic molecules of phospholipids (grey) and free cholesterol (yellow).

What It Is

Alpha-helix consists of a spiral structure made up by an amino acid chain. The inner part of the spiral is formed by the backbone, and the outer part by the side chains. The helix is stabilized by internal hydrogen bonds. Alpha-helixes can be found in membrane proteins or in areas of proteins which need to remain rigid.

Beta-sheets are made up of multiple amino acid chains or segments that are aligned side by side. The backbone is zigzagged, and each strand is connected to the next by hydrogen bonds. The beta-sheets may be parallel (with strands running in the same direction) or antiparallel (with strands running in opposite directions). The beta-sheets in spherical proteins provide structural stability.

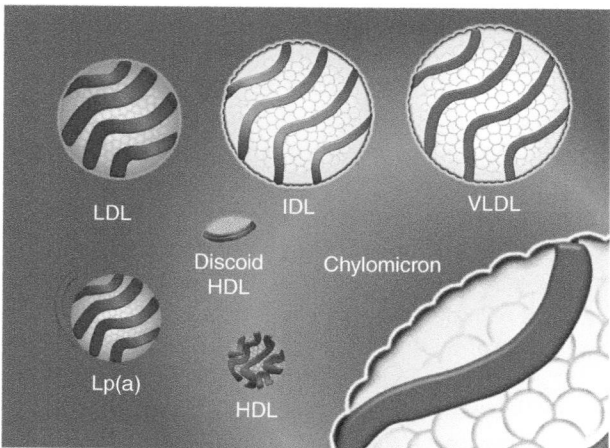

The main types of lipoproteins found in human plasma are chylomicrons, very low-density lipoprotein (VLDL), intermediate-density lipoprotein (IDL), low-density lipoprotein (LDL), lipoprotein (a) (Lp(a)), and high-density lipoprotein (HDL). These particles are arranged clockwise from discoid HDL to chylomicrons based on their size.

Lipoproteins have been separated historically based on their density differences. Fats and oils are typically less dense than water. This is why oil can be seen at the surface of water, where it looks like a rainbow. By contrast, proteins are much denser. Lipoproteins are characterized by their density based on the presence of the lipids. The more the lipids, the lighter the lipoprotein. Lipoproteins containing predominantly lipids are therefore less dense than those containing primarily proteins.

Most commonly, plasma lipoproteins can be classified according to density. This classification is based on ultracentrifugal separation, which is the operational process for obtaining lipoproteins. In order of decreasing density, the major lipoproteins found in plasma have very low density, intermediate density, low density, and high density. These are abbreviated VLDL, IDL, LDL, and HDL. Lipoproteins which are also found in the plasma of humans include chylomicrons (which are lighter than VLDL) and lipoprotein (a), which is abbreviated Lp(a), a specialized lipoprotein with a density between HDL and LDL.

Chylomicrons, VLDL, IDL, LDL, and Lp(a) contain more than half of their total mass in lipids. These lipoproteins, therefore, are commonly thought of as being light particles that are lipid rich but low in protein. HDL, on the other hand, is composed of approximately half of proteins. It is therefore a dense lipoprotein that is rich in protein. Note that albumin, a major protein in blood plasma, is the main carrier of some simple lipids like free fatty acids. Albumin–fatty acid complexes are not called lipoproteins. These include more complex lipids like phospholipids and triglycerides, as well as cholesteryl ester.

The albumin is the main carrier of fatty acids in blood plasma. Each principal lipoprotein acts as a major carrier for a specific lipid class. Chylomicrons, VLDL, and IDL are the main carriers of triglycerides. LDL and HDL are primarily responsible for transporting cholesterol.

The increased density of proteins in comparison to lipids means that the proteins take up less space. For this reason, the elevated density of lipoproteins results in a decrease of size. It is important to remember that in this nanoworld, "large" starts at 25 nm and "small" equals 10 nm and less. Lipoproteins are therefore natural nanoparticles.

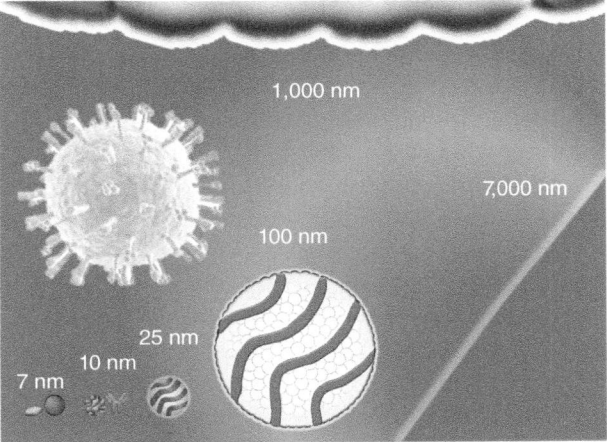

The size of human plasma lipoproteins is shown in relation to the red blood cell, influenza virus, antibody molecule, and gold dot. From left to right, discoid HDL, gold dot, spherical HDL, an antibody, LDL, VLDL, and a fragment of the red blood cell. The influenza virus and a fragment of a chylomicron are shown on top.

The size of lipoproteins decreases as the density increases, from chylomicrons through VLDL, IDL, and LDL up to HDL. The largest lipoprotein particle is the chylomicron, which can be as large as 1 μm. This is a huge dimension in the world of lipoproteins. VLDL is the second largest lipoprotein and has an average size of 80 nm. IDL is a heterogeneous collection of particles that are between the sizes of LDL and VLDL. LDL and Lp(a) are between 20 and 30 nm in diameter, whereas HDL has a diameter of 7–14 nm. Human cells, for example, are an order of magnitude bigger – the average diameter of an erythrocyte, for instance, is around 7 μm.

Lipoproteins are present in the blood due to the concerted efforts of the liver and the intestines, which produce many different types of lipoproteins. The liver can synthesize different sizes of particles based on the amount of lipids available. Certain proteins and enzymes can also affect the production of these particles. Understanding where lipoproteins are produced in the body is important to know their roles.

	Density, g/ml	Size, nm	Molecular mass, MDa	Lipid to protein ratio
Chylomicrons	<0.930	75–1200	Up to 100,000	50 to 1
VLDL	0.930–1.006	30–80	6–27	10 to 1
IDL	1.006–1.019	25–35	3.5–6	4 to 1
LDL	1.019–1.063	18–25	2.4–3.5	3 to 1
Lp (a)	1.055–1.085	~30	2.7–4.2	2.5 to 1
HDL	1.063–1.210	5–12	0.07–0.50	1 to 1

The lipoprotein classes vary in density, size, and molecular mass, which is directly linked to the differences in the proportion of lipid and protein content.

The liver and the intestine are the primary organs that produce blood lipoproteins in humans. In this process, lipids in hepatic and intestinal cells are usually associated with the appropriate proteins before being secreted into the bloodstream. Cells secrete lipoproteins in a highly regulated, complex, and energy-consuming process. This serves an important physiological function – to supply other organs and tissues with lipids for energy. The liver and the intestine can therefore be considered to be factories that produce lipoproteins. Peripheral tissues, on the other hand, can be viewed as consumers. Lipoproteins are seen as transporters in this view. The secretion and production of lipoproteins is a way to move lipids through the aqueous phase in living organisms.

After secretion, the lipoproteins move freely through the arteries and veins. There is no difference between lipoprotein particles in these blood vessels. In the blood, lipoproteins travel between innumerable blood cells under the pressure exerted by the heart upon contraction. The average person's blood cell count is estimated to be higher than that of the stars in the Milky Way. The number of lipoprotein particles in the blood of this individual is even higher than the number of blood cells. It exceeds the total number of cells in their body. There are more HDL than LDL particles in circulation. Their number is only surpassed by the stars in the entire universe!

LDL particles in the blood are specialized carriers of cholesterol. The LDL particles are formed during an endogenous metabolic cascade that originates from the hepatic release of VLDL particles, which is the vehicle to transport triglycerides. The cascade is a metabolic continuum that occurs in the blood. Lipoprotein particle transformations occur through the action of lipolytic enzymes and lipid transfer proteins. This process produces remnant lipoproteins such as IDL. It is completed by the production of LDL.

A fascinating magic trick enables lipids to be transported through the body, even though they are insoluble in water. This trick involves the use of special proteins called apolipoproteins, which help form lipoproteins.

Like any complex technological process, lipid delivery into peripheral tissues can be vulnerable to several issues. So, lipoprotein secretion by cells can be either pathologically enhanced or reduced, resulting in an improper lipid transport. The transport of lipids from the site of production to that of consumption through the blood can be altered by many factors, reflected by altered levels of both apolipoproteins and lipids. Lipoprotein uptake by tissues can be impaired, resulting in an inadequate lipid supply. Dyslipidemias are abnormal metabolic states that result from altered lipid metabolism. They are often a result of problems with lipoprotein production or transport.

The liver produces VLDL and HDL in a manner akin to industrial production of goods.

Lipoproteins can be found in the blood, but they are also present in other fluids such as follicular, cerebrospinal, and amniotic. The human brain produces lipoproteins that are different from those secreted by the liver and the intestine on the other side of the blood–brain barrier.

Brain lipoproteins secreted in the intracellular area of the brain serve as important components for the nervous systems. Further, it appears that disturbances in brain lipoprotein metabolism are involved in major pathologies such as Alzheimer's disease and Parkinson's disease. This controversial, fast-developing, and highly fascinating topic is beyond the scope of this book which focuses primarily on blood lipoproteins.

The story begins now that the setting has been defined.

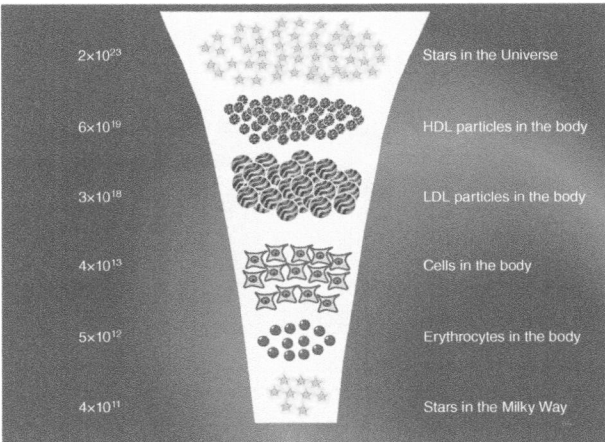

2×10^{23}	Stars in the Universe
6×10^{19}	HDL particles in the body
3×10^{18}	LDL particles in the body
4×10^{13}	Cells in the body
5×10^{12}	Erythrocytes in the body
4×10^{11}	Stars in the Milky Way

The number of lipoproteins in the blood is similar to other large numbers. Since erythrocytes make up the majority of human blood cells, their total count is very similar to that of erythrocytes.

Open Questions

Can we eradicate cardiovascular diseases and stop atherosclerosis? Can we do it through normalization of what happens to lipoproteins in the body?

History

Isolation and Purification of Plasma Lipoproteins

Who	When
Michel Macheboeuf	**1929**

Where	What
Paris, France	Macheboeuf, M. (1929) "Recherches sur les phosphoaminolipides et les sterids du serum et du plasma sanguins II. Etude physiochimique de la fration proteidique la plus riche en phospholipids et in sterides." Bulletin de la Société de chimie biologique 11: 485-503.

In the early 20th century, the concept of lipoproteins appeared.

Since a long time, it was known that organic solvents like ether could not remove all the fat from the blood serum. Joseph Nerking, at the University of Bonn in 1901, observed that much more fat could be removed from horse plasma after the breakdown of proteins. The precipitation was carried out by pepsin and hydrochloric acid. This experiment led him to conclude that specific proteins in blood plasma were responsible for binding fats.

Henry Haslam, at Cambridge University, and Harriette Chick, at the Lister Institute of London, found independently that globulins are a subset of plasma proteins which carry fats. They also identified the fats to be phospholipids, called lecithin in those days. Hans Handovsky, at the University of Gottingen in 1925, showed that the cholesterol in the serum is tightly bound to globulins. Hermann Bennhold, at the University of Hamburg, confirmed these results in the early 1930s.

A key part of this research was conducted in France. Michel Macheboeuf, along with his colleagues, discovered plasma lipoproteins at the Institut Pasteur in Paris. Researchers were intrigued by the mystery of how lipids, which are insoluble in water, could be dissolved in the bloodstream.

Lipoproteins were separated from horse serum and stained after precipitation by salt solutions. The lipoproteins precipitated by lowering the pH to 3.8 were isolated from neutral ammonium extracts of serum. The isolated compound was stable and could easily be redissolved into water to produce a clear solution.

The studies were conducted in the 1920s as part of Michel Macheboeuf's doctoral thesis at the *Université de Paris*. In 1928, his thesis was delivered at the *Faculté des Sciences de Paris* under the title *Recherches sur les lipides, les stérols et les protéides du sérum et du plasma sanguin* – "Research on lipids, sterols and proteids of serum and blood plasma." Macheboeuf first published his findings in a French magazine called *Bulletin de la Société de la Chimie Biologique* in 1929. He also presented them at an international conference in the same year.

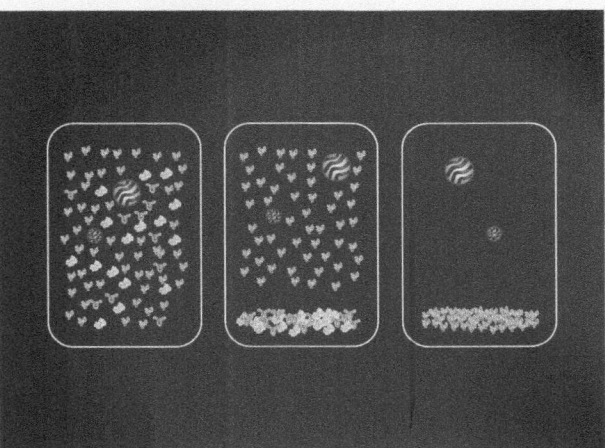

Two-step separation of LDL and HDL from plasma proteins which do not contain lipids.

A protein fraction rich in lipids, consisting of 59% protein, 23% phospholipid, and 18% cholesterol, was isolated by precipitation. This composition is very similar to the HDL we know today. The fraction was prepared from an albumin-rich precursor, which was obtained by removing the globulins from horse plasma. The lipid-rich fraction had peculiar chemical

(Continued)

(Continued)

properties – it was a clear aqueous solution containing 10% lipid that could not be extracted using ether. The globulin fraction also contained some lipids, but it was much less resistant to extraction with ether. This discovery revealed the heterogeneity in lipid–protein interactions present in plasma.

During his postdoctoral, as we now say, phase, Macheboeuf first stayed at Paris, then moved on to Lille, and, finally, was appointed full professor of biological and medicinal chemistry at Bordeaux. In his ongoing research, he discovered that the albumin fraction could be incubated with ether and alcohol to extract lipids. The addition of soap made the extraction of all the lipids from globulins, and a significant portion of the albumin-containing fraction even easier.

These observations indicated that lipids were linked with proteins in the lipid–protein complexes by a weak interaction between hydrophobic molecules hidden from the aqueous environments. These moieties, at first believed to be hydrocarbon chain-like structures, were later replaced by hydrocarbon cycles to explain the existence of cholesterol. The polar lipid moieties were guided to the outermost part, close to the polar protein moieties. This created the surface. The molecules were united by mutual affinity. The complexes were very stable. Lipids and proteins did not separate under the influence of an electrical field or gravitational force during electrophoresis or centrifugation. This structure's basic principles are very similar to what we now think of as plasma lipoproteins.

Macheboeuf named the lipidoprotein complexes "lipoproteic cenapses" in 1937, from the Greek word "cenapse," which means "joining" or "union". He did not choose the term "complexes," as it meant a poorly defined biochemical system at that time. He also characterized the chemical composition of these complexes. To achieve this remarkable feat, he had to develop analytical micromethods that could evaluate the low concentrations of phospholipids, cholesterol, and proteins in his samples. In the 1940s, he also used ultracentrifugation to determine the sedimentation constants for the cenapses. This allowed him to calculate their approximate size.

After the composition of the lipid–protein cenapses was well understood, Macheboeuf's lab undertook the brave experiment of producing them on its own. The lab added a detergent, a bile salt, to a mixture containing a protein and fatty acid. They then removed the detergent using dialysis. It is remarkable that, almost a century later, the same steps are used to produce artificial HDL particles.

Macheboeuf, in the final years of his life that ended prematurely at the age of 52 in 1953, showed that the proteins of "lipoproteic cenapses" were not identical with albumin or other plasma proteins. On the basis of this finding, Macheboeuf proposed a hypothesis for the specificity in the protein fraction of lipoproteins. It was suggested that the lipid affinity of this fraction is a defining characteristic. These proteins, which we now call apolipoproteins, were further identified and characterized by additional research. This discovery finally resolved the mystery of why lipids are insoluble in aqueous solution and solubilized within the bloodstream.

Macheboeuf's work allowed the isolation of two major types of lipid–protein complexes from horse plasma. These are now designated LDL and HDL. This research was not well known outside France. Macheboeuf was invited to speak at many scientific and academic societies and universities. He also participated in several congresses. It is curious that almost all his lectures were given in France. Only one notable exception was a lecture he gave at a Birmingham satellite meeting of the First International Congress of Biochemistry, in 1949. He also only published in French journals. These peculiarities hindered dissemination of his findings and development of new ideas.

Edwin Cohn, a Harvard University professor in Boston, confirmed the findings of Macheboeuf. Cohn's lab separated alpha- and beta-lipoproteins from the serum in the 1940s using electrophoresis, followed by precipitation. The Cohn lab used a gradual change in pH, ionic concentration, and ethanol at temperatures between 0 and −5°C. These studies focused primarily on the purification and concentration of plasma proteins. They were especially important during World War II. This research identified lipid-containing protein fractions as a side product.

Cohn's work had a greater impact on the public than Macheboeuf's. Cohn's and his colleagues' efforts to rediscover lipoproteins led to a greater recognition of the work by the scientific community.

Practical Aspects: How to Isolate Lipoproteins by Precipitation

Through the interaction of added chemicals, lipoproteins from blood serum and plasma can be precipitated selectively. These methods were widely used in the early research of lipoproteins when ultracentrifugal and electrophoretic approaches were not yet developed.

Later research found that heparin, which is a negatively charged polymer consisting mainly of sugar residues, precipitates VLDL and LDL selectively when it is added to serum. As apolipoprotein B is only present in VLDL and LDL in fasting plasma, this method is called apolipoprotein B depletion. This process requires the presence of divalent metal ions with a positive charge, such as magnesium. Dextrane sulfate can replace the combination of heparin with metal ions. These mixtures form insoluble complexes containing VLDL, LDL, metal ions, and heparin. The precipitated complexes can be collected by a quick centrifugation. Only HDL is left in the solution as in the experiments of Michel Macheboeuf. Measurement of cholesterol in this solution allows determining HDL-cholesterol concentration.

How to isolate lipoproteins
Precipitation

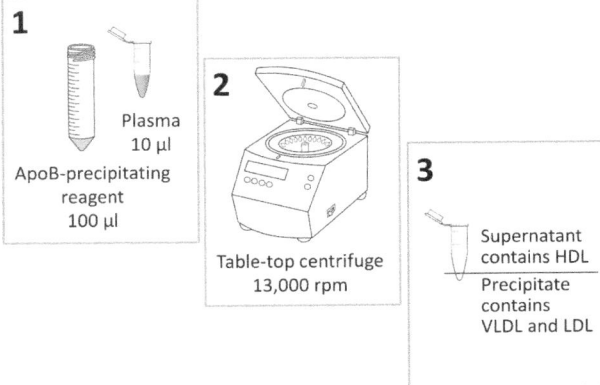

This method is less expensive and less complex than ultracentrifugation. It is for this reason that it has been used to measure HDL-cholesterol in plasma until now.

Test Your Knowledge

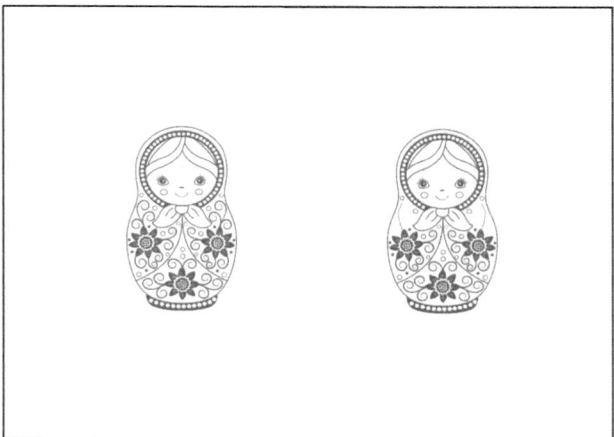

1 Which matryoshka has a higher risk of developing atherosclerosis?
   A on the left;
   B on the right.

Cholesterol, Lipoproteins, and Cardiovascular Health: Separating the Good (HDL), the Bad (LDL), and the Remnant,
First Edition. Anatol Kontush.
© 2025 John Wiley & Sons, Inc. Published 2025 by John Wiley & Sons, Inc.

2   What is the time required for a heart to pump one barrel of blood?
   A 32 minutes;
   B 300 minutes;
   C 122 minutes.

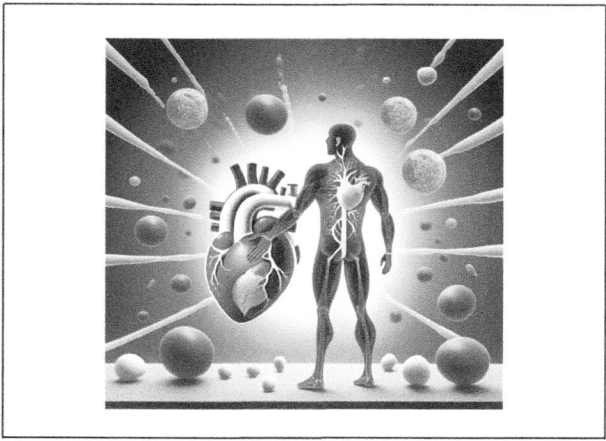

3   How much cholesterol is in the average human body?
   A 3.5 kilograms;
   B 350 grams;
   C 35 grams.

4 What do Elvis Presley, Christian Dior, Federico Fellini, and Mao Zedong have in common?
   A they all lived exclusively in the 20th century;
   B they all liked the Beatles;
   C they all died of cardiovascular causes.

5 What is the probability that a man will develop heart disease if he has all risk factors?
   A almost 20%;
   B almost 50%;
   C almost 70%.

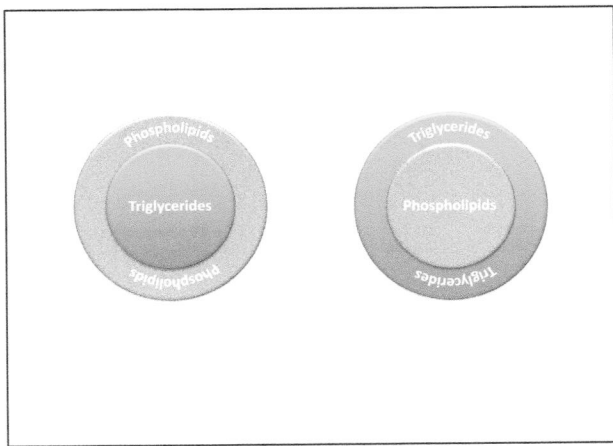

6   What is the best way to pack lipids in lipoproteins for transporting them across the human body?
    A triglycerides coated with phospholipids;
    B phospholipids coated with triglycerides.

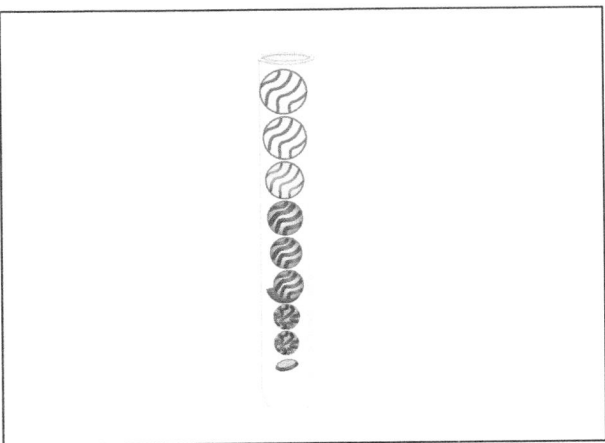

7   What lipoproteins (chylomicrons, VLDL, IDL, LDL, Lp(a), or HDL) float in seawater if they are ultracentrifuged? Seawater has a density of 1.020–1.029 g/ml.

Answers

1, A, because of the turbulences in the blood flow at the level of the carotid arteries; 2, A; 3, C; 4, C; 5, C; 6, A; 7, chylomicrons with a density of 0.93 g/ml or less, VLDL with a density of 0.93–1.006 g/ml, and IDL with a density of 1.006–1.019 g/ml will float.

Main Character One: LDL, Carrier of "Bad" Cholesterol

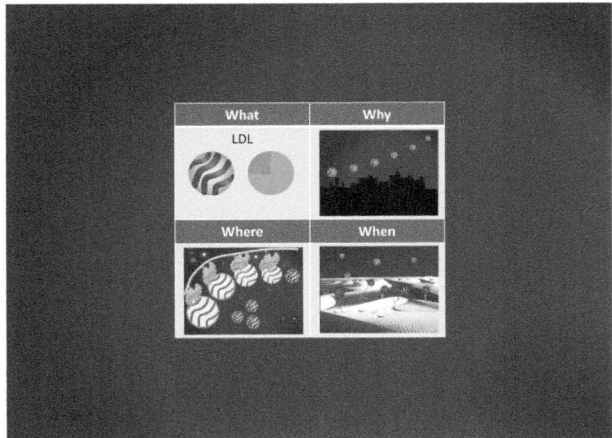

4

LDL – Why It Is Important

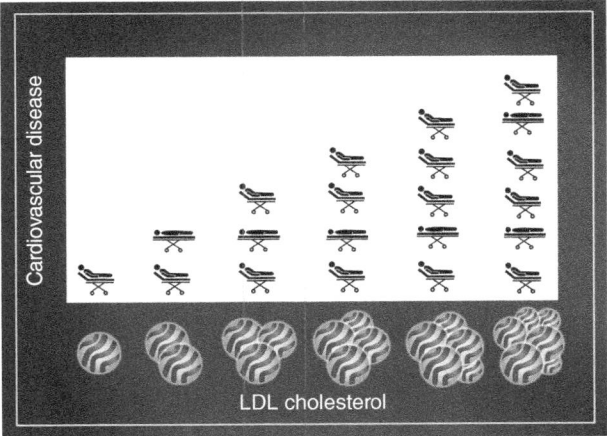

The relationship between LDL-cholesterol (plotted horizontally) and mortality due to cardiovascular disease (plotted vertically) shows how, as LDL-cholesterol levels rise, so does the risk of developing the disease. LDL-cholesterol is harmful. It is worth noting that higher LDL-cholesterol levels can lead to smaller LDL particle sizes.

The Bad is the first of our three main characters. This protagonist is also known as Bad Cholesterol. This is the name of the cholesterol that is found in LDL, or low-density lipoprotein. It is well known that having high levels of LDL-cholesterol in the bloodstream can increase the risk of heart disease and stroke. As both conditions are fatal, elevated LDL-cholesterol levels increase the risk of mortality – hence, the "bad" label. LDL is often viewed as a dangerous particle that causes heart disease and stroke. It is not often remembered that LDL can also serve a positive purpose, supplying cholesterol to the tissues. Indeed, cholesterol plays a vital role in the functioning of human cell membranes and is also essential for other cell functions. This example shows that lipoproteins are not simple characters – they are all complex and ambivalent.

LDL is by far the most known of all the characters in the story. LDL-cholesterol is the main risk factor of cardiovascular disease. There are also numerous therapies that can reduce LDL-cholesterol levels and lower cardiovascular death. The most effective of them are statins. Researchers and clinicians have made great strides in the fight against heart disease using statins. However, the disease persists despite the advances and new treatments are needed urgently.

Cholesterol, Lipoproteins, and Cardiovascular Health: Separating the Good (HDL), the Bad (LDL), and the Remnant, First Edition. Anatol Kontush.
© 2025 John Wiley & Sons, Inc. Published 2025 by John Wiley & Sons, Inc.

Retrospectively, it is clear that the path of LDL began with biomarker discoveries. This was followed by basic science, and, finally, a successful therapy. A detailed analysis reveals many twists, contradictions, and problems that seemed unsolvable for a very long time.

What It Is

LDL-cholesterol is cholesterol that LDL carries in the blood. The concentration of LDL-cholesterol is measured in milligrams per deciliters (mg/dl), or millimoles per liter (mmol/l) plasma. LDL-cholesterol levels above 130 mg/dl are associated with an increased risk of heart disease and are considered high. Treatment is required for high levels of LDL-cholesterol to reduce cardiovascular risks.

A low LDL-cholesterol can be defined as less than 50 mg/dl, and an extremely low as below 20 mg/dl.

Heart disease develops when cholesterol-rich lipoproteins build up in the arterial wall. LDL particles can easily penetrate the subcellular space in the arterial walls because they are small. Low LDL-cholesterol levels also mean low LDL particle concentrations in the arterial walls and low risk of heart disease. As LDL-cholesterol levels rise, so does the concentration of LDL particles in the arterial walls and the risk of heart disease.

Epidemiology

The amount of LDL-cholesterol in the blood is closely linked to the incidence of coronary heart disease and the progression of atherosclerosis. Numerous studies have repeatedly demonstrated that these connections exist and are strong. For example, the Emerging Risk Factors Collaboration, a study that looked at over 300,000 individuals who had no heart disease when the study began, found that high levels of LDL-cholesterol were linked to a greater risk of heart attacks and death from heart disease. Another study of 170,000 participants observed that low LDL-cholesterol can reduce the risk for heart attacks and strokes. High LDL-cholesterol levels are associated with a doubled risk of heart disease in comparison to lower levels. According to these studies, in the general population, cardiovascular events are increased by approximately 1% for every 1% increase of LDL-cholesterol. This confirms the importance of LDL-cholesterol in determining heart disease risk.

The Prospective Studies Collaboration Trial examined over 892,000 individuals who were free of heart disease when the trial began. The researchers followed the people closely for many years and discovered that high cholesterol levels in the blood are linked with a greater risk of heart disease. The same was true of LDL-cholesterol. Other studies have also shown that higher LDL-cholesterol levels are associated with a greater risk of heart disease. The longer people were followed, the stronger this link became. It is clear not only that LDL-cholesterol contributes to heart disease but that its contribution also accumulates over time. In fact, having high LDL-cholesterol levels over time increases the risk of developing heart disease. The risk was also higher if LDL-cholesterol levels rose quickly at young age. It is therefore important to manage LDL-cholesterol at a young age in order to reduce heart disease risk later in life.

Meta-analyses, which combine the results of several studies to create a single analysis, demonstrate that high LDL-cholesterol is a significant problem. It is directly linked to the incidence of cardiovascular disease in large populations and across various ethnicities. The discovery that high LDL-cholesterol is a big risk for heart disease in Caucasians was also found to be true in other groups like those from the Middle East and Asia. High LDL-cholesterol is also a risk factor for heart disease in both healthy individuals and those with diabetes, hypertension, or other diseases. This risk factor applies to older people too.

High LDL-cholesterol levels are connected not only to heart disease but also to stroke, carotid disease, and blood clots in the veins. They are also linked to signs of heart disease like calcium buildup in the arteries and thickening of the carotid arterial wall. It is interesting that high LDL-cholesterol differently affects different types of strokes. The risk is only increased for ischemic strokes, whereas the risk for hemorrhagic ones is reduced.

Even after being watched for up to 30 years, high LDL-cholesterol levels are still good at predicting heart disease. Because of how important they are, LDL-cholesterol levels are used in many tools to predict heart disease risk.

Interesting Numbers

LDL-cholesterol can increase heart disease risk by as much as 2.75 times (175%) for a 40-year-old man with elevated LDL-cholesterol of 190 mg/dl compared to a peer who has a normal LDL-cholesterol of 120 mg/dl. The risk difference can be up to 3% for every 1% rise in LDL-cholesterol and it is more noticeable in younger people.

To reduce the risk for heart disease, LDL-cholesterol should be kept as low as is possible. Recent data, however, show that LDL-cholesterol levels at extreme lows can be detrimental to health. In a Danish study, researchers discovered that people with extremely low or high LDL-cholesterol levels were at a higher risk of death from any cause. It is interesting to note that low LDL-cholesterol was not associated with an increased risk of heart attack. LDL-cholesterol levels associated with the lowest mortality risk were 140 mg/dl for all people and 90 mg/dl for those taking cholesterol-lowering medication. The findings were the same for men and women of different ages and for different causes of death, except for those related to heart disease. The relationship between LDL-cholesterol and mortality was thus U-shaped. Researchers in Korea and China also obtained similar results. This association can be explained by reverse causation, where the outcome comes before the cause. It could be that being weak or sick can lower LDL-cholesterol levels. The Danish study found that people with very low LDL-cholesterol had more health issues. This confirms the notion that very low LDL-cholesterol could be a warning sign for serious illness.

> **Open Question**
> Are very low LDL-cholesterol levels simply an indicator of poor health?

According to the newest guidelines, it is still best for high-risk people to have really low LDL-cholesterol levels. With new treatments, it is now possible to achieve low levels of LDL-cholesterol in the clinic.

> **What It Is**
> **Reverse causation** occurs when two things are connected in a way that is unexpected. Instead of one thing causing the other, the opposite happens. For example, people might believe that being overweight is the cause of depression, but find that it is actually depression that causes being overweight. In reverse causation, the cause comes before the effect, which is not how we normally think of cause and effects.

According to studies, having very low LDL-cholesterol levels can reduce heart disease risk even more than lowering LDL-cholesterol a bit. This approach does not appear to pose any safety issues. In other studies, people with naturally low LDL-cholesterol levels did not face a greater risk of heart disease. More research is, however, required to ensure that this approach is safe over the long term.

Even if we were to say that high LDL-cholesterol is the main factor, the conclusion could still be biased. People with chronic diseases tend to be older than people who are healthy. LDL-cholesterol increases with age. In addition, men are more likely to suffer from heart disease than women. Men also have higher LDL-cholesterol than women. Theoretically, the observation of higher cholesterol in LDL from patients with heart disease may, therefore, be an indication of older age and male sex. When we compare the patients with healthy people who are the same age and have the same gender, the concentrations of LDL-cholesterol in the blood still remain higher in the patients. These "age and gender-matched" comparisons are important in epidemiology because they can exclude some obvious causes for observed differences. Overall, the data clearly show that high LDL-cholesterol is associated with heart disease, diabetes, and other chronic diseases, regardless of gender or age.

> **History**
> *Separation of Lipoproteins by Ultracentrifugation*
> After World War II, the majority of routine methods for isolating lipoproteins were developed in the United States. The work was first performed at the Donner Laboratory of Medical Physics, University of California Berkeley under the direction of John Gofman. His Nobel Prize-winning PhD mentor, chemist Glenn Seaborg, described him as "very bright." The lab used centrifugation, which was done at high speed to separate biological particles based on their density.
>
> Other researchers made important contributions to the field before Gofman's studies. Kai Pedersen, a Swedish researcher at the University of Uppsala, isolated beta-lipoproteins in 1947. This was simultaneously with Edwin Cohn's work in Boston. He used ultracentrifugation to isolate the beta-lipoprotein at a density of 45% saturated magnesium sulfate. The particle identified as protein X had a density of 1.03 g/mL.

Who	When
John Gofman	**1949**

Where	What
Berkeley CA, USA	Gofman, J. W., F. T. Lindgren and H. Elliott (1949) "Ultracentrifugal studies of lipoproteins of human serum. "Journal of Biological Chemistry

Gofman began studying lipoproteins in 1947 at the Donner Laboratory, the year the Framingham Heart Study was launched. During World War II, he worked on the chemistry of plutonium at the University of California Berkeley, where he received his PhD in Nuclear Chemistry, in 1943. Gofman codiscovered isotopes such as protactinium and uranium-232 in these studies and was then recruited to isolate large quantities of plutonium by the Manhattan Project at request of Robert Oppenheimer. He became interested in the medical sciences in Berkeley and graduated from medical school in 1946 at the University of California, San Francisco. He chose coronary heart disease and the transport of cholesterol in the bloodstream to be his research area based on the work of Cohn and Svedberg.

John Gofman developed the ultracentrifugation method to separate lipoprotein fractions based on his experiences in separating radioactive isotopes. In the 1940s, Jesse Beams' and Edward Pickels' ultracentrifuges were a huge success in the United States. Gofman purchased the second model manufactured by these two men. The laboratory's radioactive studies used the same equipment for their analytical method, mainly preparative and analytic ultracentrifuges.

The first applications of ultracentrifugation in biological research were sedimentation. When the blood plasma is centrifuged, gravity forces cause proteins and viruses to slowly settle at the bottom. Both proteins and viruses are heavier (denser) than water. This allowed for the sedimentation. Arthur McFarlane, who was working in Uppsala with Theodor Svedberg in 1935, designated one of the proteins – protein X. The protein X appeared to contain lipid and was deposited near albumin, a major human plasma protein. However, its behavior inside the ultracentrifugal tube was unusual.

John Gofman started working on lipoproteins together with Frank Lindgren and Harold Elliott who were his first doctoral candidates. They studied lipoproteins by ultracentrifugation. In 1949, the first publication on this technique appeared when researchers successfully applied it to human serum. Gofman Lindgren and Elliott, after years of experimentation, showed that protein X was actually a lipid–protein complex. Its density was lower than albumin (1.33 g/ml), but higher than plasma (1.006 g/ml). As lipids are lighter than proteins, the component must contain lipids. The lipoprotein floated at the top of the tube when the density of plasma increased

(Continued)

(Continued)

by adding salt to 1.3 g/ml. The complex was identified by Pedersen as beta-lipoprotein, which he had isolated earlier. The researchers were surprised to find that HDL, another lipoprotein complex separated in their studies, was the same as alpha-lipoprotein discovered 20 years ago by Michel Macheboeuf in Paris.

The research was transformed when the idea of flotation of lipoproteins, instead of sedimentation, was introduced. Separation was achieved by flotation. Particles of different densities floated at different speeds, so lighter particles had lower densities and floated faster. They could also be harvested sooner than heavier particles. The surrounding solution density must be greater than the densest particle to be separated.

The novel method allowed the separation of several lipoprotein classes. Schlieren photography was used to detect lipoproteins that were separated. Schlieren is a technique for detecting differences in the light passing through transparent material, like water, thereby allowing the detection of irregularities in its composition. It works by shining a specific type of light on the object and then allowing the fluid surrounding the object to distort the light. This distortion causes some areas of the image to be brighter, while others are darker. It helps determine if the fluid is not uniform. This method allowed to visualize and separate lipoproteins, but not to isolate them.

Gofman classified lipoproteins according to their flotation properties. Svedberg units of sedimentation were used to describe rates of sedimentation. Gofman, along with his colleagues, proposed using reverse flotation units called Svedberg flotation units (Sf). Lipoproteins were divided into three main categories of particles with Sf more than 20, between 12 and 20, and less than 12. The differences in their flotation rates were due to the differences in density. Particles with a higher Sf value were less dense and floated faster, and vice versa. These fractions had densities ranging from less than 1.006 to more than 1.019 g/ml.

The particles which floated very quickly were identified as chylomicrons. Then, came lipoproteins with increasing densities, which were called LDL (low-density), IDL (intermediate-density), and HDL (high-density). At the end of these studies, the Donner Lab described nine lipoprotein classes, including very low-density lipoprotein (VLDL) and other particles. These lipoproteins contained all of the serum cholesterol and other components such as triglycerides, fatty acids, and phospholipids.

Parallel to this, preparative ultracentrifugation for the isolation of lipoproteins from plasma was developed. The method, however, allowed only an approximate analysis of the lipids using infrared adsorption. The preparation technique was also not easy. VLDL was first isolated using ultracentrifugation at a very low density. LDL was isolated by a second ultracentrifugation at a density of 1.06 g/ml, in a concentrated NaCl solution. HDL was then separated at a density 1.21 g/ml of D_2O containing sodium nitrate. This method was complicated by its intricacies and not suitable for routine lipoprotein analysis.

In 1950, J. Lawrence Oncley from Harvard Medical School and his colleagues isolated a lipid-containing alpha-globulin in fraction III from Cohn. The researchers floated the particle in a 6% solution of NaCl at a density of 1.03 g/ml. The chemical analysis revealed that this lipoprotein contained 23% protein and 30% phospholipid. It also had 8% free cholesterol and 39% cholesteryl ester. This composition is very similar to what we currently know as a consensus for LDL. In addition, the fraction IV of Cohn contained the alpha-lipoprotein that Macheboeuf first isolated. These studies, however, used sophisticated methodology.

   Lipoproteins isolated at the Donner Lab could be purified and studied by a variety of methods including compositional analysis, electrophoresis, electron microscopy, and others. The molecular mass and round shape of the major lipoproteins' classes were quickly established. This method was so successful that it was adopted by many laboratories around the world to study lipoproteins – a new class of biological complexes.

What It Is

Svedberg flotation unit is defined as a flotation speed of 100 femtometer/s measured at 26°C in NaCl solution of a density of 1.063 g/ml at a gravitational force of 1 dyne per acceleration of 1 g. This unit abbreviated as Sf is used to characterize the flotation rates of plasma lipoproteins that vary according to their physical properties (density and size) and chemical composition (lipid to protein ratio). John Gofman introduced the unit in 1951.

   In 1951 and 1952, the first clinical data were obtained by using this experimental method. Healthy individuals were compared with patients suffering from a variety of diseases. Patients with diabetes and atherosclerosis had higher levels of IDL in their plasma than healthy people. These patients also had lower HDL. Experiments performed on rabbits quickly confirmed these results. The plasma levels of IDL, VLDL, and LDL were greatly increased when the animals were given cholesterol according to the studies of Anitschkow. In humans, fat intake sharply increased IDL and LDL concentrations, while diets restricted in fat and cholesterol reduced them. IDL and LDL levels increased with age. These concentrations were also elevated in obesity. The concept of "bad" cholesterol was born at this moment.

   Gofman, along with his colleagues, proposed a diet low in fat and low in cholesterol to fight heart disease. Furthermore, they introduced in 1954 the term "hyperlipoproteinemia" to cover several clinical conditions in which concentrations of LDL were elevated. They also found that men had higher levels of IDL, LDL, and HDL than women. These differences, however, vanished in subjects older than 45.

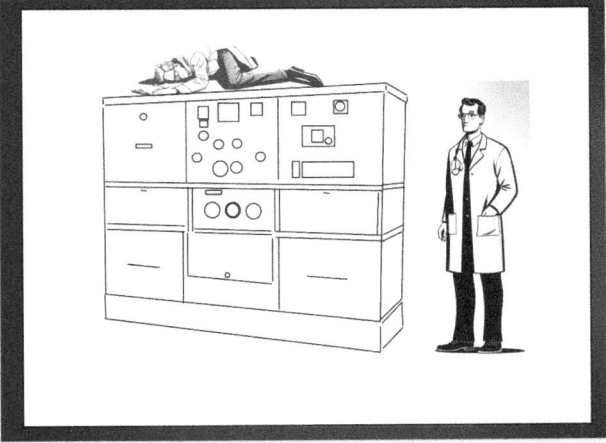

Specialized Instruments Corporation, Spinco, produced an analytical ultracentrifuge in 1948. It was designed by Edward Pickels. Thanks to this first practical, commercially developed ultracentrifuge, this important technique became widely available. Beckman Instruments acquired the company in 1955.

(Continued)

(Continued)

Gofman's findings were groundbreaking, but he was determined to do more. His main goal was to find a cure for heart disease. The team at the Donner Lab was incredibly excited about this possibility and worked tirelessly, even sleeping on the ultracentrifuge which ran around the clock. Gofman's proposed that the lipoprotein profile, particularly the IDL concentrations, could better predict the risk of heart attacks than plasma cholesterol alone. At the time, this hypothesis sparked a heated debate.

Gofman, to test his hypothesis, launched the Cooperative Study of Lipoproteins and Atherosclerosis in 1952. Four US laboratories recruited initially 15,000 subjects for this early cooperative project on cardiovascular epidemiology. This was the first large-scale biomedical longitudinal study. In the study, 4,914 middle-aged men aged 40–59 who were clinically normal were recruited.

It was necessary to organize a large number of people in order to control the quality of ultracentrifugal analysis at each laboratory and then centrally analyze the data. The study had several objectives, including evaluating potential associations between lipoprotein levels and coronary events. These associations were compared to those of total serum cholesterol levels. The other goals were to determine the normal range for plasma concentrations of IDL (called Sf 12–20 lipoprotein then) in healthy men, women, and children, measure the effects of a low fat, low cholesterol diet on lipoproteins, and characterize the distribution of lipoproteins among patients with diabetes and hypertension.

Circulation published the results of this study in 1956. The article compared 82 validated cases of coronary heart disease with the rest of the population. The authors concluded there was a predictive value to the lipid measurements. They disagreed, however, on how to interpret their data. This was a highly unusual situation. The report included a majority and minority statement by the investigators in two sections titled Discussion A and Discussion B. The disagreement was caused by several factors, including changes in the study objectives, some methods, and the diagnostic criteria of cardiovascular disease during the study.

John Gofman and his colleagues at the Donner Lab wrote Discussion A exclusively. It was a minority opinion. They developed an atherogenic index based on the concentrations of LDL, IDL, and VLDL. The index had a significant association with cardiac events. They also found that IDL concentration discriminated coronary events more accurately than total cholesterol. The study showed for the first time that elevated blood lipids predicted coronary heart disease and were not just a consequence of the disease.

All the other authors, including the chief statistician, contributed to Discussion B. They concluded, contrary to Gofman and his colleagues' evaluation, that the current study did not confirm that lipid levels could be used to predict who would develop coronary heart disease.

The discussion continued to state that while this does not exclude the atherogenic characteristics of lipid fractions, or their use for characterizing population variations, the study of individual predictions might be better carried out on more diverse populations or youth. The majority of authors also did not believe that the measurement of IDL (Sf 12–20) or VLDL (Sf 20–100) concentrations and the atherogenic index would be superior to the simple measurement of plasma cholesterol. They did not accept Gofman's statistical approach, which derived a new index from available data. They were also afraid of the complexity of the measurements, which at the time was not something that could be reliably done in clinical laboratories. Indeed, the technique of analytical ultracentrifugation used by Gofman required a lot of time and skilled personnel and was difficult to implement in clinical practice.

Unfortunately, the report led to the official rejection of Gofman's conclusions. The study had a major impact on Gofman's decision to leave the field of lipoproteins in 1956. His lab was now a leading research center in this emerging field and a worldwide referral center for lipoprotein studies. Ultracentrifugation and electrophoresis were successfully combined to characterize lipoproteins. Trudy Fortes used electron microscopy to obtain the first pictures of lipoproteins. However, the results of the Donner Lab, as well as the very concept of lipoproteins, were not widely accepted by other researchers. Gofman then turned attention to his previous research, studying the noxious effects that irradiation has on human populations, cells, and individuals in relation to cancer risk and cardiovascular disease. This unfortunate outcome had painful consequences for the entire field of cardiovascular medicine. It took several decades before the clinical significance of LDL and HDL levels in predicting risk was discovered. Anyway, this work did create the base of information linking biochemistry to atherosclerosis.

If you go in the same direction as everyone **else**, it does not necessarily mean that you are going in the right direction

Time has shown John Gofman right. The Framingham Heart Study, a large-scale epidemiological study, and subsequent work have fully validated his findings and concepts. In 1966, he released epidemiological data that revealed a relationship between HDL concentrations at baseline and future heart disease. HDL levels were lower among those who developed heart disease than in the rest of the population of 2,200 subjects followed over 12 years.

(Continued)

(Continued)

In the end, measuring LDL-cholesterol and HDL-cholesterol became a standard procedure in preventive cardiology. Gofman's pioneering method of lipoprotein separation using ultra-centrifugation led to the discovery of lipoprotein receptors, as well as other breakthroughs regarding lipid metabolism and atherosclerosis. The emphasis of Gofman on the importance of prevention of heart disease based on lipoproteins rather than treating the disease was revolutionary. His work was recognized by several awards within 10–15 years of the publication of his findings. He was among the first researchers to successfully work at the boundaries of physics and chemistry with biology and medicine, in a field that was later called biomedicine. His work exemplified the principle formulated in an excellent way by Claude Bernard: "Every advance made in science begins with a technological advance." Remarkably, this work was done in a small research lab without a huge budget. Frank Lindgren later said that the work was carried out by a visionary and dedicated scientist, who was in his prime, and two excited, enthusiastic half-time graduate students.

Gofman's successors consider him to be The Father of Clinical Lipidology, who began the shift away from lipids and toward lipoproteins. His studies opened up a new research field on the structure of the protein-containing fractions which carry lipids through the blood. Furthermore, he contributed to the concept of a risk factor long before the Framingham Heart Study results were published.

Practical Aspects: How to Isolate Lipoproteins by Ultracentrifugation

Ultracentrifugation is the classical method used to isolate and fractionate lipoproteins. Theodor Svedberg, the future Nobel Prize-winning scientist from Sweden's University of Uppsala, developed ultracentrifugation in the 1920s. The first ultracentrifuge in 1923 was constructed here. John Gofman, along with his colleagues in the 1940s and 1950s, used Schlieren optical analysis to separate plasma lipoproteins. Richard Havel, a researcher at the National Institute of Health (NIH) in Bethesda (USA), and his associates developed preparative ultracentrifugation, which allowed the isolation of lipoproteins rather than their separation by the analytical method. These methods were used to separate lipoproteins by ultracentrifugation, usually at speeds exceeding 105,000 g and for a period of 20 hours.

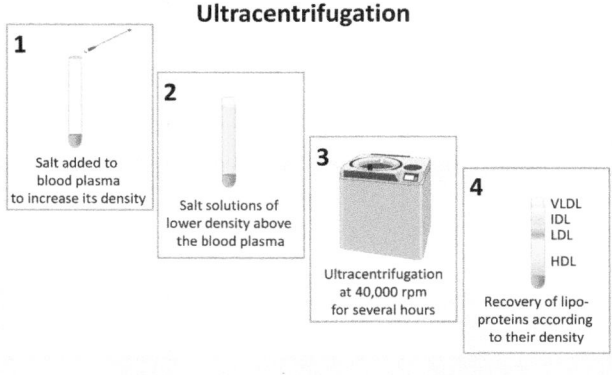

How to isolate lipoproteins

Ultracentrifugation

1 Salt added to blood plasma to increase its density

2 Salt solutions of lower density above the blood plasma

3 Ultracentrifugation at 40,000 rpm for several hours

4 VLDL IDL LDL HDL — Recovery of lipoproteins according to their density

The ultracentrifugal flotation rate in high-salt KBr solutions was used to differentiate between the three main lipoprotein classes: VLDL, LDL, and HDL. Each lipoprotein was isolated by a separate ultracentrifugation performed overnight. The whole procedure involved a sequence of three steps and was therefore termed "sequential ultracentrifugation." This method quickly became the standard for all laboratories that study lipoproteins.

Separation can be accelerated using a gradient of salt concentration within each test tube. This allows for the isolation of major lipoproteins by a single ultracentrifugation. The density gradient procedure has several advantages over sequential ultracentrifugation. These include a shorter centrifugation time and the use of a swing-out rotor. Ultracentrifugally isolated lipoproteins have a low level of artefactual products of lipid and protein degradation, which are attenuated by the procedure. The lipoproteins can be analyzed in a physical, chemical, and biological manner.

However, all ultracentrifugation-based methods of lipoprotein fractionation have a major drawback: they subject lipoproteins with high centrifugal forces and ionic strengths. These methods are also too expensive and labor-intensive for routine clinical use. Deuterium oxide and sucrose buffers can be used to achieve physiological pH and ionic strength. When lipoproteins are fractionated by vertical analytical profiling, centrifugation times can be reduced significantly. This method allows for a much quicker separation than other ultracentrifugation methods. It is, however, hampered by an incomplete separation of plasma proteins and lipoproteins. Finally, the use of iodixanol instead of KBr can reduce lipoprotein modifications during the procedure. However, separation between subfractions could be affected.

Genetics

Genetic studies have shown repeatedly that genes such as *LDLR* and *PCSK9*, which are associated with lower levels of LDL-cholesterol, are linked with a reduced risk of coronary artery disease. This is strong evidence that LDL-cholesterol causes heart disease. The relationship between the effects of each gene on LDL-cholesterol and heart disease is clear and consistent. The genetic variants linked to LDL-cholesterol have a major impact on heart disease through their effects on LDL-cholesterol levels. Indeed, it is highly unlikely that multiple genes associated with different biological pathways relating to lowering LDL-cholesterol would all have the same effect on heart disease risk.

The walls of arteries are home to many of the genes that cause heart disease. These genes could affect the way that the arteries respond to LDL-cholesterol or how they retain it. Researchers continue to study how genes and other factors affect the retention of LDL-cholesterol within the arterial wall.

It becomes evident when we examine all the evidence, including genetic studies, large-scale studies of groups of people, research using a method called Mendelian randomization, and experimental studies with interventions, that LDL does not only indicate an increased risk but also causes issues in the arteries which lead to cardiovascular diseases. This strong correlation between high LDL-cholesterol levels in the blood and heart disease risk meets all of the criteria of causality. Numerous studies, including those that examined people's genetics, have consistently shown that higher LDL-cholesterol levels increase the risk of developing heart disease. Moreover, long-term high LDL-cholesterol further increases the risk of heart disease. It is crucial to establish causality, as the prevalence of high LDL-cholesterol in a population is typically around 25% and can be even higher. This means that at least a quarter of the global population is at risk of heart disease and stroke as a result of high LDL-cholesterol concentrations.

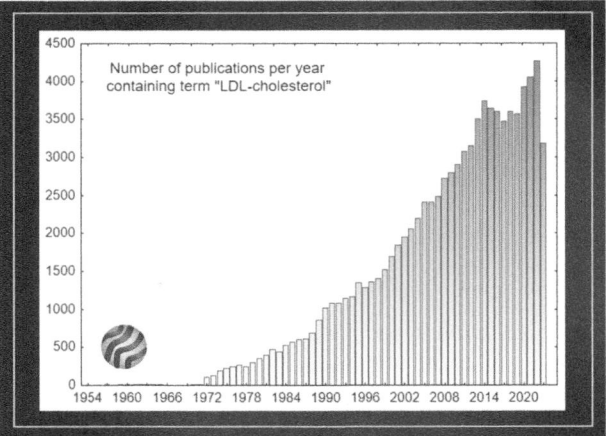

In the 1970s, studies of LDL-cholesterol began. They have continued to grow ever since.

Interesting Numbers

In the United States, 33.5% or 71 million adults have high LDL-cholesterol.

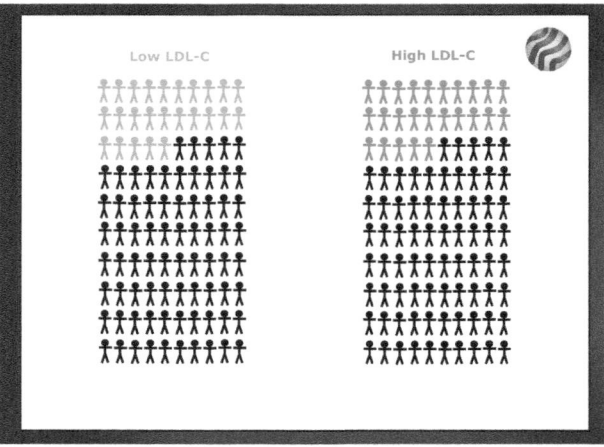

About half of the population has either high or low levels of LDL-cholesterol. The precise proportion depends on the definition of these conditions.

Practical Aspects: How to Measure LDL-Cholesterol

LDL-cholesterol is a key factor in the evaluation of risk for heart disease and stroke. This makes it important to accurately measure the LDL-cholesterol concentrations of a person. A standardized method is available to measure LDL-cholesterol in a small amount of blood plasma. Using a commercially available kit of reagents and a photometer, the concentrations of total cholesterol and triglycerides are first measured in plasma. In a similar manner, HDL-cholesterol is measured in plasma after removing other lipoproteins from plasma by precipitation with a specific reagent. Third, VLDL-cholesterol is calculated either by dividing the plasma concentration of triglycerides by five using a formula devised by William Friedewald and coworkers in 1972 or by other methods. Fourth, LDL-cholesterol is calculated by subtracting HDL-cholesterol and VLDL-cholesterol from total cholesterol. The entire measurement is usually completed in one hour.

How to calculate LDL-C

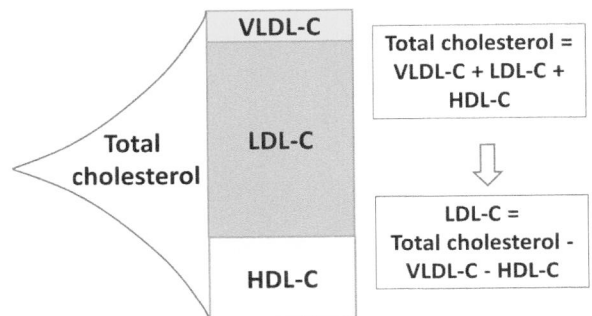

A lipid test measures concentrations of two important lipids, specifically cholesterol and triglycerides. Lipoproteins are divided into three types: HDL, LDL, and VLDL. The latter represents a large portion of remnant lipoproteins when fasting. When we speak of cholesterol in lipoproteins, we mean HDL-cholesterol, LDL-cholesterol, and VLDL-cholesterol. LDL-cholesterol is calculated by subtracting HDL-cholesterol and VLDL-cholesterol from total cholesterol.

Friedewald's formula was proposed as a replacement for the need to separate VLDL particles by ultracentrifugation to calculate LDL-cholesterol. Friedewald and coworkers found that when they studied the composition of VLDL particles from hundreds of patients, VLDL-cholesterol was roughly equal to plasma triglycerides divided by five. Notably, the relationship stayed for patients without marked hypertriglyceridemia, i.e. those with plasma triglycerides below 400 mg/dl. Patients with marked hypertriglyceridemia being rare, it was concluded that the formula could be used for a large population. The method of calculating LDL-cholesterol is called "beta-estimation" because LDL is beta-lipoprotein. In patients with marked hypertriglyceridemia, ultracentrifugation is used to isolate LDL and to determine its cholesterol content. This method is more accurate and is therefore called "beta-quantification." The beta-estimation has another limitation: Friedewald's LDL-cholesterol calculation includes both LDL-cholesterol and IDL-cholesterol. This formula is therefore an approximation.

Recently, it has been suggested that more complex formulae be used to improve the accuracy of LDL-cholesterol calculations. The Martin–Hopkins formula, which was introduced in 2013, as well as the Sampson NIH equation developed in 2020 are two examples. These equations are still used relatively little, despite their advantages over Friedewald's formula.

Direct methods for determining LDL-cholesterol were developed to improve accuracy. This is achieved by masking or consuming the cholesterol in the other lipoproteins fractions. However, these methods are not commonly used because they can often give inaccurate results.

It is important for clinicians to remember that LDL-cholesterol levels are not always stable and can change based on several factors, even if lipid-lowering treatments are not used. These factors include the ambient temperature, the season, and certain medications.

The terms "cholesterol," "LDL," and "LDL-cholesterol" are often used interchangeably, which can cause confusion. Cholesterol plays a role in the production of bile acid and hormones. It is also an important component of cell membranes. LDL is the main carrier of cholesterol. In medical practice, we estimate LDL levels by using its cholesterol concentration, abbreviated LDL-cholesterol. This is used to assess the risk of heart disease and to evaluate the effectiveness of treatments in clinical trials.

Interesting Numbers

LDL-cholesterol rises by 1.9% (with two-day lag) for every +5°C increase in the mean ambient temperature.

LDL and LDL-cholesterol are not the same. LDL is a type of lipoprotein composed of cholesterol (both esterified and nonesterified), phospholipids, proteins, and a small amount of triglycerides. LDL-cholesterol refers specifically to the cholesterol contained within LDL. Measuring the concentration of LDL-cholesterol in the blood can help determine the risk of developing heart disease. In laboratory settings, LDL-cholesterol includes all subtypes of LDL (LDL1 through LDL5) as well as Lp(a) cholesterol. However, in clinical jargon, "LDL" is often used instead of "LDL-cholesterol," which is technically incorrect.

Practical Aspects: How to Measure Lipoprotein Particle Concentrations by NMR

The NMR spectroscopy is based on the fact that the signal from certain parts of lipoprotein particles, called terminal lipid methyl group protons, changes with the particle size. This is due to differences in the local magnetic fields in the particles. By analyzing the NMR signals, this technique can be used to measure different lipoprotein subgroup sizes. This allows the determination of concentrations of VLDL, LDL and HDL particles in the blood. This technique has some limitations, including assuming that each subtype contains a fixed number of methyl protons, making arbitrary distinctions between large, medium, and small subtypes, and dealing with the interference of other lipoproteins.

**How to measure concentrations
of lipoprotein particles by NMR**

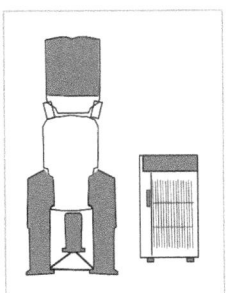 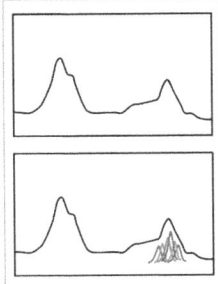

The amount of LDL-cholesterol in the blood is usually closely related to the LDL particles' count. It follows that measuring LDL-cholesterol levels will give a good indication of the number of LDL particles. In certain conditions, such as metabolic syndrome, high triglyceride, or diabetes, the relationship between LDL-cholesterol and LDL particles can, however, be inconsistent. It is possible that there are more small, dense LDL particles with less cholesterol. In such cases, determining LDL-cholesterol may not accurately reflect the number of LDL particles or the impact of LDL on heart disease. Directly measuring the particle number or concentration of apolipoprotein B (one molecule of which is present in every LDL particle) may provide a better reflection of how LDL impacts heart disease risk. Most studies use LDL-cholesterol as an estimation for measuring the concentration of LDL.

What It Is

Nuclear magnetic resonance (NMR) is a physical phenomenon when the atom nuclei in a strong magnetic field are disturbed by a weak magnetic field and produce a signal with a frequency that shows the strength of the magnetic field at the nucleus. NMR spectroscopy is a technique that measures this signal.

History

Plasma Cholesterol and Heart Disease

Heart disease became the leading cause of death for Americans in the mid-20th century. The focus of public health shifted from fighting infectious diseases to controlling heart disease. After several decades of debate, the American Heart Association, founded in 1924 and set to mark its centennial in 2024, began to address public health issues in addition to its original scientific and clinical goals. After two World Wars, the death rate among Americans was still high and heart disease was their main enemy. It was necessary to conduct scientific research in order to determine the causal factors that led to the disease. This project was launched with the goal of curing and preventing heart disease. In Bethesda, Maryland, the National Institute of Health was founded in 1946. Its mission was to promote health research. The National Heart

(Continued)

(Continued)

Institute, which was later renamed to the National Heart, Lung, and Blood Institute, focused on heart disease in 1949.

Who	When
The Framingham Heart Study	**1961**

Where	What
Framingham MA, USA	Kannel, W. B.,T. R. Dawber, A. Kagan, N. Revotskie and J. Stokes, 3rd (1961) "Factors of risk in the development of coronary heart disease-six year follow-up experience." Annals of Internal Medicine 55: 33-50.

   In the early 20th century, it became evident that the infectious disease model could not be applied to chronic diseases such as coronary heart illness. Heart disease, unlike infectious diseases, has no single cause. Hippocrates first proposed the idea of miasma, or "bad air," as a cause of many diseases including chronic disorders in the 4th century BCE. The concept was abandoned only at the end of the 19th century after the pioneering work of Robert Koch and Louis Pasteur. Multiple factors were thought to be responsible for heart disease and other chronic illnesses, like diabetes. The concept of risk factors was born out of a novel approach that involved associations, rather than causes. In order to identify these factors, it was important to eliminate confounding variables that could distort results. In the new paradigm, questions about uncertainty, causality, and probability were central. In fact, the concept of a "risk factor" implies that it does not cause the disease but increases its likelihood. As an example, elevated LDL-cholesterol does not cause heart disease in everyone. However, it increases the likelihood that it will. Causation cannot be determined with certainty. To address these issues, and to study the multiple factors that lead to heart disease, new statistical methods were developed. These methods were based largely on correlations that suggested causal relationships but did not prove their existence. This was a turning point in the evolution of epidemiology from the old (causative), to the new (probabilistic).

   The Framingham Heart Study was the basis of this new framework. In the 1940s, Framingham was a typical American small town with approximately 25,000 residents, located 33 km east of Boston. In the early 20th century, the town had a long history of participating in epidemiological research. It was also close to prestigious institutions such as Harvard Medical School. The experts chose this town because of these circumstances.

The method was simple, but very original at the time. The National Heart, Lung, and Blood Institute designed the study, which was commissioned by Congress. Researchers wanted to know if there were any factors that explain why some people are more likely to develop heart disease than others. This study was unique in that it did not use real patients. Researchers were more interested in future patients who might develop the disease. This approach, called prospective, was quickly adopted in epidemiology. An innovation was the understanding of how atherosclerosis develops and causes heart disease. The doctors of the day viewed it as part of natural aging and did not acknowledge lifestyle or other factors. It was revolutionary to even consider such factors.

The longitudinal study began in 1948, with a randomly selected group of 5,209 people aged 30–59 without any signs of heart disease. The participants agreed to provide free consultations for a 20-year period. Only 43 myocardial ischemic events were detected after four years. However, even this short period of time allowed for the identification of three conditions that are more common in those who experience these events, including elevated blood pressure, elevated blood cholesterol, and obesity. Men and women have different cardiovascular risks, with more men having events.

The initial observations were confirmed by further follow-up. The data revealed, among other findings, that high plasma cholesterol concentrations were a predictor of myocardial infarction. In addition, those who had an event were more likely to smoke and be inactive. The investigators were able to derive from the analysis of these data the concept of coronary risk factors that William Kannel, along with his colleagues, proposed in 1961. They showed, among other things, that risk factors were associated with myocardial infarction and preceded its occurrence. The researchers also differentiated between positively and negatively associated risk factors, i.e. those that were linked to a higher (positive) or a lower (negative) probability of developing the disease. They also showed that risk factors can act synergistically, amplifying one another, so that their combined effects are greater than their individual effects.

In fact, the concept of risk factors has been around since the 19th century. It was used in the insurance industry to identify factors that reduce life expectancy, such as elevated blood pressure or body weight. The Framingham Heart Study was the first to provide a scientific answer. The term "risk factor," as used by doctors and the public, was first introduced at this time. They accepted the concept with surprising ease, perhaps because of their traditional belief that a person's health is a combination of genetics, environment, and behavior. Further research has added social stress as a risk factor for coronary heart disease. Thus, the concept of multifactoriality in coronary heart disease became firmly established.

The Framingham Heart Study was originally intended to run for 20 years. In 1968, the study committee concluded that it was time to end the study, as the hypothesis of the study had been thoroughly evaluated. The Congress did not accept this recommendation and instead voted to continue with the study. In 1971, the Framingham Cohort was monitored for an additional year. Not only did plasma cholesterol but LDL-cholesterol also appear as a risk factor for coronary heart disease. In the 1970s, elevated blood pressure became a stroke risk factor. In the 1980s, low HDL-cholesterol was added to the list. The results of a further follow-up study showed that postmenopausal women are at heightened risk for heart disease in comparison to women who were premenopausal. The research identified menopause in women as a significant risk factor.

(Continued)

(Continued)

It is hard to believe the study continues and that it is now in the fourth generation of participants. Members of the first cohort, as well as the children and grandchildren of those members, volunteered to submit to a medical examination, a physical exam, and medical testing every three to five years. This informed consent produced a wealth of data on the subjects studied, including their psychological and physical characteristics. Framingham Heart Study became the gold standard for a longitudinal cardiovascular epidemiology study because of its scope, quality, and duration. The study was a catalyst for a number of similar projects including the Busselton Health Study in Australia, the Copenhagen City Heart Study in Denmark, and the Caerphilly Heart Disease Study in the United Kingdom.

The results of this study have allowed for the development of a calculator that can estimate the 10-year cardiovascular risks of individuals without or with diagnosed heart disease, provided they know their age, gender, LDL-cholesterol, HDL-cholesterol, blood pressure, and hypertension treatment. They also need to report if they smoke and have diabetes. The Framingham Score can be used to estimate the risk. Low risk is defined by the probability of 10% or less that coronary artery diseases will develop in 10 years. Intermediate risk is 10–20% and high risk is 20% or more.

Age is the strongest predictor of cardiovascular risks in any risk equation. The Framingham Risk Score is an important tool that can be used to determine the benefits of prevention. It also helps to choose lifestyle changes and medical treatments in order to prevent heart disease.

Practical Aspects: How to Measure Blood Lipid Values

How to measure blood lipids
Total cholesterol

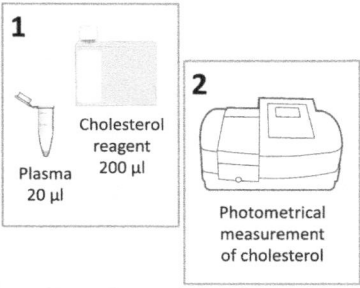
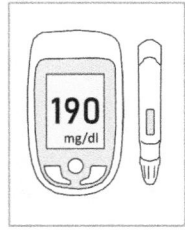

Evaluating lipoprotein levels in the blood is crucial for assessing the risk of cardiovascular disease. It is important for everyone to have these levels measured, as it could potentially save lives. LDL-cholesterol, HDL-cholesterol, and triglyceride levels are typically measured when analyzing blood lipids. Additionally, measurements of apolipoprotein B, apolipoprotein A-I, and lipoprotein(a) can be performed if requested by a doctor.

5

LDL – What It Is

This image shows an LDL particle in the circulation. The molecule of the apolipoprotein B-100 is shown by brown ribbons surrounding the particle. Phospholipids are represented by pink circles. Red blood cells (erythrocytes) and white blood cells (leucocytes) are depicted as red and white stars in the background. The size of LDL is magnified 400 times larger than that of the cells.

LDL particles, like other lipoproteins, help lipids dissolve outside of cells in water and move them through the body. All LDLs are spherical, and they function as vehicles that transport lipids in the blood. LDL is rich in cholesterol, and it is the major carrier of this substance in plasma.

LDLs are complex structures that include lipids, proteins, and other components. Visual representations of LDL show this characteristic, as they always combine lipids with proteins. Even artistic representations created by artificial intelligence software that imitate famous artists like Leonardo da Vinci and Juan Miro depict the dual composition. The algorithm creates images that allow the separation of the two main components of LDL.

Approximately 80% of the mass of an LDL particle is composed of lipids. Each particle contains around 3,000 lipid molecules, with the majority (2,000–2,700) being cholesterol molecules, and most of those (about 1,600) being esterified.

Apolipoprotein B-100 is the main protein found in these particles. It helps maintain the particle structure. Apolipoprotein B-100 is found in each LDL particle, along with several other proteins. Apolipoprotein B-100 is one of the biggest proteins in mammals. Although other proteins may be smaller, their numbers can be large. In fact, the average LDL particle contains more than 20 additional proteins.

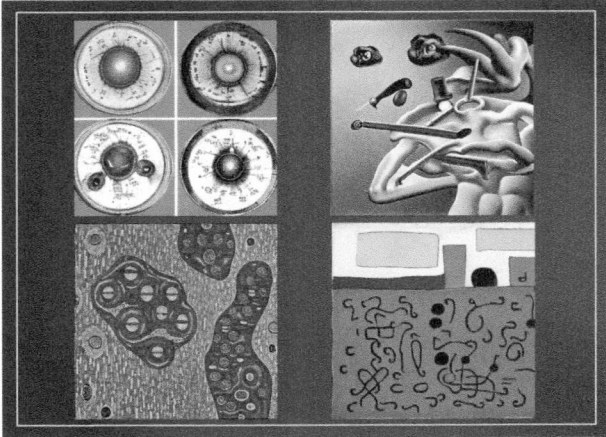

Stable Diffusion AI can generate artistic representations of LDL (clockwise starting at the top left), in the styles of Leonardo da Vinci, Salvador Dali, Juan Miro, and Gustav Klimt, according to a prompt that included the artist's name and the term "low-density lipoprotein." Source: Black Technology LTD/https:// stablediffusionweb.com/ (accessed 9 February 2024).

LDL particles, like other lipoproteins, have an outer shell and an inner core. The core is made up of hydrophobic lipid molecules, mostly cholesteryl ester with a small amount of triglycerides. In one LDL particle, there are on average 170 molecules of triglycerides, which is approximately 10 times lower than the number of cholesteryl esters. The core is surrounded by a shell of phospholipids and nonesterified cholesterol, with around 700 and 600 molecules, respectively, in each LDL particle.

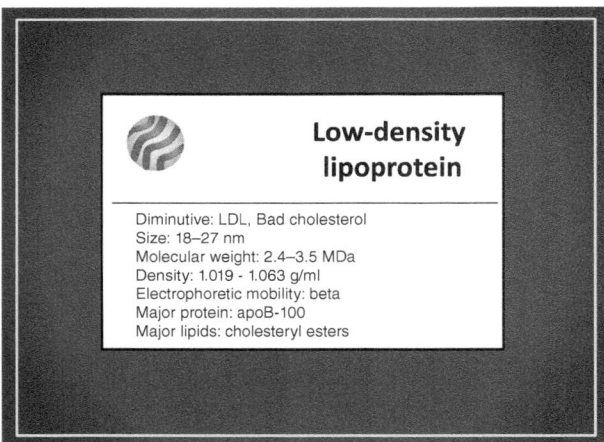

The apolipoprotein B-100 is wrapped around the LDL particle's surface as a belt, and it interacts with a few lipids both on the outside and the inside. The protein is composed of domains which can move around the surface and contribute to the flexibility of the particle. Indeed, studies found that apolipoprotein B-100 in LDL is made of flexible and ordered parts.

Apolipoprotein B-100's flexibility results in a variety of sizes and compositions for LDL particles. These particles can vary in mass and size due to differences in lipid content, ranging from 22 to 27 nm with an average size of 25 nm and an average mass of 3 MDa. Additionally, their density can range from 1.019 to 1.063 g/ml. The composition of the particles is constantly changing based on the lipids they contain, making it challenging to study their specific details due to their dynamic nature and complex structure.

Cryoelectron microscopy is a sophisticated technique that can examine LDL particles at their natural state. The measurements reveal that the particles are made of different rigid and flexible parts, which are composed of lipids. Analysis allows for the creation of 3D models that reveal rigid and flexible domains of lipids and apolipoprotein B-100. The models show that the particles are composed of a core with layers of lower density and an outer layer with a higher density. The knob-shaped protrusion is also visible. At a lower temperature, the particles are more compact and have more ordered lipid molecules, but the mobility of apolipoprotein B-100 remains largely intact. The particle structure is more visible at higher temperatures. Analyzing the models allows for the approximate location of each domain of apolipoprotein B-100 on the LDL surface.

Proteins

Apolipoprotein B-100

Apolipoprotein B-100 is the main protein found in LDL. It plays a crucial role in transporting lipids, such as cholesterol, to all cells in the body. The protein is present not only in LDL but also in VLDL, lipoprotein (a) (Lp(a)), and IDL. Apolipoprotein B-100 is responsible for organizing these particles and facilitating their formation. Additionally, it helps bind lipoprotein particles to cells in order to deliver the lipids they carry. This process is facilitated by specialized proteins on the cell surface known as receptors. However, due to its large size and tight association with lipids, isolating apolipoprotein B from plasma is challenging. As a result, the structure and organization of the protein in LDL remains insufficiently studied.

Apolipoprotein B-100 is one of the two main types of apolipoprotein B, which is made by the liver. The other type is called apolipoprotein B-48 and is only produced by the small intestine.

Apolipoprotein B-100 is 550 kDa in size and contains 4,563 amino acids as building blocks. Both types of apolipoproteins B are made by a gene named *APOB*, with a major role played by messenger RNA. A special process known as RNA editing is used to create a stop sign at position 2,153 of the genetic material. This corresponds with 48% apolipoprotein B-100. A special gene determines the type of apolipoprotein that will be produced. Due to RNA editing, apolipoprotein B-48 and apolipoprotein B-100 share a similar starting part but have different endings. The name "apolipoprotein B-48" is derived from the fact that 48% of full apolipoprotein B-100's sequence is made up by apolipoprotein B-48. Apolipoprotein B-48 can only be found in the chylomicrons produced by the small intestine.

It is important to understand that each VLDL, IDL, LDL, or Lp(a) particle from the liver contains only one apolipoprotein B-100 molecule, while each chylomicron from the intestine contains only one apolipoprotein B-48 molecule. This means we can determine the number of lipoprotein particles by looking at the total amount of apolipoprotein B-100 and apolipoprotein B-48 in the blood. Since each particle has one molecule of apolipoprotein B, the concentration of apolipoprotein B tells us how many particles there are.

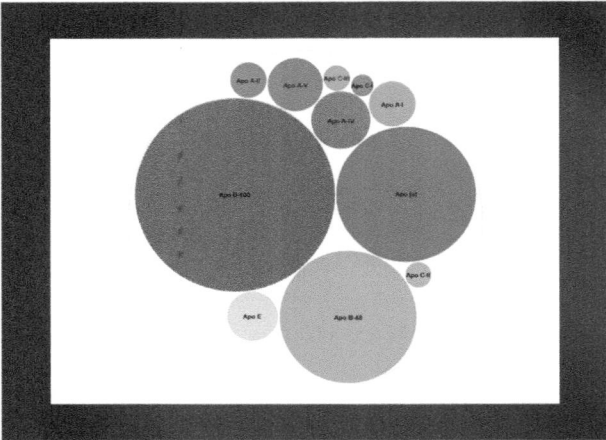

Size matters. The molecular mass and size of apolipoproteins are vastly different. The size is linked to what they can do and how. The following are the masses (in kDa), in decreasing order: apolipoprotein B-100, 512; apolipoprotein (a), 250 (the lowest possible); apolipoprotein B-48, 241; apolipoprotein A-IV, 45; apolipoprotein A-V, 39; apolipoprotein E, 34; apolipoprotein A-I, 28; apolipoprotein A-II, 17; apolipoprotein C-II, 8.8; apolipoprotein C-III, 8.8; apolipoprotein C-I, 6.6. The size of the circle is proportional to the molecular mass of a protein.

Dalton, or Da, is a unit used to measure the mass of atoms and molecules. It is defined as 1/12 of the mass of a carbon atom. The mass of a carbon atom is 12 Daltons. Big molecules can have masses of thousands or even millions of Daltons, which are often written as kDa (kilodalton) or MDa (megadalton).

Apolipoprotein (a)

Apolipoprotein (a) is only found in the subtype of LDL particles known as lipoprotein (a), in short Lp(a). This lipoprotein floats in a density between 1.050 and 1.080 g/ml and migrates more quickly on electrophoresis than LDL. Apolipoprotein (a) is available in different sizes. Some people have larger versions of the protein and others smaller. This variation is due to a variable number of fragments known as kringle 4 repeats within the *LPA* gene that codes for apolipoprotein (a). The protein may contain 10–50 fragments, each of which is 114 amino acids long. It is interesting that apolipoprotein (a) is mostly found in primates. However, it has also been discovered in hedgehogs, where it looks different from the human version but still shares some similarities.

Apolipoprotein (a) is similar in structure to tissue plasminogen activator and plasminogen, which are both important for the breakdown of blood clots. It competes with plasminogen which decreases the breakdown rate of blood clots. It can also lead to blood clots by stimulating the production of the protein plasminogen activator inhibitor-1. It can also make blood clotting more easy by preventing another substance, tissue factor pathway inhibitor, from working properly.

Apolipoprotein (a) binds also to certain oxidized lipids which cause inflammation, attract immune cells, and cause muscle cells to grow within the blood vessels. It can also be involved in the healing of wounds and repair of tissues.

Some people have low apolipoprotein (a) levels and still appear to be healthy. Apolipoprotein (a) in the blood does not therefore seem to be required for health under normal circumstances. It is mostly found in humans and monkeys, so it may not be necessary for survival. However, in some situations such as exposure to certain diseases, this protein could be beneficial.

LDL is unique in that it contains only trace quantities of other apolipoproteins besides apolipoproteins B-100 and (a). Thereby, LDL is distinguished from other lipoproteins that contain a variety of apolipoproteins including apolipoproteins E, C, and others.

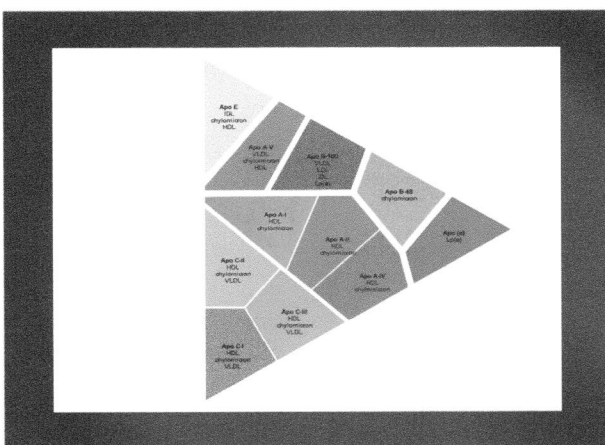

Apolipoproteins are not distributed equally across lipoproteins. The LDL fraction contains only apolipoproteins B-100 and (a), while other lipoproteins contain a variety of apolipoproteins ranging from apolipoprotein A-I to apolipoprotein E.

Enzymes and Other Proteins

LDL also contains enzymes and proteins other than apolipoproteins.

Lecithin-cholesterol acyltransferase (LCAT) is a well-known enzyme in LDL. It has a molecular mass of 63 kDa. LCAT catalyzes esterification of cholesterol to cholesteryl esters in plasma lipoproteins. This occurs primarily in HDL but can also occur within particles containing apolipoprotein B. Apolipoprotein B-containing lipoproteins account for approximately 25% of the plasma LCAT activity. LCAT's three-dimensional structure is maintained by special disulfide S-S bonds and is similar to that of other hydrolases and lipases. The enzyme transfers the fatty acid residue of a phospholipid molecule to a cholesterol molecule. The lipid core of lipoproteins accumulates cholesteryl esters that are formed by the LCAT-catalyzed reaction between phospholipids, cholesterol, and fatty acids.

What It Is

Lecithin-cholesterol acyltransferase is an enzyme that plays a central role in the fate of cholesterol within blood plasma. LCAT catalyzes the esterification of cholesterol into cholesteryl ester within plasma lipoproteins.

Another enzyme found in LDL is platelet-activating factor-acetyl hydrolase (PAF-AH), also known as lipoprotein-associated phospholipase A_2 (LpPLA$_2$). This 53 kDa enzyme breaks down platelet-activating factor, which is a bioactive lipid with many functions in the cell. PAF-AH thereby regulates platelet-activating factor's biological action. The enzyme also cleaves highly pro-inflammatory oxidized short-chain phospholipids but does not affect long-chain, non-oxidized phospholipids. The primary source of circulating PAF-AH is the macrophage. Plasma PAF-AH is primarily bound to atherogenic LDL particles that are small and dense, and to Lp(a). The active site of the enzyme is near the surface of the lipoprotein while still being accessible to the aqueous solution.

What It Is

Platelet activating factor-acetyl hydrolase (PAF-AH) is an enzyme found on LDL or HDL. It breaks down platelet-activating factor, which is a bioactive lipid with many functions in the cell. PAF-AH also cleaves bioactive lipids like oxidized phospholipids.

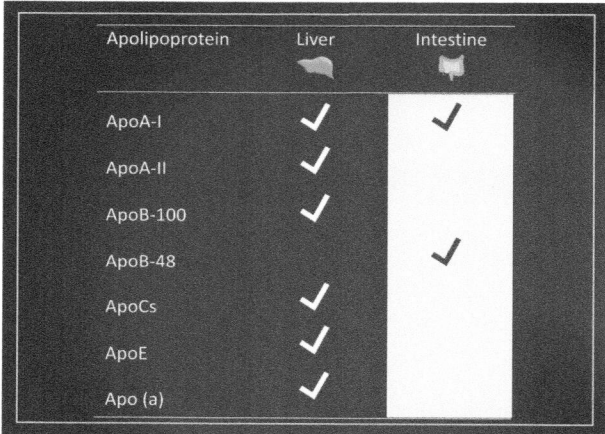

The liver is the winner! The liver produces the majority of the apolipoproteins, while only a small amount is produced in the intestine.

Practical Aspects: Difficulties of Lipoprotein Isolation from Blood

Ultracentrifugation of lipoproteins in highly concentrated solutions of salt can change the distribution and amount of certain proteins. Proteins that are loosely attached to lipoprotein surfaces can be transferred from the lighter LDL and VLDL fractions to the denser HDL fractions. Other methods of lipoprotein separation, such as immunoaffinity chromatography and precipitation, can result in particles heavily contaminated by plasma proteins, or exposed to harsh conditions which may alter their structure and composition.

Lipids

LDL, like other lipoproteins in plasma, contains four main lipid classes: phospholipids, cholesteryl esters, free cholesterol, and triglycerides. Plasma LDL typically contains 35–40% of cholesteryl ester, 20–25% phospholipids, and 8–12% triglycerides. It also contains 5–10% free cholesterol. Recent mass-spectrometric analyses called lipidomics identified hundreds of molecular lipid species in

LDL and other plasma lipoproteins. The number of individual molecules of lipid in the human body has been estimated to be several hundred thousand. It appears that at present, the only limit on the number of lipid molecules detected in lipoproteins comes from the sensitivity available in technologies.

Cholesteryl ester is the main lipid class in LDL. It accounts for around half of all lipid mass. Cholesteryl linoleate is responsible for the majority of cholesteryl esters. It is interesting that the majority of plasma cholesteryl ester are formed by LCAT in HDL, not LDL. The cholesteryl esters produced in this reaction are first accumulated in HDL's lipid core and then transferred to LDL, and other large apolipoprotein B-containing lipoproteins, under the influence of the cholesteryl ester transfer protein (CETP), which exchanges the molecules for triglycerides.

The second largest lipid class in LDL is phospholipids. They are essential for maintaining the structure and function of LDL, as well as other lipoproteins. Phosphatidylcholine, which makes up approximately 70% of LDL's phospholipid, is a structural lipid. Sphingomyelin, the second major surface lipid in LDL, is an important factor for enhancing surface rigidity. LDL is an important plasma carrier of sphingolipids including sphingomyelin, ceramide, and others. Ceramides are signaling molecules that play a role in cell death, growth, and differentiation. Their accumulation in LDL may be harmful.

Lysophosphatidylcholines, low-molecular-weight lipolytic products of the LCAT reaction, constitute a minor subclass of LDL phospholipids. They are pro-inflammatory. Phosphatidylinositol, phosphatidylserine, and phosphatidic acid are negatively charged minor phospholipids that may, however, significantly impact on the net surface charge of LDL particles, thereby modulating interactions of LDL with biomolecules and cells. In addition, LDL contains considerable amounts of phosphatidylethanolamine and plasmalogens.

Sterols like cholesterol are found in the monolayer of lipids at the surface of LDL and help regulate their fluidity. Cholesterol is the dominant sterol in LDL. This reflects the important role played by this lipoprotein for cholesterol transport throughout the body. LDL contains other sterols at lower levels, as shown by the minor amounts of phytosterols and oxysterols.

Triglycerides in LDL are the remnants of the triglyceride-rich core of VLDL particles. Triglycerides are more fluid than cholesteryl ester and increase the fluidity of the particles. Other minor bioactive LDL lipids include diacylglycerides, monoacylglicerides, free fatty acids, glycosphingolipids, gangliosides, sulfatides, and lysosphingolipids.

Multiple oxidants in the human body can oxidize many of the lipid molecules that are carried by LDL. These oxidation-prone lipid molecules are primarily those with readily oxidable polyunsaturated moieties. Oxidized fats have pro-inflammatory properties that can increase cardiovascular disease. To protect lipids from oxidation, LDL is rich in low-molecular-weight antioxidants such as coenzyme Q10, and tocopherols (vitamin E).

LDL also contains multiple sugar components as a part of glycosylated proteins and lipids. It additionally contains microRNAs that can regulate gene transcription in various tissues.

Practical Aspects: How to Measure Molecular Lipoprotein Composition by Mass Spectrometry

Lipoproteins are primarily composed of lipids and proteins. Each is made up of a multitude of molecules. Mass spectrometry, a highly sensitive and precise analytical technique, has recently been developed for the identification and quantification of individual molecular species of proteins and lipids. Proteomics and lipidomics are two of the omics techniques. These methods can reliably detect hundreds of proteins and lipids in lipoproteins. Mass spectrometry also

(Continued)

(Continued)

allows measurements of sugar residues in proteins and lipids, which are called glycoproteins and glycolipids. This glycomics approach allows for the glycomic profile of lipoprotein particles.

How to detect proteins and lipids in lipoproteins

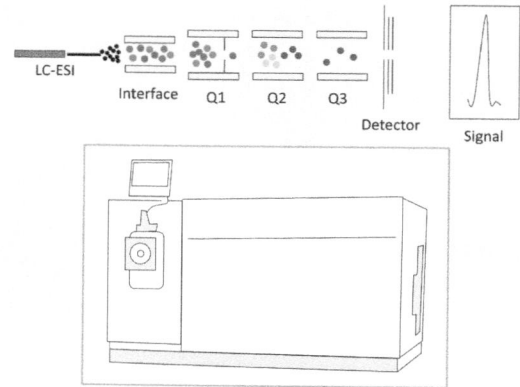

When studying proteins and lipids using mass spectrometric methods, the molecules of interest are first dissolved in a suitable solvent and then chemically modified if necessary. They are then separated by class using liquid chromatography (LC) before being injected into a mass spectrometer as a spray of electrically charged droplets created through electrospray ionization (ESI). The molecules are then separated from each other under vacuum in a first quadrupole device (Q1) and broken into smaller ions in a second quadrupole (Q2). The ions produced are further separated in a third quadrupole (Q3) and quantitated in a detector, resulting in a specific signal for each molecule.

Heterogeneity

Large, Light, and Small, Dense LDL

LDL particles are composed of subtypes with different characteristics.

LDL can be subdivided into different subtypes depending on their size, using methods such as NMR, size exclusion chromatography, and ion mobility. There are three main subtypes of LDL: large, medium, and small. Healthy individuals with normal levels of lipids have a higher proportion of medium-sized LDL particles than large or small particles.

A special technique, called non-denaturing gradient gel electrophoresis, can also be used to separate LDL subtypes according to their size and charge. These subtypes range from LDL1 through LDL7, and each has a different size range. LDL1 is the largest while LDL7 is the smallest.

Ultracentrifugation separates lipoproteins by density. The LDL pool can be subdivided into five different types using this technique. The subtypes of LDL particles isolated by ultracentrifugation range from large, light LDL1 and LDL2 to intermediate LDL3 and small, dense LDL4 and LDL5 particles. The intermediate LDL3 subtype is most prevalent in healthy people with normal cholesterol levels. The density of LDL particles ranges between 1.013 and 1.063 g/ml. This technique shows large LDL particles to be less dense and smaller particles denser. As the density of the particles increases, their size decreases. This is accompanied by an increase in protein content. Parallel to this, the particle's surface-to-core ratio also increases.

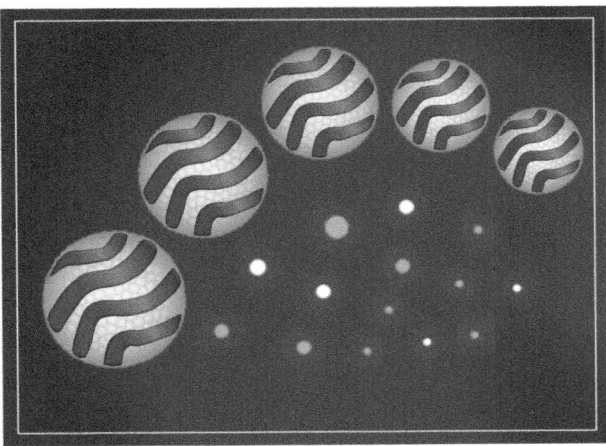

There are five major subtypes of LDL: large LDL1 and LDL2, intermediate LDL3, and small LDL4 and LDL5. Red blood cells (erythrocytes) and white blood cells (leucocytes) are depicted as red and white stars in the background. The size of LDL is magnified 400 times larger than that of the cells.

The profile of LDL particles can be altered by disease. Pattern A is a type of LDL particle profile where large particles are concentrated more than small particles. Pattern B is a pattern with elevated levels of small LDL. Researchers believe the pattern B increases the risk of heart disease because smaller particles are more likely to enter the arterial walls. Pattern B, which involves higher levels of small, dense LDL, is more common in people who have higher triglyceride, like those with diabetes. It is possible that the pattern B is simply a reflection of hypertriglyceridemia. The concentration of triglycerides within the blood plays a major role in determining the profile of LDL particles. As triglyceride concentrations increase, the LDL profile shifts from large, light to small, dense particles. The gender also plays a role; men tend to have more of small, dense LDL than women.

Several studies suggest that heart disease is more closely linked to pattern B than it is with LDL-cholesterol levels as measured by a standard lipid test. The tests of the LDL particle profile are, however, more expensive and less widely available. Therefore, the measurement of LDL-cholesterol remains more popular.

Apolipoprotein B-100 is the main reason for differences between LDL particles. The unique properties of this protein define the shape of LDL. Apolipoprotein B-100 binds lipids, and its shape changes depending on the amount of lipids bound. The protein can create differently sized LDL particles. Apolipoprotein B-100 is very flexible, and it can bind to lipids in order to form stable particles.

> **Practical Aspects: How to Evaluate LDL Heterogeneity**
>
> Different methods can be used to study differences between lipoproteins. The results are affected by the choice of method. Among the most common methods are gel electrophoresis, analytical, sequential, zonal and density gradient ultracentrifugation, NMR, size exclusion chromatography, gel filtration, ion mobility, and affinity chromatography. All of these methods show that LDL particles can be classified into three main types: large, intermediate, and small. Intermediate LDL3 is the most common form of LDL. People with heart disease often have higher levels of small, dense LDL particles.
>
> While these methods can provide useful information about LDL types, they are only able to give an overview of the situation at a particular time-point. The study of how certain components are transported between the different types of LDL particles during the entire lifetime of LDL can give valuable insight into health and disease. Unfortunately, these methods are not very well developed at this time.

Lipoprotein (a)

Lipoprotein (a) is a subtype of LDL found in the blood. Lp(a) is very similar in composition to LDL, but it additionally contains apolipoprotein (a). The size of this protein can vary, with some individuals having larger proteins and others having smaller ones. Those with larger proteins typically have lower levels of lipoprotein (a) in their blood, possibly because it takes longer for the larger proteins to be produced and released into the bloodstream by cells.

LDL and Lp(a), which additionally contains apolipoprotein (a).

Increased concentrations of Lp(a) can increase the risk of stroke, heart disease, aortic stenosis and blood clots. This effect is independent of LDL-cholesterol levels. Genes are largely responsible for the amount of Lp(a) in the blood.

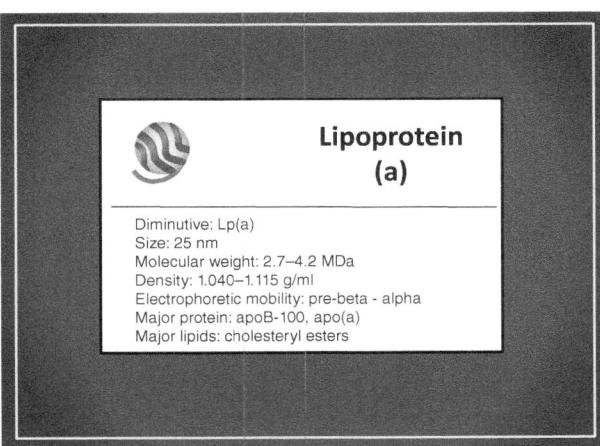

Lipoprotein (a)

Diminutive: Lp(a)
Size: 25 nm
Molecular weight: 2.7–4.2 MDa
Density: 1.040–1.115 g/ml
Electrophoretic mobility: pre-beta - alpha
Major protein: apoB-100, apo(a)
Major lipids: cholesteryl esters

6

LDL – Where It Comes From

LDL particles are produced in the body through a complex process that involves several steps. The process begins in the liver and then continues in the blood where proteins and enzymes help convert other lipoproteins into LDL. This latter is pivotal in cholesterol metabolism, as it delivers cholesterol to various tissues within the body. To maintain a healthy LDL level, the liver then removes LDL from the blood.

The processing of LDL involves several steps, which are illustrated in a clockwise direction, starting from the top left. Initially, the liver releases VLDL, which is subsequently broken down by lipoprotein lipase into IDL and LDL. The LDL then attaches to LDL receptors and is removed from the blood. However, LDL can also penetrate the arterial wall and cause atherosclerosis.

Lipolysis

LDL is arranged around a molecule of apolipoprotein B-100. This protein is produced in the liver where it is mixed with lipids, first of all with triglycerides. This combination, however, is not secreted into the blood as LDL. Instead, larger VLDL particles with a high content of triglycerides are formed. The VLDL particles when in the bloodstream are first broken down into remnants or IDL. Only then are they converted to LDL. IDL is an intermediate form in the breakdown of VLDL. LDL, on the other hand, is its terminal form. Lipases, which are lipolytic enzymes, cause these transformations. These enzymes are responsible for breaking down the triglycerides in the lipoprotein particles. Lipoprotein lipase is the most important lipase found in the bloodstream.

Cholesterol, Lipoproteins, and Cardiovascular Health: Separating the Good (HDL), the Bad (LDL), and the Remnant, First Edition. Anatol Kontush.
© 2025 John Wiley & Sons, Inc. Published 2025 by John Wiley & Sons, Inc.

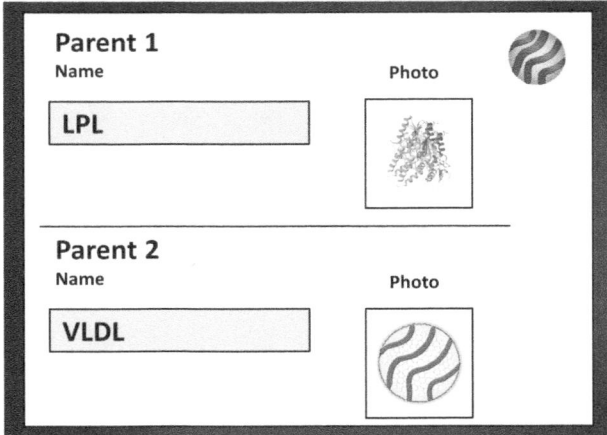

Parent 1

Name

LPL

Photo

Parent 2

Name

VLDL

Photo

Lipoprotein lipase (LPL) and VLDL are key factors in the formation of LDL.

What It Is

Lipases help break down lipids within the body. Transformations of lipoproteins occur under the action of several lipases, such as lipoprotein lipase, hepatic lipase, endothelial lipase, PAF-AH, and LCAT.

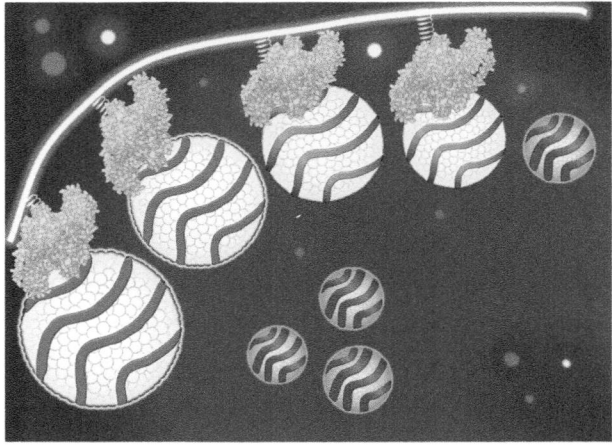

Lipolysis of VLDL into IDL and further into LDL is performed by lipoprotein lipase. This enzyme breaks down triglycerides present in the lipoprotein particles. To be active, the lipase must be in a dimer form, comprising two identical molecules.

Lipoprotein Lipase

Lipoprotein lipase is an enzyme that breaks down triglycerides in VLDL, IDL, and chylomicrons. This process releases fatty acids which are taken up by the muscles and adipose (fat) tissues where they are used for energy production. Lipoprotein lipase can be found primarily in the muscles, heart, and adipose tissues. Lipoprotein lipase is active in dimers composed of two identical molecules. Normally, the enzyme is bound to a thin layer of heparin-sulfate proteoglycans at the surface of blood vessel walls.

The process of breaking down triglycerides with lipoprotein lipase is complex and involves other proteins like glycosylphosphatidylinositol-anchored high-density lipoprotein binding protein 1 (GPIHBP1). This protein binds VLDL, chylomicrons, and other lipoproteins rich in triglycerides and facilitates their interaction with lipoprotein lipase.

What It Is

Lipoprotein lipase (LPL) is the most important enzyme for breaking down triglycerides in triglyceride-rich lipoproteins like chylomicrons, VLDL, and IDL.

Hepatic lipase is produced primarily in liver cells. It is released into the bloodstream, where it attaches to the surface cells of the liver and the blood vessel lining. Hepatic lipase is more efficient in breaking down phospholipids on the surface of LDL and HDL than triglycerides. Hepatic lipase action leads to lower LDL particle levels in the blood because they are cleared by the kidneys faster.

Two ways of making these LDL particles are possible depending on the amount of triglyceride in the liver. The liver will produce larger VLDL particles which are richer in triglyceride if there is lots of triglyceride available. These particles are known as VLDL1. They are broken down in the circulation to produce smaller LDL4 or LDL5. The liver will, however, produce smaller VLDL particles if there is not enough triglyceride to synthesize VLDL. They are destroyed, resulting in large LDL1 or LDL2.

When triglycerides are low or normal, large LDL1 and LDL2 are formed, which have a strong affinity for LDL receptors. These particles are not as stable in the blood as smaller LDL4 or LDL5 that come from VLDL1.

It is of note that the names of the particles are not entirely logical. Why should VLDL1 not be converted to LDL1 instead of LDL4 or LDL5, while VLDL2 not become LDL2 but rather LDL1 or LDL2? It is because the terms used to describe the subclasses for VLDL and LDL were developed separately. Researchers initially separated LDL into five subtypes and VLDL into two.

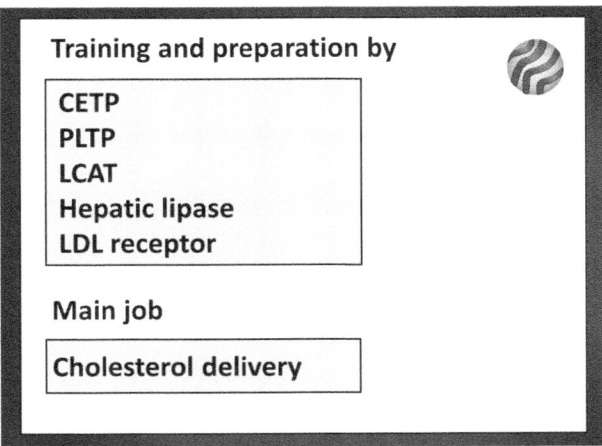

LDL processing in the blood is carried out by various proteins and enzymes, such as CETP, PLTP, LCAT, hepatic lipase, and LDL receptor. The primary role of LDL is to deliver cholesterol to cells.

Lipid Transfer

CETP

Other proteins, besides lipases are also involved in the formation LDL. CETP is one of them. This protein is involved in moving lipids from one type of lipoprotein particle to another. CETP is more specifically involved with the lipids that are found in the cores of lipoproteins. These lipids are triglycerides and cholesteryl esters. They are not water soluble and cannot leave lipoproteins by themselves. CETP transfers a triglyceride or a cholesteryl ester molecule from the lipoprotein core to another lipoprotein by moving it through water.

What It Is

Cholesteryl ester transfer protein (CETP) is an exchange protein in plasma that transfers triglyceride and cholesteryl ester molecules between different lipoproteins. It retrieves cholesteryl ester from HDL to transfer it to other lipoproteins such as VLDL in exchange for triglyceride.

CETP normally moves molecules from lipoproteins that are high in these molecules to lipoproteins that do not have many. The protein does this by transferring triglycerides from lipoproteins that are rich in triglycerides (such as VLDL, chylomicrons, and IDL) to those that are low in triglycerides. CETP also transfers cholesteryl ester from lipoproteins rich in cholesterol (such as LDL and HDL) to those that are low in cholesterol (such as VLDL and chylomicrons). This activity is a result of a molecular exchange between lipoproteins. CETP helps to make LDL particles richer in triglycerides. This pathway shows how all lipoproteins in the blood are constantly interacting with each other.

The transfer of cholesteryl ester molecules from HDL particles and the transfer of triglycerides from VLDL and its remnants influence the composition of LDL subtypes. CETP transfers triglyceride molecules from VLDL to the core of LDL particles in exchange for cholesteryl ester, resulting in LDL rich in triglycerides. Hepatic lipase can further break down these LDL particles to remove triglycerides and phospholipids on the surface. This results in smaller, denser, and more stable LDL particles. This pathway is more active in those with high blood triglyceride concentrations. It is for this reason that such people tend to have more of small, dense LDL. Men are more likely to have small, dense LDL than women because they have a higher level of hepatic lipase activity.

PLTP

Lipids on the surface of lipoprotein particles, like phospholipids and cholesterol that has not been esterified, can leave the particle without the help of other proteins. A special protein also exists to facilitate the transfer of phospholipid molecule across the aqueous phases. This protein is known as PLTP, or phospholipid transfer protein. This protein is able to exchange phospholipids among lipoproteins.

What It Is

Phospholipid transfer protein (PLTP) is an exchange protein that transfers phospholipid molecules between plasma lipoproteins.

7

LDL – What It Does

Supply of Cholesterol

LDL is essential for transporting cholesterol throughout the body. LDL receptors, which are proteins present on the surface of the cells to capture LDL particles, deliver the cholesterol to the cells. This is how cholesterol is brought to the body tissues. The liver catches two-thirds of the LDL-cholesterol, and the rest is transported to other tissues. Large LDL particles that are loaded with cholesterol make excellent vehicles for cholesterol in tissues.

The liver receives cholesteryl esters primarily from LDL. This ester is produced preferentially by LCAT in HDL and then transferred to VLDL or LDL by CETP. LDL transports 70% of the cholesteryl ester produced by LCAT in HDL to the liver via the LDL receptors. This helps remove cholesterol from the tissues via HDL. CETP is also involved in the optimization of LDL particle structure and their ability to attach to LDL receptors.

Endocytosis is a method by which animal cells obtain cholesterol from LDL. Cells make LDL receptors when they need more cholesterol. The receptors wait on the cell surface for LDL particles and go inside the cells when they catch them. The LDL particles enter the cells and are sent to subcellular structures called lysosomes, where they are broken.

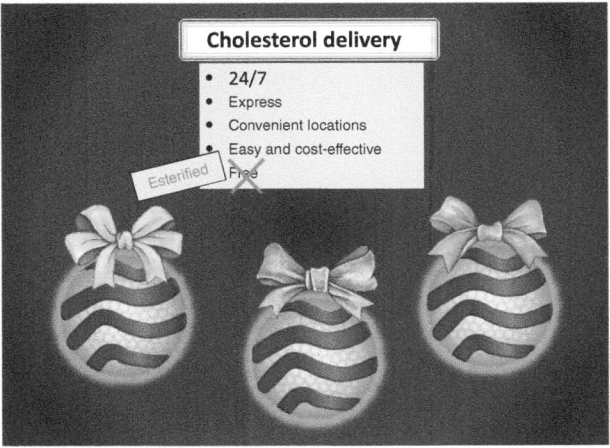

LDL particles ensure continuous delivery of cholesterol to cells in the form of cholesteryl ester.

Cholesterol, Lipoproteins, and Cardiovascular Health: Separating the Good (HDL), the Bad (LDL), and the Remnant, First Edition. Anatol Kontush.
© 2025 John Wiley & Sons, Inc. Published 2025 by John Wiley & Sons, Inc.

The lysosome's internal space is acidic. The LDL receptor changes shape when it is exposed to the acidic lysosome. The receptor's complex with LDL is broken apart by this change, allowing the receptor to return to the cell membrane. In lysosomes, cholesterol is released from LDL and moved to the the outer plasma membrane and the endoplasmic reticulum inside the cells. The cholesterol in the endoplasmic reticulum creates a feedback loop by stopping the production of receptors for LDL. Additionally, cholesterol is transformed into cholesteryl ester to be stored as lipid droplets. Animal cells regulate cholesterol by transferring it between membranes. The LDL receptors return to the cell's outer plasma membrane and the cycle begins again.

LDL particles bind to LDL receptors at the surface of hepatocytes.

Under some circumstances, LDL can actually help cells stay alive by providing them with cholesterol. It is surprising that LDL can also remove cholesterol from cells, although it is not as good at it as HDL.

LDL is also important in protecting against dangerous germs. *Staphylococcus* can be a germ which lives in the body. It can also cause serious infections. How and why can this germ go from living in the body to causing infection? In this respect, it is crucial that LDL stops this change because it can block activation of the germ. Apolipoprotein B-100 in LDL is responsible for this blocking and for protecting against *Staphylococcus*.

LDL Receptor

The LDL receptor located on the surface of the liver removes more than a half of all LDL particles. This protein interacts specifically with LDL. The receptor binds to apolipoprotein B-100, which helps remove a significant amount of LDL from the blood. Other receptors located in the liver and some in the blood vessel walls clear the rest of LDLs.

The LDL receptor is responsible for helping cells to take up cholesterol from LDL. This helps maintain the normal level of LDL-cholesterol in the blood. The process occurs in all cells but mainly in the liver which has a lot of the LDL receptors and removes approximately 70% of LDL. This happens because the LDL receptors are saturated when there is only a small amount of LDL in the blood. They cannot take in more LDL-cholesterol, especially in tissues outside of the liver. The receptors form clusters of coated pits, which are small pockets on the surface of the cells. These pockets separate to allow LDL into the cells. When exposed to acidic environments, the receptors dissociate LDL from them, fold themselves back and return to the cell surface. This rapid recycling of receptors delivers cholesterol efficiently to cells. The amount of cholesterol in the cell controls the production of receptors. If there is too much, less receptors are made. The receptors are produced by subcellular structures called ribosomes and then modified by the Golgi complex before being transported onto the surface of the cell in small sacs known as vesicles.

What It Is

LDL receptor is present on the surface of the liver and other tissues. The protein interacts specifically with the apolipoprotein B-100 on LDL and helps remove a large portion of LDL from the blood.

The LDL receptor can recognize certain parts of apolipoprotein B-100 on the surface of LDL particles. The receptor is made up of different parts, each with a different function. The terminal part which is composed of seven sections similar to each other is responsible for attaching LDL. Each section contains certain amino acids, and they form bonds with the other sections. The way that the sections retain calcium and form these bonds is important for maintaining the structure of the receptor as it enters the cell with the LDL particle bound to it.

A propeller-like part is located next to the LDL-binding part of the proteins. The propeller is what helps the protein to change shape inside cells and release bound LDL particles. A part of the protein crosses through the cell membrane and helps it travel to specific locations inside cells. This part can also be found in other proteins of similar structure.

What we eat can affect the amount of LDL receptors in the blood. A high intake of saturated fat and cholesterol can reduce the number of LDL receptors and therefore the removal of LDL. LDL levels in the blood rise as a result. Eating less fat and cholesterol, on the other hand, can increase LDL receptors. This will decrease blood levels of LDL.

The liver can tell if it needs to produce more cholesterol or less based on the amount of cholesterol in hepatocytes. The cells can take two actions if there is a shortage. They can most often increase the number of LDL receptors on their surface to import more cholesterol. They can also increase their own cholesterol production inside the cell if this does not work. This process is complex and involves a series of chemical reactions. This chain has several key steps that are controlled by enzymes.

HMG-CoA Reductase

One of the key enzymes that regulate cellular production of cholesterol is called HMG-CoA reductase, short for hydroxymethylglutaryl coenzyme A reductase. It is found in a part of cells known as the endoplasmic reticulum and has eight transmembrane domains. In normal mammalian cells, there is typically enough cholesterol from the breakdown of LDL to suppress this enzyme and limit its effects, especially when the diet contains a lot of cholesterol. However, the enzyme can be activated when there is an insufficient supply of cholesterol through the LDL receptor.

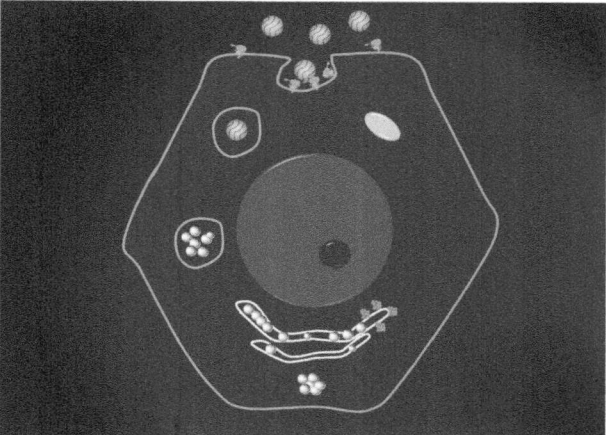

The LDL receptor on the hepatocyte binds LDL particles, which then move to the endosome and lysosome to be degraded and recycled. LDL particles are broken down in the lysosome and their content is either stored or used for cell needs.

PCSK9

Cells can break down the LDL receptor when they already have enough cholesterol and do not need to take more from the blood. The breakdown process is complex and involves several proteins. One of these is proprotein convertase subtilisin/kexin type 9, in short PCSK9. The liver secretes this enzyme, which is found in a variety of tissues and cells throughout the body. PCSK9 binds to the LDL receptor on the surface of the cell and helps it break down. Activated PCSK9 breaks down the LDL receptors and prevents them from being recycled back to cell surfaces. This means they cannot remove any more LDL particles. This results in less cholesterol being delivered to cells and LDL-cholesterol levels increasing. When PCSK9 is inhibited, however, LDL receptors can be recycled back to the cell surface so that they can remove more LDL particles from the blood. When the enzyme is inhibited, LDL-cholesterol levels will decrease.

When PCSK9 binds to the LDL receptor on the cell surface, it also makes the receptor go inside the cell. But, unlike when LDL binds, PCSK9 stops the LDL receptor from changing shape. The receptor is then sent to a lysosome, where it is broken down and does not return to the cell surface. PCSK9 lowers the number of LDL receptors at the cell surface and therefore the amount of LDL processed.

What It Is

Proprotein convertase subtilisin/kexin type 9 (PCSK9) is an enzyme that attaches to the LDL receptor in cells and helps it break down.

PCSK9 plays a role in a number of processes that occur at different stages of the atherosclerosis. It regulates the expression of important receptors in macrophages, which play a part in creating foam cells and plaques in the arteries. PCSK9 regulates inflammation in the blood vessel. PCSK9 also affects blood clotting, by altering platelet function, recruiting white cells, and forming blood clots.

Removal of Cholesterol

Maintaining cholesterol balance is dependent on the removal of cholesterol from the blood via the LDL receptor. The body gets its cholesterol from food and the liver, with the liver being the primary source. The body will produce more cholesterol if the diet does not provide enough. If the body has enough cholesterol from the diet, then it will stop producing as much. This adjustment is due to changes in liver activity, which are influenced by overall cholesterol production in the body. The liver produces more cholesterol in humans than what is delivered from the diet, even though a large amount of cholesterol is delivered. Therefore, reducing liver production of cholesterol could help lower overall cholesterol levels.

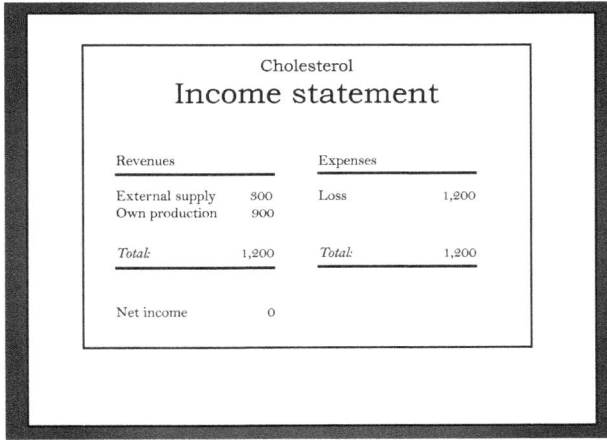

Maintaining a balance between the intake and expenditure of cholesterol is crucial for the body. The values are measured in milligrams per day.

Normal cholesterol balance is achieved by the body producing cholesterol and absorbing the dietary cholesterol from the intestine. The human body receives only 300 milligrams of cholesterol per day from food. Own body production is three times higher. This amounts to an extra 1,200 milligrams of cholesterol in the body each day. Due to the toxicity of excess cholesterol, the same amount needs to be eliminated every day.

Both the amount of cholesterol in the diet and the production of cholesterol in the body affect its levels in the blood. Because cholesterol cannot dissolve in blood, it must be transported to the liver for removal from the body. LDL particles transport 70% of the cholesterol that is eliminated from the body to the liver. HDL particles make up the remaining 30%.

The liver eliminates excess cholesterol by turning it into bile. The neutral cholesteryl ester hydrolase is the first enzyme to release cholesterol derived from lipoproteins. In the liver, cholesteryl esters are broken down by the enzyme. The cholesterol released from the liver can be either used to produce bile acids or sent directly into bile.

The liver releases a mixture of these acids along with the cholesterol into the bile when cholesterol is converted into different bile acids. The majority (95%) are absorbed into the intestines while the remainder is excreted in the feces. The process of bile acid release and reabsorption is crucial for the digestion and absorption of fats in food. In total, the intestines reabsorb two-thirds of the intestinal cholesterol. The liver removes cholesterol from other tissues through this process, which is also known as reverse cholesterol transport.

As a result, LDL slowly leaves the blood, where it remains for three to four days. Similarly, Lp(a) stays in the blood for about three to four days. How Lp(a) is removed from the blood is not well understood. Contrary to LDL, it is not removed primarily through the LDL receptor. The kidneys help remove Lp(a).

8

LDL – How It Stops Working

Cholesterol plays a vital role in the proper function of animal cells. Over time, special biochemical pathways were therefore developed to save and reuse cholesterol. However, these pathways were not prepared for the modern conditions that are characterized by excessive food intake. The complex biological machinery designed to deal with a deficiency in cholesterol can become detrimental in modern times, leading to elevated cholesterol levels in the blood and a buildup of cholesterol in the arterial walls.

Evolutionary genetics indicates that coronary heart disease is not directly affected by natural selection, as it usually occurs later in life and does not affect reproductive success. However, a theory called the ancestral-susceptibility model proposes that common complex diseases like coronary heart disease may be caused by a mismatch between ancient genes and current environment. According to this theory, certain genes that may have been beneficial in the past could be harmful today. High cholesterol levels, for example, may have helped to produce hormones during the evolution of humans, but they now can cause heart problems.

In a similar way, the "thrifty gene" hypothesis states that certain genetic variations have evolved in humans to maximize food storage and energy efficiency, but they may also contribute to diabetes in the current environment. Blood clotting genes may have been selected in the past to help with survival following injuries. However, they now can lead to artery blockages. Our immune response to pathogens was also shaped by evolution, but an immune system that is overactive to combat infections can increase the risk for coronary heart disease.

Causes

Dyslipidemias

Both genetics and lifestyle affect plasma LDL-cholesterol. Heritability, or the amount of LDL-cholesterol that is genetically determined, is thought to account for 40–50%. LDL-cholesterol levels are governed by multiple genes, environmental factors, and their interactions. LDL metabolism is affected by genes that produce LDL receptors and lipid transfer proteins, as well as other proteins that regulate LDL processing. Individual genes can have dramatic effects on LDL-cholesterol levels.

About one-third of the mass of LDL is cholesterol. In individuals with too much LDL, the cholesterol level in the blood is higher than normal. This hypercholesterolemia is a form of dyslipidemia. Lipid disorders are classified according to the concentration of lipids that are affected, whether it is cholesterol, triglycerides, or both. Dyslipidemias are classified according to the lipoproteins that contain the affected lipids.

Cholesterol, Lipoproteins, and Cardiovascular Health: Separating the Good (HDL), the Bad (LDL), and the Remnant, First Edition. Anatol Kontush.
© 2025 John Wiley & Sons, Inc. Published 2025 by John Wiley & Sons, Inc.

> **What It Is**
>
> **Dyslipidemia** is a medical condition characterized by abnormal levels of lipids within the blood plasma.

According to the system developed by Donald Fredrickson in 1967, dyslipoproteinemias are classified in five major types from I to V. Within the types, further distinction can be made as reflected by small letters, such Types Ia, b, and c. Type I dyslipoproteinemia involves elevated concentrations of chylomicrons, the largest and the lightest lipoprotein. In Type II dyslipoproteinemia, concentrations of LDL are elevated, without (Type IIa) or with (Type IIb) those of VLDL. Dyslipoproteinemias of Types III, IV, and V respectively involve elevated plasma levels of IDL, VLDL, and VLDL together with chylomicrons. In Types I and V, patient's plasma appears milky, while it is turbid in Types III and IV, and clear in Type II.

Dyslipidemias are also classified according to the cause. They can be primary, secondary, or a mixture of both. Primary dyslipidemias, also known as familial dyslipidemias, are caused by genetic disorders which can lead to abnormal lipid concentrations without any obvious risk factors.

Familial Hypercholesterolemia

The most well-known form of dyslipidemia is probably familial hypercholesterolemia. Its direct link to heart disease is the reason for its fame. It is a hereditary disease characterized by high levels of cholesterol in blood plasma. This condition is characterized by elevated plasma concentrations of LDL-cholesterol and apolipoprotein B. Type IIa dyslipoproteinemia has particularly high concentrations of them, while Type IIb also has higher levels of apolipoprotein C-III. As every genetic condition, it can exist in two forms, homozygous and heterozygous.

> **What It Is**
>
> **Homozygous** and **heterozygous** are used to indicate if a living organism possesses two identical copies of the same gene for a specific trait (homozygous) or two different copies (heterozygous).

In the condition of homozygous hypercholesterolemia, individuals have only faulty LDL receptors. This prevents them from binding LDL particles in the bloodstream.

Carl Muller was a Norwegian professor in internal medicine who discovered the homozygous condition in 1938. The patients with this condition have high cholesterol levels since birth. As LDL is responsible for the majority of cholesterol found in plasma, LDL-cholesterol and LDL particle concentrations are also elevated. Patients with homozygous familial hypercholesterolemia develop coronary heart disease, atherosclerosis, and xanthomas quickly.

In the condition of heterozygous hypercholesterolemia, half of the LDL receptors are faulty, while the other half functions normally. As a result, there is only a partial ability to bind LDL particles.

Heart disease is common in these patients and can occur as early as in childhood or early adolescence. They usually present with a myocardial infarction in their 20s if they do not receive treatment. The association between familial hypercholesterolemia, atherosclerosis, and heart disease is the strongest evidence in support of the key role of cholesterol in this disorder. Homozygous disease is rare, affecting only 1 in 100,000 to 1 in 1,000,000 individuals worldwide.

The heterozygous condition is milder, with LDL-cholesterol levels not exceeding 50% of the homozygous patient's. This is the most common form of familial hypercholesterolemia, which affects about 1 in 200 to 1 in 300 people. Untreated, this condition may also cause high LDL-cholesterol levels and increase heart disease risk. The severity of heart disease and the risk of developing it in both forms of familial high cholesterol are determined by levels and duration of elevated LDL-cholesterol.

A mutation in the *LDLR* gene (coding for the LDL receptor) or, less frequently, the *APOB* or *PCSK9* genes, is the most common cause of familial hypercholesterolemia. All of these mutations lead to high LDL-cholesterol. LDL is not removed from circulation if there is a problem with the interaction of LDL with the LDL receptor. This can lead to higher LDL-cholesterol levels and an increased risk of heart disease. People with certain mutations (called nonsense or splice sites, or frameshifts) in the *LDLR* have an LDL-cholesterol average of 280 mg/dl. Those without these mutations, however, have a level of LDL-cholesterol of around 135 mg/dl. The disruptive mutations are 13 times more prevalent in those who have had a heart disease or a recent heart attack compared to people without these conditions.

In any family with familial hypercholesterolemia, each child has an equal chance of inheriting the mutation that causes the disease. Siblings who inherit the mutation have higher LDL-cholesterol levels in their blood and are at a higher risk of heart disease than their unaffected siblings. This is another piece of evidence that LDL is a cause of heart disease. This conclusion is also supported by the fact that mutations in PCSK9 may increase or decrease LDL-cholesterol levels. The effects depend on whether or not the mutations result in a gain or loss of function for the enzyme. This will increase or decrease the risk of cardiovascular disease. The gain of function for PCSK9 results in elevated LDL-cholesterol and increased heart disease risk, while people with rare loss-of-function variants have hypocholesterolemia and are protected. This was discovered by Helen Hobbs and Jonathan Cohen at University of Texas in Dallas.

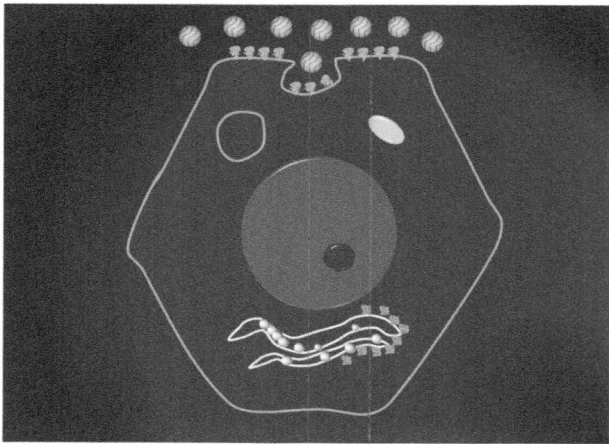

A lack of LDL binding to the LDL receptors prevents cells from taking in cholesterol.

Secondary Dyslipidemias
Secondary dyslipidemias are distinct from primary (genetic) dyslipidemias. These are the ones that can be altered, such as by lifestyle or environment. Diabetes, liver disease, and kidney disease can all increase the risk of dyslipidemia. The diet can also affect the risk – too much alcohol or carbohydrates can raise it. Saturated fats are known to delay the clearance of LDL-cholesterol from the blood and increase its level.

Secondary dyslipidemia can also be caused by certain medications such as beta-blockers and diuretics, birth control pills, and antipsychotics. Smoking, being pregnant, and being overweight are also factors that increase risk.

Atherogenic Dyslipidemia
LDL is the end product of a modification process on VLDL. The body's ability to process VLDL may affect how LDL particles are formed, what they consist of, their function, and the way they are broken down.

VLDL is the main carrier of triglycerides. The amount of plasma triglycerides will affect the size and composition of LDL. Too much triglycerides can cause the formation of small, dense LDL particles, which may increase heart disease risk. Atherogenic dyslipidemia is a type of dyslipidemia that includes high levels of both small, dense LDL and large VLDL1. This is a common feature of diabetes and metabolic syndrome. Apolipoprotein B-100 levels are higher in people with a high number of small, dense LDL particles than they are for normal people. In this case, the LDL particles have more protein and less cholesterol.

Certain properties of small, dense LDL can increase the risk for coronary heart disease. It appears to enter the arterial wall quicker than larger LDL. Small, dense LDL may also have a more difficult time binding to LDL-receptors and stay in the blood longer than larger LDL particles. Small, dense LDL particles also have a greater affinity for the proteoglycans found in arterial walls. This helps them to accumulate there more effectively. LDL particle composition can influence their toxic effects on the human body. This lipoprotein's lipids are susceptible to oxidative modifications, which can lead to toxic oxidized molecules. The small, dense LDL contains substances such as oxidized phospholipids which can cause blood vessel problems.

It is hard to know the exact contribution that small, dense LDL makes to heart disease, as it is closely linked to other lipoproteins. Recent studies have found that levels of small, dense LDL, rather than those of large, light LDL, can predict heart disease.

Other Dyslipidemias

Lp(a)   Other types of lipoproteins similar to LDL can also make the process of developing artery blockages worse. These lipoproteins primarily include Lp(a). Unlike LDL, Lp(a) can aggravate thrombosis without needing any further modifications. The lipoprotein carries oxidized phospholipids which are deleterious for the arteries. Recent data show that the risk associated with high Lp(a) can even be greater than that associated with high LDL-cholesterol.

Lipoprotein-X   Lipoprotein-X is an abnormal form of lipoproteins. This unusual particle is found in the blood when there is a blockage in the flow of bile, a condition known as cholestasis. This can happen with jaundice, which is called obstructive. The biliary or pancreatic canal may become narrowed or blocked. This stops the normal flow of the bile from the blood into the intestines.

Lipoprotein-X has a diameter of 30–70 nm, which is comparable to VLDL and IDL. The density of the lipoprotein is, however, similar to LDL. Lipoprotein-X is composed of lamellar layers. It contains a large amount of phospholipids and cholesterol that has not been esterified, but little protein, cholesteryl esters, or triglycerides. Lipoprotein-X also contains a large amount of albumin and apolipoprotein C. Its particle core is aqueous, and it contains albumin.

Lipoprotein-X is formed in cholestasis for unknown reasons. This may occur when bile is not excreted through the liver, but instead returns to the blood. In the lab, when the bile and the blood or the albumin are mixed together, particles that resemble lipoprotein-X form. It is possible that the physical and chemical properties of mixing the bile with the blood are what cause lipoprotein-X to be formed. Lipoprotein-X is also found in the blood of people who have a condition known as familial LCAT deficiency. It could be because there is not enough LCAT enzyme to convert phospholipids into cholesteryl esters.

Lipoprotein-Y and Lipoprotein-Z   Lipoprotein-Y is a large particle, similar to LDL. It is 30–40 nm in diameter and carries a high amount of triglycerides. It contains apolipoprotein B-100. Lipoprotein-Y particles are formed when the LCAT or hepatic lipase is not working properly. These lipoproteins resemble normal IDL particles, or remnants of VLDL.

Lipoprotein-Z is another abnormal particle. It has a diameter of 18–20 nm, close to that of LDL. Lipoprotein-Z is similar to LDL, but it has less cholesteryl ester and more triglycerides. This lipoprotein can be found in those with alcohol-related liver disease.

Extremely Low LDL-cholesterol Genetic conditions can be associated with both very high and very low LDL-cholesterol levels in the blood. Hypobetalipoproteinemia is a group of disorders that affect how the body processes fats. Mutations in the *APOB* gene are usually responsible for these disorders. They are characterized by very low levels of LDL-cholesterol and apolipoprotein B. Most common symptoms are a swollen or fatty liver, problems with growth, and serious developmental issues which result from fat malabsorption.

Abetalipoproteinemia, another rare disorder, affects the way the body absorbs vitamins and fats. Low levels of vitamins and other important nutrients are the result. This can lead to problems with the eyes, brain, and muscles. Extreme cases of this disorder can lead to ataxia, which is a lack of muscle control. This may result in difficulty with walking, balance and coordination, as well as difficulties with speech, swallowing, eye movements, and speaking. This disorder is caused by changes in the microsomal triglyceride transfer protein (MTTP), which is necessary for producing apolipoprotein B-containing lipoproteins such as VLDL and chylomicrons. The condition is characterized by extremely low levels of LDL-cholesterol, apolipoprotein B, and triglycerides. Beta-lipoproteins are also undetectable. Only 100 cases of abetalipoproteinemia have been reported in the world.

In the both hypobetalipoproteinemia and abetalipoproteinemia, most problems can be avoided by eating a low-fat diet and taking vitamins. Eating enough calories is essential for growth.

History

Classification of Dyslipidemias

Who	When
Donald Fredrickson	**1965**

Where	What
Bethesda MD, USA	Fredrickson, D. S. and R. S. Lees (1965) "A system for Phenotyping Hyperlipoproteinemia." Circulation 31: 321–327.

Donald Fredrickson was a young researcher when he joined the National Heart Institute in 1953. Christian Anfinsen, who won the Nobel Prize in Chemistry in later years for his research on ribonuclease, was the head of the lab where he worked. Anfinsen was appointed a few years earlier to put together a team of leading investigators that could help set the national agenda in cardiovascular research.

Fredrickson, along with his colleagues in 1964, analyzed the data of plasma lipoprotein concentrations for a variety of clinical conditions. They proposed a classification of dyslipoproteinemias. These data were obtained using ultracentrifugation and electrophoresis.

However, the former technique was time consuming and costly. It was not accessible in clinics. Fredrickson's classification was therefore based upon paper electrophoresis.

Electrophoresis was applied to lipoproteins in the 1950s. Lipoproteins were separated by a migration in a constant electric field applied to a paper or a gel that contained a salt solution at a constant pH termed buffer. The "lipidogram" was the name given to the profile of plasma lipoproteins that electrophoresis produced. Bengt Swahn, a Swedish scientist in 1952, created the first lipidogram.

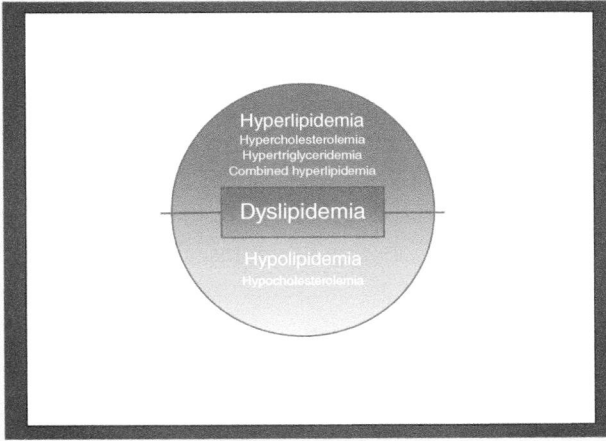

The dangerous realm of dyslipidemia is cut in two halves of elevated (hyperlipidemia) and reduced (hypolipidemia) plasma concentrations of lipids. It should be remembered that hyperlipidemia occurs more often than hypolipidemia.

In 1963, Robert Lees and Frederick Hatch developed this approach at the Massachusetts General Hospital in Boston. Albumin was added to the buffer and this improved the separation of lipoproteins. It was easy and inexpensive to use the method in routine clinical practice. In 1965, Fredrickson described the details of this technique which was called a system for phenotyping hyperlipoproteinemia. Fredrickson, Lees, and Robert Levy published the classification in *New England Journal of Medicine* together with an overview of lipoproteins. This approach quickly gained wide acceptance and was recommended by the World Health Organization in 1970.

What It Is

Dyslipidemia refers to a medical condition characterized by an abnormal concentration of lipids (triglycerides or cholesterol) in blood plasma. Hyperlipidemias, which are abnormally high concentrations of lipids within plasma, constitute the majority of dyslipidemias. Conditions involving elevated concentrations of cholesterol or triglycerides are, respectively, termed hypercholesterolemia or hypertriglyceridemia. Combined hyperlipidemia is a condition in which both triglycerides and cholesterol are elevated.

(Continued)

(Continued)

Hypolipidemia is a condition in which lipid concentrations are abnormally low. Hypocholesterolemia is a condition that involves low cholesterol concentrations. Hyperlipidemia thereby represents a subset of dyslipidemia and a superset of hypercholesterolemia, while hypolipidemia is a subset of dyslipidemia and a superset of hypocholesterolemia.

Dyslipidemia also affects lipoprotein concentrations because lipoproteins transport plasma lipids. Dyslipidemia is also known as dyslipoproteinemia. When one class of lipoproteins is specifically affected, this condition can be referred to as abetalipoproteinemia (reduced LDL-cholesterol), hypoalphalipoproteinemia (reduced HDL-cholesterol), or hyperalphalipoproteinemia (elevated HDL-cholesterol).

According to Fredrickson, dyslipoproteinemias should be classified in five major types from I to V. Within the types, further distinction can be made as reflected by small letters, such Types Ia, b, and c.

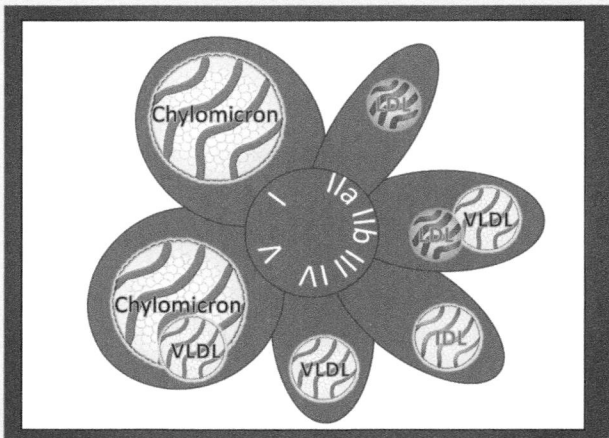

The flower of dyslipoproteinemia, according to Fredrickson's classification. There are five major types, ranging from I to V, of this condition: Type I, with elevated chylomicrons; Type II, with elevated LDL concentrations without (Type IIa) or with (Type IIb) those of VLDL; Type III, with elevated IDL concentrations; Type IV, with elevated VLDL concentrations; and Type V, with elevated concentrations of VLDL and chylomicrons.

In the 1960s, genetic changes associated with dyslipoproteinemias were not known. This is why Fredrickson's approach did not include them. This classification is based on the relative concentrations of lipoproteins in plasma, and it is strictly phenotypic. Fredrickson's phenotypes do not always reflect a change in genotype. In this system, the relationship between an abnormal lipid profile and the underlying pathology is not always clear. Now we understand that only two of the phenotypes, specifically Type I and Type IIa, are caused by a single gene, while the rest are influenced by multiple genes.

It is interesting to note that the relative positions of lipoproteins upon electrophoresis are reflected in the order of the classification. In fact, chylomicrons are not able to migrate and remain at start, while LDL is slow. IDL and VLDL, on the other hand, migrate faster.

After the Fredrickson studies, clinical lipidology experienced a rapid acceleration. It was recommended that the classification be applied to patients with abnormal lipid levels. Diagnostics of dyslipidemia were greatly simplified and standardized. Kits containing the reagents needed for diagnosis were produced. The production of electrophoresis equipment followed. This breakthrough was the catalyst for lipoproteinology. In the United States, lipids became a focus of national research. In 1970, 12 lipid research centers were opened in the United States as part of the Lipid Research Clinics Program. This program became known as LRCP. The early 1970s saw the opening of lipid research centers in Canada, Germany, Israel, and Russia. This effort brought together the interests of science, industry, and politics. The system was then quickly included in biochemistry textbooks. Fredrickson's classification was a great success, and he gained fame as a result. He became director of the National Institutes of Health in 1975, and the personal physician to the King of Morocco that same year.

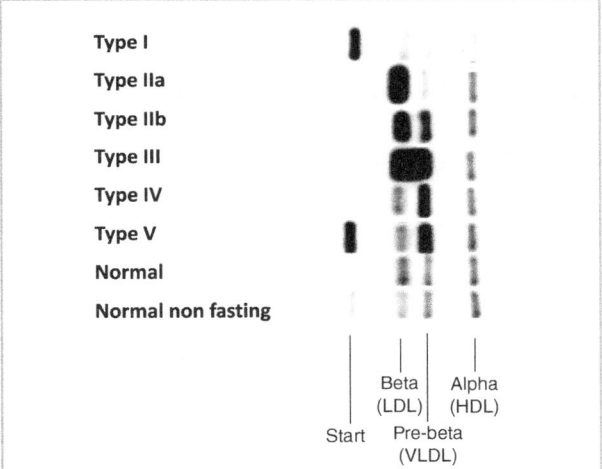

How to diagnose dyslipidemia
according to Fredrickson

A lipidogram is a major tool for distinguishing between the main types of dyslipidemia.

In the 1980s, the NIH created the National Cholesterol Education Program in order to raise awareness among clinicians and patients regarding the link between cholesterol and cardiovascular disease. Dewitt Goodman was the leader of the program, which developed the first national guidelines to treat high cholesterol in order to prevent heart diseases. These guidelines were published by the American Heart Association in 1988. They were a major step forward in preventive cardiology. The program published a series of expert panel reports called Adult Treatment Panel I, II, and III. Adult Treatment Panel II recommends that patients be treated to reduce their LDL-cholesterol levels to less than 100 mg/dl as a secondary prevention measure and to less than 130 mg/dl as a primary prevention measure. These guidelines have helped to reduce heart disease and preventable deaths over the last few decades.

To say that Fredrickson's approach was easily accepted would, however, be exaggerated. It was not always simple to distinguish between Types II and III. Hypolipoproteinemias and

(Continued)

(Continued)

hyperalphalipoproteinemia were not included in the system. The classification did not mention Lp(a) which was discovered in 1963 by Kare Berg in Norway. It was unclear how to distinguish between primary and secondary dyslipidemias.

Overall, these studies revealed that lipoprotein analysis was superior to plasma lipid analysis. The debate was settled by this, which fully supported John Gofman's conclusions from the 1950s. The Fredrickson classification confirmed and extended the observations made by Gofman based on the Schlieren profiles.

Practical Aspects: How to Separate Lipoproteins Using Electrophoresis

Electrophoresis can separate lipoproteins produced from exogenous lipids (chylomicrons) and endogenous particles, such as beta-, pre-beta, or alpha-lipoproteins. These particles are similar to LDLs, VLDLs, and HDLs obtained by ultracentrifugation.

Electrophoresis was the primary method used to detect dyslipidemias for a long period of time. This is due to its simplicity and speed compared to ultracentrifugation. Remember that the lipidogram cannot measure lipoprotein levels in plasma. The nonlinear and complex relationships between the concentrations of lipoproteins and the intensity with which a dye stains the lipoproteins are the reason for this. Their relative concentrations are estimated by comparing them to control plasma from a healthy person with normal lipoprotein levels. The electrophoretic lipogram was finally replaced with direct measurements of the lipoprotein levels.

How to separate lipoproteins
Electrophoresis

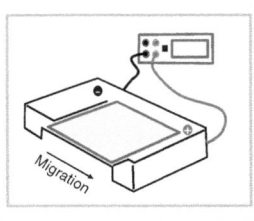
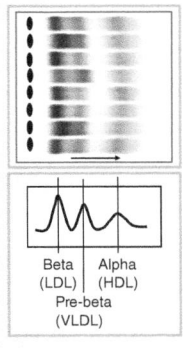

Beta (LDL) | Alpha (HDL)
Pre-beta (VLDL)

What It Is

Electrophoresis separates biomolecules and their complexes based on their mobility in an electric field. The mobility of molecules is determined by their size and charge.

Consequences

Cholesterol Buildup in Arterial Walls

All individuals have LDL particles in their blood. A small proportion of LDL can pass through the inner surface of the arteries and stay there. LDL particles bind with proteoglycans, which are branched molecules on the surface of cells. Endothelium is not a barrier for many lipoproteins including chylomicron remnants, VLDL, IDL, and HDL. They are mostly cholesterol rich and can remain under the endothelium of the arterial wall in order to cause cholesterol buildup.

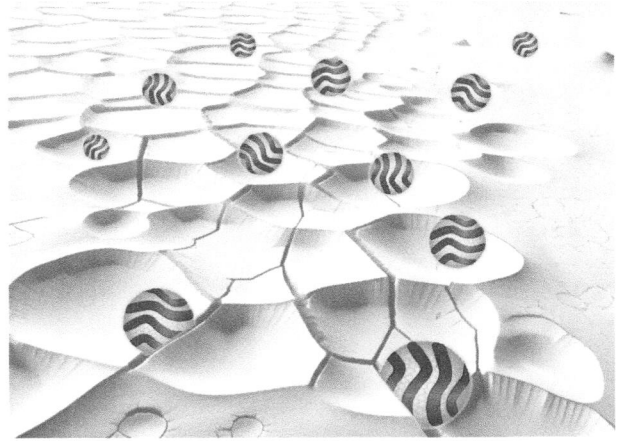

LDL is the most common harmful type of lipoprotein found in the blood, responsible for carrying cholesterol to the arterial wall. Several risk factors can impact the ease with which LDL and other harmful lipoproteins can pass through the endothelium and enter the arterial wall.

Understanding how LDL travels through the endothelium during atherosclerosis is crucial, but we still have much to learn about the molecular control of this process. It has been believed for a long time that LDL simply moves passively through a compromised endothelium based on its size and concentration. However, recent evidence challenges this idea, suggesting that LDL actually moves through specialized transport vesicles. This pathway involves specific membrane structures like caveolae, as well as receptors such as scavenger receptor class B type I (SR-BI) and the LDL receptor. New evidence indicates that SR-BI interacts with cytoplasmic proteins to facilitate LDL transport across endothelial cell layers. Interestingly, the levels of SR-BI are higher in arteries affected by atherosclerosis compared to normal arteries. This is consistent with accelerated transport of LDL across the endothelium.

> **What It Is**
>
> **Scavenger receptor class B type I (SR-BI)** is a cellular receptor which binds HDL as well as many other partners, such as modified LDL. The receptor is involved in the movement of cholesterol from cells into HDL and back.

Contrariwise, estrogens (hormones present in high concentrations in women before menopause) can reduce the movement of LDL across the inner surface of the artery. This occurs because estrogens reduce the activity of SR-BI within the cells lining arteries. Estrogens also have many other protective effects on arterial wall cells. These findings may in part explain why women are at a lower heart disease risk before menopause. Lowering blood cholesterol can also improve endothelial cells' barrier function. Other factors, such as inflammation and high sugar levels in the blood, can increase LDL's movement across the endothelial cell layer.

The amount of time that LDL stays in the blood can be important for the risk of developing atherosclerosis. This is because it determines how much LDL is exposed to the arteries and how likely it is to undergo changes that can lead to heart disease. When LDL cannot be cleared by the liver and stays in the blood for a longer time, it can accumulate in the blood and also in the arterial walls. There LDL can be taken up by macrophages located just beneath the arterial surface. These cells turn into lipid-loaded cells, also called foam cells, and start the buildup of lipids in the arteries.

Modifications of LDL

LDL particles are capable of causing atherosclerosis in different degrees. This can be affected by various factors, such as their ability to clump, bind to proteoglycans, and undergo oxidation. LDL's harmful effects on arteries are multiple. First of all, these include formation of foam cells by macrophages after they absorb LDL. LDL particles must be chemically altered or clumped together to build foam cells. Lipids such as phospholipids or cholesterol, or the apolipoprotein B-100 can be oxidized by enzymes (e.g. by myeloperoxidase, an enzyme secreted by immune cells) or without them. The clumping can be caused by natural processes or enzymes. Cells in the arterial walls can cause this.

LDL particles can be retained by proteoglycans on arterial wall cells. When this happens, the LDLs are exposed to reactive oxygen species and other active molecules released by the cells. As a result, modified LDL is formed, which is easily taken up by macrophages. This uptake leads to the appearance of foam cells that are filled with cholesterol.

The atherosclerotic plaques contain a variety of products from lipid and protein oxidation. This process can involve different enzymes, such as myeloperoxidase and nitric oxide synthase. LDL is likely to be oxidized by reactive chlorine, nitrogen, and oxygen species. Local oxidative stress is responsible for lipoprotein oxidation within the arterial wall. It means that the balance between harmful prooxidants and protective antioxidants is skewed in favor of prooxidants. The development of atherosclerosis is closely related to oxidative stress within the arterial wall, which leads to oxidized LDL.

There is a small portion of LDL that has a strong negative charge. It is often regarded as oxidized LDL. This subclass is associated with the development of cardiovascular disease. The electronegative LDL may cause inflammation, just like LDL oxidized in vitro. It is more dense, has higher content of several apolipoproteins and can clump together easier. This clumping may cause it to stick to the proteoglycans on the arterial wall and cause heart disease.

Recent research has suggested that the accumulation of LDL in the arterial walls can be a response to infection. This buildup protects vulnerable areas of the artery against microbe invasion. The antimicrobial properties in lipids and fibrin structures within the arterial walls as well as the contraction and movement of smooth muscle cells all combine to combat these infections. When lipoproteins are oxidized near infection sites, they can also travel through the bloodstream, promoting the accumulation of lipids in vulnerable areas of artery tissues. These oxidized lipoproteins can also trigger an immune response by acting as signals.

Its ability to accumulate within cells of the arterial wall, especially macrophages, is what makes oxidized LDL so harmful. This occurs through scavenger–receptor pathways that are not suppressed as cells accumulate cholesterol. This leads to the formation of lipid-loaded foam cells. Modified lipoproteins trigger biological reactions that also contribute to the progression of atherosclerosis. This includes endothelial dysfunction and the formation of lipid-rich plaques prone to rupture. Modified LDL contains oxidized phospholipids that are powerful agents for causing inflammation.

Local Inflammation

The immune system responds when certain parts of the apolipoprotein B-100 are modified or destroyed by harmful molecules produced by activated immune cells. This response includes specific cells and antibodies that fight harmful substances. The immune system recognizes modified apolipoprotein B-100 as a type of harmful foreign agent.

Inflammation occurs in the arterial wall when cholesterol accumulates in the arteries. This inflammation is caused by different types of blood cells, but certain early steps are crucial. Monocytes, which are white blood cells, adhere to the walls of arteries by interacting on their surface with specialized proteins. The cells then pass through the walls and move into the interior space, where they become macrophages. Lymphocytes, another type of white cells, also accumulate in this area. They can make inflammation worse by interacting and causing the macrophages to produce more inflammatory molecules.

Highly active lipids like oxidized phospholipids can also cause inflammation. When lipoproteins are oxidized, these molecules are released. They have an effect both locally and throughout the body. The lipid molecules can attract immune cells such as monocytes and macrophages. They also attract lymphocytes and dendritic cells. This inflammation can lead to chronic cell death, which will further aggravate the disease. When foam cells accumulate too much cholesterol, they die. Cholesterol crystals may form when the cells die. The crystals can damage cells and trigger an immune reaction involving molecules that are associated with the cell damage.

History

Discovery of the LDL Receptor

In the 1950s, there was a significant interest in cholesterol, leading to a major effort to understand how the body produces this molecule. Four biochemists, Konrad Bloch, Feodor Lynen, John Cornforth, and George Popják, were instrumental in clarifying most of the steps in this process. The scientists showed that cholesterol is made in four steps. These include combining three acetates to form a six-carbon mevalonate intermediate, changing the mevalonate to activated isoprenes, combining six 5-carbon isoprenes to create a linear 30-carbon squalene and finally changing squalene to the steroid nucleus and then to cholesterol. The third step controls the speed of the whole process. This step is controlled by the HMG-CoA reductase enzyme.

Who	When
Michael Brown Joe Goldstein	**1974**

Where	What
Dallas TX, USA	Brown, M. S. and J. L. Goldstein (1974) "Expression of the familial hypercholesterolemia gene in heterozygotes: Mechanism for a dominant disorder in man." Science 185: 61–63.

By 1960, the major contours of this pathway had been completed. Bloch and Lynen were awarded the Nobel Prize for this work in 1964. Sven Friberg, Rector of the Karolinska Institute, said at the Nobel Banquet that these discoveries may provide weapons against some of mankind's most serious diseases, particularly cardiovascular diseases. Achievements like these make it not unrealistic to look forward to a time when mankind will not only live under vastly improved conditions but will itself be better, he added.

This time came sooner than expected. Michael Brown and Joe Goldstein first met in Boston, at the end 1960s. The young interns were interested in the molecular basis for human diseases. Then, they went to the NIH, where Donald Fredrickson was their collaborator. In the early 1970s, they moved from Bethesda to the University of Texas at Dallas. Their intention was to perform studies of the genetics of hyperlipoproteinemias. They wanted to clarify biochemical pathways by which genetic alterations increased cholesterol concentrations in plasma of people with hypercholesterolemia. They started studying patients with the rarest and deadliest form of this disease, which was homozygous hypercholesterolemia. They wanted to find the cause of this disorder, whose carriers were common among myocardial victims.

First, they isolated skin cells called fibroblasts from healthy people and studied how the cells handled cholesterol. They quickly discovered that cells produced little cholesterol when cholesterol-containing lipoproteins were present outside the cell. When cholesterol was removed, the cells began producing more cholesterol. HMG-CoA reductase, which controlled the production of cholesterol, was weakly active if cholesterol was present in the surrounding environment but highly active if cholesterol was removed. Very small amounts of cholesterol were sufficient to suppress the activity of the enzyme. The researchers also noted that while LDL and VLDL produced these effects, HDL did not. This suggested the role of apolipoprotein B which was present in LDL and VLDL but absent from HDL.

When they turned to the fibroblasts obtained from the patients with hypercholesterolemia, they were puzzled by the absence of these effects. The cells from the patients did not react to changes in cholesterol concentrations in the outside environment. Brown and Goldstein first thought the intracellular HMG-CoA reductase enzyme could be defective in patients. When they added the cholesterol solution to the cells, not as LDL, but in ethanol instead, the enzyme activity was, however, suppressed. This observation led them hypothesize that there was a protein at the membrane surface that had a defect in the patients. They called this protein the LDL receptor. The hypothesis went that when the receptor was defective, the binding of LDL did not occur and there was no signal sent into the cells to the enzyme to decrease its activity. The binding had to involve apolipoprotein B.

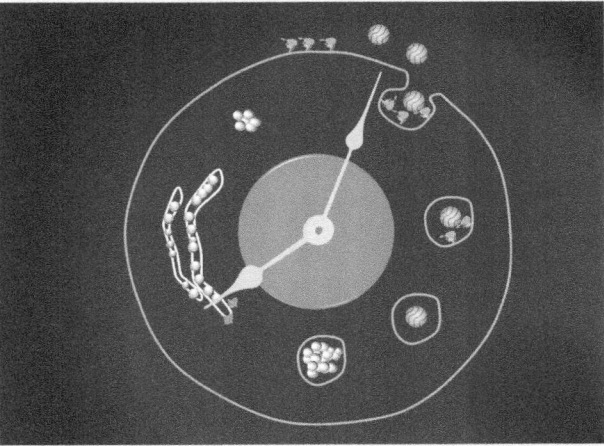

In 1974, Brown and Goldstein discovered the LDL receptor using in vitro experiments. LDL with a radioactive tag was used to establish the existence of the receptor in normal fibroblasts. After incubating radioactively labeled LDL with cells, the label was discovered in one of their proteins. Researchers also observed that LDL bound with the receptor entered the cells where it was digested into lipids, amino acids, and other compounds. Cholesterol was released from LDL and then transferred to the cell membrane where it produced a signal which reduced the activity of the enzyme. The cholesterol also reduced the number of LDL receptors on the

(Continued)

(Continued)

surface of the cells. Using this pathway, the cells could control the uptake and utilization of cholesterol, which is essential for cell viability.

Further research revealed that individuals with familial hypercholesterolemia had defective forms of this receptor, had less of it, or did not have it at all. In some homozygous cases, the LDL receptors were absent and, in heterozygous cases, they were half the number of normal. As a result, LDL particles were less rapidly removed from the blood, their concentrations were increased and they more readily penetrated the arterial wall to induce atherosclerosis and heart disease. In 1984, the existence of the LDL receptor was confirmed in vivo both in normal individuals and in patients with familial hypercholesterolemia.

Subsequently, studies revealed the existence of a subset of HDL that was able to bind to the LDL receptor. These particles obligatory contained apolipoprotein E, which was shown to be a partner for the receptor in addition to apolipoprotein B. As a consequence, the LDL receptor was renamed to the apolipoprotein B/E-receptor. However, this term is rarely used at present. Further refinement of this pathway led to the discovery by Nabil Seidah and Michel Chrétien in France of PCSK9 which breaks down the LDL receptors.

Together, these studies delivered an outstanding example of how cell biology and molecular genetics could be combined to elucidate the molecular basis of a disease. Clinical and fundamental research came together to understand familial hypercholesterolemia. This breakthrough revolutionized studies of lipids and lipoproteins. Moreover, the mechanism of receptor-mediated endocytosis described by Brown and Goldstein was soon shown to be a general process by which cells pick up extracellular material.

The Nobel Prize was awarded to Brown and Goldstein for this discovery in 1985.

9

LDL – What to Do to Correct It

In light of the discovery that LDL is linked to heart disease, the research made it important to lower LDL-cholesterol to prevent heart problems. LDL-cholesterol became a key target for the prevention of heart disease and stroke.

Two main approaches are used to lower LDL-cholesterol: lifestyle changes and therapies. The first step in treatment is to change your lifestyle. Dietary factors and lifestyle modifications have no side effects, which makes them an ideal first line of defense against heart disease. Dietary changes are the easiest way to reduce LDL-cholesterol, but they may not be the most effective. Maintaining a healthy body weight is also important to keep LDL-cholesterol low and prevent cardiovascular disease. Finally, exercise improves circulation, lowers blood pressure, and decreases LDL-cholesterol while reducing heart stress.

It's always time to target LDL.

Lifestyle

Nutrition

The levels of LDL-cholesterol in the blood can be affected by nutrition. LDL-cholesterol is influenced by the types of carbohydrates and fats. A number of nutrients can reduce LDL-cholesterol.

Cholesterol, Lipoproteins, and Cardiovascular Health: Separating the Good (HDL), the Bad (LDL), and the Remnant,
First Edition. Anatol Kontush.
© 2025 John Wiley & Sons, Inc. Published 2025 by John Wiley & Sons, Inc.

Soluble fiber is found in foods like oats, barley, kidney beans, apples, pears, and Brussels sprouts. This fiber lowers LDL-cholesterol. Soluble fiber works by reducing cholesterol absorption in the intestine.

Foods containing monounsaturated and polyunsaturated fatty acids, such as soybeans, olive oil, rapeseed oil, colza, and fish oil, can also help lower LDL-cholesterol levels. Low heart disease rate in Crete and other Greek islands is believed to be due to olive oil consumption. The low heart disease rate in Japan may be due to the consumption of rapeseed and fish oil. Moreover, soybean protein can block cholesterol production in the liver.

Plant sterols and stanols are important compounds in fruits, vegetables, and a healthy diet. In fact, phytosterols reduce the absorption of cholesterol by the intestine. Another food that can reduce LDL-cholesterol is red yeast rice. China is a major consumer of this natural product which may contain small amounts of lovastatin. However, red yeast rice can also contain toxins. Almonds, soy protein, flaxseeds, and tomatoes can help reduce LDL-cholesterol a bit. Avocados and turmeric can help it drop a lot. Pulses, hazelnuts, walnuts, wholegrain foods, and green tea can also help lower LDL-cholesterol.

On the flip side, consuming foods that are high in unhealthy fats, first of all trans-unsaturated fats, can increase LDL-cholesterol. These unhealthy fats can be found in fatty meats, deli-style meats, butter, cream, ice cream, coconut oil, palm oil, as well as most deep-fried takeaway foods and commercially baked products like pies, biscuits, buns, and pastries.

By blocking the LDL receptor, saturated fatty acids increase LDL-cholesterol levels. In contrast, monounsaturated fatty acids can lower LDL-cholesterol levels when saturated fats are substituted in the diet. This happens because the body is able to eliminate LDL faster. To reduce the risk of cardiovascular disease, The National Cholesterol Education Program suggests that saturated fats should not exceed 7% of total energy intake. Dietary recommendations to prevent and treat cardiovascular disease include replacing trans and saturated fats with healthier foods in order to reduce LDL-cholesterol.

What It Is

Trans-unsaturated fats, also known as **trans-fats**, are a type fat that is found in many foods. Trans fats can be produced both naturally and in factories when oils are refined.

Axel Keys developed the equation to predict the effects of different fatty acids on serum cholesterol levels. This relationship states that saturated fatty acids increase total and LDL-cholesterol twice as much as polyunsaturated fatty acids lower them. The outcome is however strongly dependent on simultaneous changes in other dietary components like carbohydrates.

Some people can have higher levels of LDL-cholesterol and total cholesterol when they consume more fat. This link is still controversial, as the amount and type of cholesterol consumed in the diet do not always influence the cholesterol levels in the blood. Further, a low-fat diet does not necessarily reduce plasma cholesterol or heart disease.

Cholesterol is present in animal products such as eggs, liver, shrimp, and dairy. It is not visible to the naked eye in foods, unlike other dietary fats such as triglycerides. It is recommended that we consume less than 200 mg of cholesterol each day to reduce LDL-cholesterol. Only 25–35% of calories per day should be fats.

Maximal amounts of fats in the diet recommeneded in the United Kingdom.

Interesting Numbers

The average intake of total fats by adults aged 19–64 is 34.6%.

The average daily intake of saturated fats is 12.6%.

The average intake of trans fats is 0.7%.

The data are calculated as a percentage of daily energy intake and are from the United Kingdom in 2008.

Different forms of hyperlipidemia have different dietary guidelines. Donald Fredrickson suggested in the 1970s that dietary fat should be restricted to 10% of energy for type I and 30% for type V hyperlipidemias. For types II, III, and IV, it was recommended to increase the ratio of polyunsaturated to saturated fats and to reduce body weight.

Indigenous people's levels of LDL-cholesterol and heart disease rates are a perfect illustration of the importance of a healthy lifestyle. Those communities that are still isolated and hunter–gatherers, with very low LDL-cholesterol levels between 50 and 70 mg/dl, show this. The rates of heart disease in these communities are also low. For example, the Tsimane, who live in Bolivia's Amazon, have much healthier hearts than people from the United States. Tsimanes also have low blood pressure, heart rate, cholesterol, and glucose levels. It could be because of their lifestyle. They live a lifestyle that suggests eating low-fat foods, consuming wild game and fish, not smoking, and being physically active throughout the day which help prevent heart artery issues. The Tsimane spend only 10% of their time inactive, while most people in the modern world are inactive more than half of the day (54%). On an average, Tsimane men are physically active for six to seven hours per day while women are active for four to six hours. The majority of their diet (72%) is made up of carbs, including healthy carbs with lots of fiber, such as rice, plantain, and manioc. They also eat fruits, nuts, and nut butters. Animal meat makes up 14% and is the main source of protein in their diet. They eat very little fat (14%), which is approximately 38 g/day. This population has very few smokers. A computer tomography scan showed that almost 9 out of 10 Tsimane residents had no heart disease risk. Only 13% of Tsimane people had a low risk, and only 3% had moderate or high risks. Tsimane has the lowest level of vascular ageing ever recorded. Comparatively, in the United States, only 14% of computer tomography scans showed no heart disease risk. The half of participants (50%) showed a moderate to high risk.

This study suggests that adopting the Paleolithic eating habits of our Stone Age forefathers can help improve LDL-cholesterol levels. Recent studies have shown that this diet is popular and includes lean meats, nuts, olive oils, fruits, and vegetables. The average American diet is composed of about 50% carbohydrates, a third fat, and 15% protein. The long-term health effects of the Paleolithic diet are, however, not yet fully understood. The Paleolithic lifestyle encourages nutrient-dense whole foods and discourages processed food containing excess sugar, salt, and unhealthy fats. It is still important to remember that excluding legumes, dairy products, and whole grains may lead to an inadequate intake of nutrients.

Losing their traditional lifestyle and diet could also be a risk factor for blood vessel aging. Modern diets and lifestyles may be associated with a higher risk of heart disease. Age, smoking, high blood pressure or cholesterol, lack of physical activity, obesity, and diabetes are the main risk factors for heart disease. It is no wonder that indigenous people from the United States, Australia, and New Zealand suffer from health issues related to obesity, diabetes, high cholesterol, and blood pressure. In addition, they have a poor lifestyle with little exercise, bad eating habits, excessive alcohol consumption, and smoking. They are at higher risk of heart disease because of these factors. Diabetes is a major cause of heart disease in American Indians and Alaska Natives. Poor American Indians also have poor heart health.

Physical Activity

Aerobic exercise can lower LDL-cholesterol. A study of 4,700 participants that examined 51 exercise programs lasting at least 12 weeks and involving aerobic exercise showed a 5% decrease in LDL-cholesterol. In other studies, both resistance and aerobic training significantly lowered LDL-cholesterol. To achieve this result, however, intense exercise is needed.

High concordance between spouses for LDL-cholesterol and other risk factors is an example of the importance of lifestyle habits. Genetics cannot explain this phenomenon as the spouses are not genetically related. Many studies have shown that spouses share similar heart disease risk factors. These include high cholesterol, blood pressure, weight, smoking, and other factors. In one study, it was found that plasma cholesterol levels are 20% higher in one spouse if the other is high. It is possible that they share similar habits. Another explanation could be that they are drawn to each other because of their similar health.

Therapy

The risk of heart attack or stroke increases when LDL-cholesterol levels are high in the bloodstream. Therefore, it is logical that medications which reduce LDL-cholesterol levels can effectively reduce the number of heart attacks.

Two types of medications can lower LDL-cholesterol – those that stop the body's production of cholesterol, and those which prevent its absorption. They can be combined in medical practice to reduce LDL-cholesterol more effectively.

Statins

Statins have been the most widely used and well known of all lipid-lowering molecules. The first statins were created in the 1980s. Later on, several companies, including Pfizer and AstraZeneka, were especially successful in bringing to the market highly efficient molecules of atorvastatin (Lipitor) and rosuvastatin (Crestor) that markedly reduced cardiovascular disease. At present, statins approved by the European Medicines Agency and the US Food and Drug Administration include atorvastatin, simvastatin, rosuvastatin, fluvastatin, pitavastatin, lovastatin, and pravastatin. Despite statin therapy, the risk of developing this disease is still substantial. Statins do not work for a large number of patients.

The main way that therapies to stop the body making cholesterol work is by increasing the clearance of LDL from the blood via the LDL receptor. The number of liver receptors that remove LDL-cholesterol can be increased. The liver must be deficient in cholesterol to increase the number of LDL receptors. When the liver cells are not producing enough cholesterol, this is what occurs.

The liver produces cholesterol with the help of several enzymes, the most important being HMG-CoA reductase. Statins target this enzyme. The molecules cause a shortage of cholesterol within the liver cells. The liver produces more LDL receptors to help it pick up more LDL particles in the blood. This causes a quicker removal of LDL particles from the blood and lowers LDL-cholesterol levels. Statins inhibit the enzyme, lowering cholesterol in LDL, and other lipoproteins, including IDL and VLDL. However, statins have no significant impact on Lp(a).

Statins enhance the presence of LDL receptors on cell surfaces, leading to a faster elimination of LDL from the bloodstream and a reduction in LDL-cholesterol levels.

Statins have many other beneficial effects, such as reducing inflammation and oxidative stress in the body. The anti-inflammatory and antioxidant effects may be important in reducing heart disease. Statins protect lesions from rupture and stabilize them. However, statins can have negative side effects, such as muscle pain and liver injury. They can also cause memory loss, fatigue, and reduced sex hormones. They may also elevate blood sugar levels.

In many large-scale studies, statins were shown to have a beneficial effect on heart disease. These trials always randomized the comparison of treatment to placebo. The studies were also double blind – neither the participants nor the doctors knew which patients received treatment and which placebo. Statins consistently lower LDL-cholesterol by 25–35%, and heart attacks are reduced by 25–30% compared to placebo. The reduction in heart issues is greater if treatment is longer and begins earlier. In these studies, lowering cholesterol had no major side effects.

Statins were shown to lower cholesterol in several populations. This is especially true for middle-aged men who have a history of heart disease. The effects of the drug are less pronounced in women and older people.

Remember that only a small percentage of patients who take statins develop heart problems, usually less than 10%. While statins don't markedly reduce the absolute risk of heart disease, it would take treating more than 50 patients for five years to prevent just one heart attack.

Clinical studies involving large-scale randomized trials, which examine the effects of lowering LDL-cholesterol on cardiovascular events, provide strong evidence that causality exists between LDL-cholesterol and heart disease. It is also important to take into account the design of the trial as factors such as small sample sizes and few events can impact the interpretation of the results. Several therapies that reduce LDL-cholesterol can, however, have negative effects, such as an increased incidence of diabetes. This can decrease the clinical benefit.

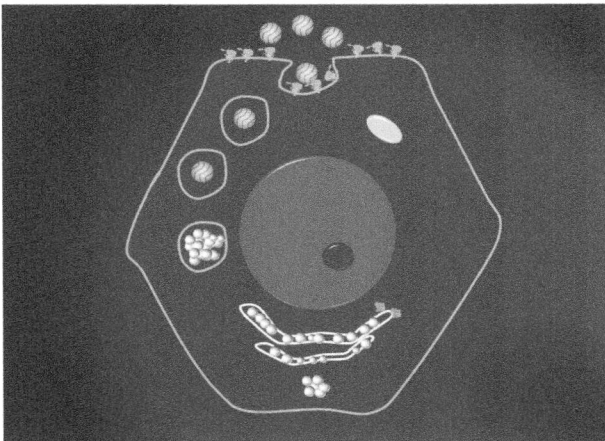

Statins block cellular cholesterol production, which increases the number of LDL receptors on cell surfaces. This, in turn, accelerates the uptake of LDL by cells and enhances the supply of cholesterol.

It is important to not overinterpret the results of individual studies and focus instead on the collective evidence from all available trials with a high quality. Meta-analysis is a gold-standard method of analyzing data from many trials. A study of this kind looked at the data from 26 large trials that involved almost 170,000 patients who were taking statins. In this meta-analysis, statins reduced the risk of heart disease by 22% per each 39 mg/dl decrease in LDL-cholesterol. The effect was constant over the course the treatment with a slight reduction in the first 12 months. No matter what a person's cholesterol level was or if they already had heart disease, the reduction in risk was similar. This analysis shows that statins reduce the risk of developing heart disease by lowering LDL-cholesterol levels. In other studies, lower LDL-cholesterol levels have similarly been linked to fewer heart attacks. Data show that the plaque buildup can be stopped when LDL-cholesterol levels are at 70 mg/dl. These data support the "lower is better" approach that many researchers advocate.

Open Question

Can LDL-cholesterol be safely lowered to the very low levels seen in hunters-gatherers?

A recent systematic review of published randomized-controlled trials revealed that statins reduced cardiovascular events by 26%, cardiovascular mortality by 15%, and overall mortality by 9%. Statins are also effective in primary cardiovascular preventive measures, with consistent benefits for cardiovascular events, cardiovascular death, and all-cause deaths. Researchers estimate that if an extra 10 million high risk people around the world had statin treatment it would save about 50,000 lives each year or 1,000 lives a week. Statins are also promising against heart disease in a form of polypill, a single drug product that combines multiple medications. A statin combined with a blood pressure-lowering drug or aspirin can be very effective.

History

Discovery of the First Statin

Who	When
Akira Endo	**1976**

Where	What
Tokyo, Japan	Endo, A., M. Kuroda and K. Tanzawa (1976) "Competitive inhibition of 3-hydrox-3-methylutarylcoenzyme A reductase by ML-236A and ML-236B fungal metabolities, having hypocholesterolemic activity." FEBS letter 72: 323–326.

Scientists in universities and companies began searching for drugs that would lower cholesterol as clinicians discovered more evidence of the connection between high cholesterol and heart disease. In the 1950s and 1960s, companies searched for chemicals that could stop the production of cholesterol. Triparanol, the first drug that lowers cholesterol by preventing its production, was developed in the 1950s. The molecule stopped the final step of making cholesterol and other sterols began to build up. In the United States, it was first used in 1959. However, serious side effects such as cataracts led to its removal from the market by the early 1960s. At that time, fibrates were the only drugs which could lower cholesterol. However, their effects were modest at best and they were not always well tolerated.

Akira Endo was a Japanese postdoctoral scientist who worked in New York in the late 1960s. He was struck by the high rate of heart disease in the United States, which he attributed to elevated plasma cholesterol. He returned to Japan and began searching for an inhibitor that would block cholesterol biosynthesis. He hoped to find a molecule that would reduce both cholesterol production and heart disease. He was inspired by Alexander Fleming's biography, who in 1928 discovered penicillin from a blue–green mold of the *Penicillium* genus.

Endo searched for inhibitors to the HMG-CoA reductase, which was just identified as a key element in the control of blood cholesterol by Brown and Goldstein. He worked for the Sankyo Pharmaceutical Company in Tokyo, where he studied fungi. Fungi are known to have a wide range of molecules that affect human biochemistry. Endo selected this strategy in 1971 because the company used fungi for searching new antibiotics. He hypothesized that if some fungi could inhibit bacterial growth, others could inhibit cholesterol synthesis. At the time, there was no evidence that fungi produced such inhibitors. This hypothesis was bold and innovative.

Endo set up a lab at Sankyo to isolate inhibitors. In April 1971, he began his project by using the culture broths of thousands of fungi. After testing 3,800 fungi strains for a year, he discovered that a mold culture broth had a strong inhibitory effect. Citrinin was found to be the active substance, and it inhibited cholesterol production in rats. Citrinin was, however, toxic to kidneys, so the research was discontinued. This setback gave him courage and hope that he could find better active ingredients in the future.

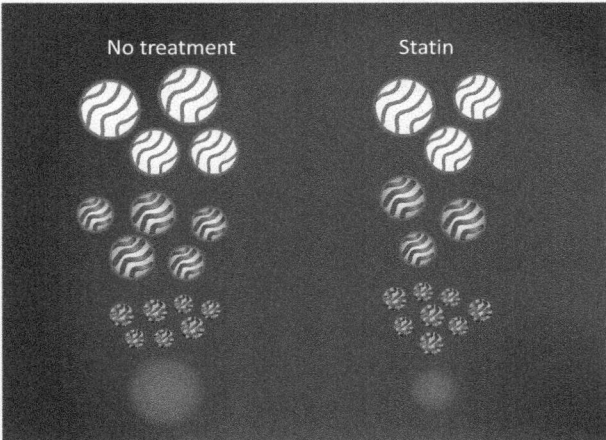

Statins lower LDL-cholesterol and triglyceride levels, while also raising HDL-cholesterol.

Endo found a second active culture of blue–green mold called *Penicillium citrinum* in the middle of the summer 1972. It was found in a rice sample from a grain store in Kyoto. This mold is the same as those that can contaminate fruits like oranges and melons. In 1973, after testing more than 6,600 species of molds, he isolated a factor from *Penicillium citrinum* which inhibited HMG-CoA reductase. The molecule was highly affine to this enzyme. Endo discovered that HMG-CoA reductase and his molecule had structural similarities. The molecule worked by competing against the enzyme. It was a strong inhibitor of HMG-CoA reductase. It stopped the production of sterols by different cell types, including those of people with high levels of cholesterol.

Unexpectedly, the inhibitor had no effects whatsoever on rats' cholesterol levels in the blood. Endo was initially disappointed but quickly realized that the lipoprotein metabolism in rats is different from humans. In particular, rats have lower LDL. Similar inconsistencies between data obtained in different animal species were observed by Nikolai Anitschkow some 60 years earlier. The compound compactin (ML-236B), when given repeatedly, actually increased the level of HMG-CoA reductase within the rat liver. This negated its ability to inhibit the enzyme. It was discovered that this was the primary reason compactin failed to work in rats.

Researchers thought that other species might have different effects. In 1976, subsequent studies on other animals, including hens, monkeys, and dogs, and humans revealed that compactin reduced plasma cholesterol.

In 1978, the first trial of compactin was started in patients with hypercholesterolemia. The study was conducted in collaboration with Akira Yamamoto, a physician from the Osaka University Hospital. Soon, a significant reduction in blood cholesterol levels was noted. Yamamoto administered compactin to eight dyslipidemic patients. The cholesterol levels of the patients dropped on average by 30%. There were no serious side effects.

(Continued)

(Continued)

Sankyo developed compactin as a result of these treatments. The company began testing compactin in November 1978 on humans as part of a phase one clinical trial. In the summer 1979, Sankyo moved to phase two and distributed compactin to patients with severe cases of high cholesterol in 12 hospitals. All hospitals reported that compactin was effective and safe. Mevastatin was the new name for compactin, which was, however, never commercialized. Sankyo halted its research on this molecule in August 1980 after it was discovered that dogs given high doses of the molecule developed cancer. Since that time, mevastatin was no longer used. Later, it was derivatized into the drug pravastatin.

Discovering compactin was akin to finding a needle in a haystack.

Pharmaceutical companies began searching for another statin after discovering the effects of compactin on dogs and monkeys. In 1978, Merck Research Laboratories isolated lovastatin from *Aspergillus terreus*. The molecule differed from compactin by a single methyl group. This was the first statin available to the public. Merck had a history in cholesterol studies, including the discovery of mevalonate. The drug was developed in the department of Roy Vagelos by Al Alberts, a brilliant researcher who never finished his PhD.

Brown and Goldstein discovered in 1981 that lovastatin can lower LDL-cholesterol in dogs. This was also observed in humans. The laboratory of Hiroshi Mabuchi at Kanazawa University in Japan reported that compactin treatment lowered LDL-cholesterol in patients with familial hypercholesterolemia. These patients' LDL-cholesterol was reduced by around 30%.

Roger Illingworth of Oregon Health Sciences University and Scott Grundy and David Bilheimer of University of Texas tested the new drug in July 1982 on patients who had very high cholesterol and were not improving with other medications. The drug was very effective at lowering LDL-cholesterol and caused few side effects. Merck conducted large-scale studies and found lovastatin to be effective at lowering cholesterol levels. It also did not cause tumors, and it was well tolerated. Merck requested approval of the drug from the US Food and Drug Administration in November 1986. It became the first statin commercially available in September 1987. To celebrate the approval, a Merck helicopter was given the lovastatin number, MK803, as its tail number.

Statins, reflecting their microbial source and clinical importance, were later referred to as "penicillins for cholesterol."

Inhibitors of Cholesterol Absorption

In large studies, it was also shown that other treatments which lower LDL-cholesterol levels reduce the risk of developing heart disease. Ezetimibe is one such treatment developed by a team led by Harry "Chip" Davis at Schering-Plough and approved by the US Food and Drug Administration in 2002. It works by preventing the absorption of cholesterol in the intestine. The drug blocks the Niemann Pick C1-like 1 transporter, a protein involved in cholesterol absorption by the enterocytes. This results in a better excretion of cholesterol from the blood.

In a study of more than 18,000 patients with acute coronary syndromes, ezetimibe combined with statin medications further reduced LDL-cholesterol and decreased major cardiac events by 6.5%. In another study of over 9,000 patients with kidney disease, combining a statin with ezetimibe additionally lowered LDL-cholesterol and reduced the risk of heart attacks by 17%. These results were comparable to those that would be expected if only statins were used. A recent systematic review showed that ezetimibe significantly reduced cardiovascular events by 7% with no effect on all-cause or cardiovascular mortality.

Cholestyramine also reduces cholesterol absorption. The molecule removes from the intestine bile acids, which are required to absorb cholesterol. Cholestyramine prevents the reabsorption in the intestine of bile. It is a resin that acts like a sponge to absorb bile acids and prevent them from being taken up by the intestine. They are instead excreted from the body via feces. As a result, the liver converts more cholesterol to bile acids to maintain the balance. The cholesterol level in the blood is lowered. In a trial called the Lipid Research Clinics Coronary Primary Prevention Trial (LRC-CPPT), cholestyramine lowered LDL-cholesterol by 27 mg/dl and reduced the risk of heart disease by 19%.

Finally, a special type of surgery known as ileal bypass reduces the amount of cholesterol in the intestine by simply shortening its length. In a study called Program on Surgical Control of Hyperlipidemia (POSCH), ileal bypass reduced LDL-cholesterol by 71 mg/dl. This helped reduce the risk of heart disease by 35%. These results were also similar to those seen with statins or statin-ezetimibe combination therapy.

> **Open Question**
>
> Can the excretion rate of cholesterol into the bile from the liver be increased to reduce both levels of LDL-cholesterol and heart disease?

PCSK9 Inhibitors

PCSK9 plays a role in maintaining blood levels of LDL-cholesterol by degrading LDL receptors. By blocking PCSK9, LDL-cholesterol can be reduced. A recent systematic review revealed that PCSK9 inhibitors significantly reduced cardiovascular events by 16%. No effect has yet been observed on cardiovascular or overall mortality.

In 2015, the US Food and Drug Administration approved two drugs, evolocumab and alirocumab, to lower LDL-cholesterol through the inhibition of PCSK9, when other medications such as statins failed or were problematic. These drugs are highly specific antibodies for PCSK9. They bind exclusively to PCSK9 and thus can interfere with its normal function. They can also decrease Lp(a) levels. The antibodies also reduce inflammation and oxidative stress.

Researchers found in a recent study that adding evolocumab to a statin reduced LDL-cholesterol levels even further among people who have heart disease. The reduction in LDL-cholesterol also reduced the risk of heart attack or stroke. The study tracked participants for approximately two years and found that statins had similar effects to evolocumab on cardiovascular risk. These drugs are however expensive. They cost more than 6,000 US dollars per year at full retail. Importantly, the antibodies can reduce LDL-cholesterol by up to 60% in patients with homozygous form of familial hypercholesterolemia. Statins are not effective in these individuals because they do not have functional LDL receptors.

Inclisiran is a small interfering RNA (siRNA), which blocks the liver's production of PCSK9. Recent studies show that this molecule can reduce LDL-cholesterol up to 50% by just a few injections. The drug is long lasting and only needs to be injected twice a year. This greatly simplifies treatment.

Adenine base editors, another type of genetic tool, were successfully used in the editing of the PCSK9 gene on experimental animals. This greatly reduced LDL-cholesterol. The tools for gene editing were quickly removed from the body, and no unintended DNA changes were detected. Finally, vaccines that cause the production of antibodies to PCSK9 are being developed.

What It Is

Small interfering RNA (siRNA) is an RNA type that has double-stranded ends and can prevent certain genes from being expressed. This is done by destroying the messenger RNA, or mRNA, that is produced after a gene has been turned on.

Bempedoic Acid

A drug called bempedoic acid gets converted into its active form in the kidney and the liver by an enzyme. This active form inhibits ATP citrate lyase which is the enzyme that helps make cholesterol upstream of HMG-CoA reductase. It lowers cholesterol, just like statins do, by increasing the number of receptors in the body that are used to eliminate LDL-cholesterol.

Bempedoic acid exemplifies another way to reduce the risk of heart diseases caused by high LDL-cholesterol levels. The molecule can reduce LDL-cholesterol by 20–25% and cardiovascular disease incidence by 14%. It has been proven that bempedoic acid is effective, particularly for patients who cannot tolerate statins. It is safe and does not cause muscle pain or high blood glucose. It is most effective when combined with ezetimibe, or a statin of high intensity.

Inhibitors of Apolipoprotein B-100 Production

It may be better to stop producing apolipoproteins B-100 and B-48 in order to reduce the harmful lipoproteins which cause heart disease. Drugs like lomitapide or mipomersen can achieve this by blocking apolipoprotein B synthesis, as well as the addition of lipids to VLDL or chylomicrons during their formation in the liver and intestines. These drugs are used only in severe cases of high blood cholesterol because they can cause fatty liver.

The liver toxicity was the main reason for stopping the development of mipomersen, an antisense oligonucleotide that targets apolipoprotein B. By contrast, lomitapide, an inhibitor of microsomal triglyceride transfer protein, can lower LDL-cholesterol by up to 60% in patients with homozygous familial hypercholesterolemia. This drug can, however, cause stomach problems, increased liver enzymes, and liver fat accumulation.

Inhibitors of Lp(a) Production

It is interesting that most drugs used to lower LDL-cholesterol have little effect on Lp(a). Only niacin taken at 1–3 g/day is able to reduce Lp(a) by up to 30%. This drug, however, does not reduce heart disease when combined with statins. Moreover, lifestyle changes can only reduce Lp(a), to a minimal extent. New treatments that target Lp(a) are currently being developed.

Pelacarsen, for example, is an antisense oligonucleotide which works by stopping the production of apolipoprotein (a). It has been proven in clinical studies that it can reduce Lp(a), at various doses, by up to 80%. The drug has also been shown to lower LDL-cholesterol and apolipoprotein B. Nevertheless, pelacarsen can cause side effects such as muscle pain, headaches, and urinary tract infection.

What It Is

Antisense oligonucleotides are short, synthetic DNA or RNA pieces that can change the way RNA functions and the way proteins are made.

Olpasiran and SLN360 are two small interfering RNAs that can decrease the production of Lp(a) by hepatocytes. Both medicines have been shown to reduce Lp(a) levels by up to 10 times in clinical studies. They also appeared to be safe. They work well if they are given every 150 days. Lepodisiran is a novel small interfering RNA that reduced Lp(a) by up to 96% in just 2 weeks. It also maintained this reduction for 48 weeks.

Open Question

Can the risk of heart disease be decreased by lowering plasma levels of Lp(a)?

Reduction of Inflammation

Inflammation and oxidation are central for cholesterol bildup in the arteries. This suggests that attenuated inflammation could reduce the risk of heart disease even without changing lipid levels. Canakinumab is a therapeutic monoclonal anti-inflammatory antibody that targets interleukin-1 beta, a pro-inflammatory molecule. The antibody was tested in a double-blind, randomized Canakinumab Anti-Inflammatory Thrombosis Outcomes Study (CANTOS) on 10,061 patients with myocardial infarction. The antibody helped decrease ongoing inflammation in the body and led to fewer heart problems coming back compared to a placebo treatment by 15%. This was true even though their cholesterol levels stayed the same. However, there was a higher chance of getting an infection.

Efficacy in Targeting LDL-Cholesterol

High blood pressure, smoking, diabetes, and high blood sugar are all factors that increase the risk of heart disease. Studies show that LDL-cholesterol levels affect cardiovascular risk regardless of any other risk factors. Moreover, individuals with a greater risk of cardiovascular disease or more risk factors will see a greater reduction in risk when LDL-cholesterol levels are lowered. People with a genetically higher risk of heart disease will benefit more by lowering LDL-cholesterol levels.

Meta-analyses from many studies revealed that lowering LDL-cholesterol in different ways was associated with a reduction of heart problems of 20–25% for each 40 mg/dl reduction of

LDL-cholesterol. This shows that lowering LDL-cholesterol is beneficial regardless of how it is done. Statins, as well as ezetimibe and evolocumab, all seem to reduce the risk of cardiovascular events in a similar way. A number of studies have shown that genetic variations which reduce LDL-cholesterol levels are as effective in reducing cardiovascular risk.

The only exception to this finding is the effect of a class of drugs called CETP inhibitors. In a study called the Assess of Clinical Effects with Evacetrapib of Cholesteryl Ester Transfer Protein Inhibition in Patients with High Risk of Vascular Outcomes Trial (ACCELERATE), it was discovered that the CETP inhibitor evacetrapib reduced LDL-cholesterol by 29 mg/dl compared with just taking a statin. It did not, however, reduce the risk for cardiovascular events. This could have happened because evacetrapib increased blood pressure slightly or negatively affected HDL. When a CETP inhibitor is added to statins, the latter seem to have less effect on lowering LDL particles, which can lead to differences in measured LDL-cholesterol and apolipoprotein B reductions. Anacetrapib, another CETP inhibitor, reduced cardiovascular disease rates by 9% in a study with 30,000 participants. However, the effect was less than expected, based on how much LDL-cholesterol had been lowered (by 31%).

The genetic data as well as the data from the majority of trials suggest that the impact of LDL-cholesterol on cardiovascular events will be the same, regardless of the method used to lower LDL-cholesterol. Changes in LDL-cholesterol are likely to be similar in magnitude to changes in the concentration of LDL particles.

The effect of reducing LDL-cholesterol on the chance of heart disease depends on how high a person's LDL-cholesterol was to start with, how much it is lowered, and how long they are treated. Studies have shown that by lowering LDL-cholesterol levels for a prolonged period of time, the risk of cardiovascular diseases can be reduced even more. This is a much larger reduction than taking statins later in life, after atherosclerosis has already developed. The amount of LDL-cholesterol and the length of exposure to it will influence the risk of cardiovascular diseases. This conclusion is supported by, for instance, the West of Scotland Coronary Prevention Study, as the researchers in this study found that the benefits of lowering LDL-cholesterol increase over time.

According to studies and clinical trials, reducing LDL-cholesterol by 40 mg/dl can reduce the risk of cardiovascular event by 10% within the first year. This increases to 16% in two years and 20% in three years. It is possible that reducing LDL-cholesterol stabilizes any plaque buildup already present in the arteries. After three years of treatment, each additional year may lead to a further reduction of 1.5% in cardiovascular events for every 40 mg/dl decrease in LDL-cholesterol. If a patient is on a lipid-lowering drug for five years, the relative risk of cardiovascular diseases could be reduced by 20–25% per 40 mg/dl decrease in LDL-cholesterol. If a patient takes the medication (or has lower LDL-cholesterol levels) for 40 years, this could reduce the risk by 50–55% for each 40 mg/dl reduction of LDL-cholesterol. It would be an important achievement, but not enough to eliminate heart disease.

It is important to determine who will be the biggest beneficiaries of treatments that reduce LDL-cholesterol in the future. It is also critical to start lowering LDL-cholesterol levels early in people who are at high risk for cardiovascular disease, especially those with familial hypercholesterolemia. The difference between primary and secondary preventive measures is crucial. Primary prevention is taking action to prevent an illness from starting. Secondary prevention is the early detection and treatment of an established disease. After a heart attack, the risk of an accident that is fatal is greater than before. LDL-cholesterol levels should be reduced more aggressively for secondary prevention than primary prevention.

Another important question is how much LDL-cholesterol should be reduced. On the one hand, studies on heart disease indicate that lower levels are preferred. On the other hand, studies on mortality however indicate that people with very low LDL-cholesterol are at higher risk for death. These results suggest caution should be used when lowering LDL-cholesterol excessively.

What It Is

Primary prevention is the prevention of an illness from occurring.
Second prevention is the early detection and treatment of an established disease.

History

Reduction of Heart Disease by Lowering Blood Cholesterol

Who	When
Terje Pedersen and Merck	**1994**

Where	What
Oslo, Norway	"Randomized trial of cholesterol lowering in 4444 patients with coronary heart disease: The Scandinavian Simvastatin Survival Study (4S)." Lancet 344: 1383–1389.

In the 1970s, a link between high cholesterol levels in plasma and heart disease was established. However, there was no effective treatment available to reduce cholesterol levels. In 1955, Canadian pathologist Rudolf Altschul discovered that niacin (also known as nicotinic

(Continued)

(Continued)

acid) had a cholesterol-lowering effect. It was at this time the only drug to lower cholesterol. The vitamin also reduced plasma triglycerides. In the 1950s, clofibrate, a drug that lowers cholesterol better than nicotinic acid, was discovered. In the 1960s, more powerful drugs from the same family called fibrates were developed. These drugs could reduce cholesterol in some patients. It was unclear how they worked. These drugs were not perfect though, in terms of either their effectiveness or safety. The effects of these drugs on heart disease were less than desirable.

A drug called cholestyramine could also reduce cholesterol by preventing the body from reabsorbing its bile acids in the intestine. These acids were used in the liver to produce cholesterol. It worked for some patients but not all. In the 1970s and 1980s, a large Lipid Clinics Coronary Primary Prevention Trial (LRC-CPPT) took place. Researchers wanted to see if lowering high cholesterol in men of middle age with cholestyramine would reduce their risk of developing heart disease. Nearly 4,000 men were randomly assigned to take either the medication or a placebo. The study lasted more than seven years. Both groups followed a low-cholesterol diet. The group that took cholestyramine saw a greater reduction of LDL-cholesterol than the placebo group at the end. The group taking cholestyramine also had a reduced risk of heart disease of 19%, which included a 24% reduction in the risk of dying of heart disease as well as a 19% decrease in the risk of a heart attack. There was no difference between the groups in terms of the risk of dying. This study was the first to show that men with high cholesterol levels can lower their risk of heart disease. However, the effects were modest.

4S, the key to success.

A large study called the Multiple Risk Factor Intervention Trial aimed to determine if men with multiple heart disease risk factors could be helped in order to reduce their mortality rate. The study included more than 12,000 male participants. The special intervention group received instructions to eat in a way that would provide 30–35% calories as fat. Herewith, 10% of the calories should come from saturated fats and 10% from polyunsaturated. The group was also told to consume around 300 mg of cholesterol and to adjust their carbohydrate consumption as necessary. The participants were also encouraged by counseling and videos to stop smoking. A special program was used for the management of high blood pressure. It included medication and weight loss. It was advised to the usual care group that they should consult their doctor or another healthcare provider as necessary for managing their risk factors. Both groups showed a decrease in the frequency of risk factors after six to eight years. However, the group that received the special intervention had slightly better results. The two groups did not, however, have a significant difference between them in terms of the number of deaths due to heart disease.

HMG-CoA inhibitors were discovered in the late 1970s. They significantly lowered cholesterol levels and opened a way to combat heart disease. The Scandinavian Simvastatin Survival Study (also known as the 4S Study) was a large-scale trial that examined the effects of simvastatin, a statin, on people who have high cholesterol and heart disease. Over 4,000 people from Scandinavia took part in the study, which was sponsored by Merck. The study was designed to determine if simvastatin would help people with high cholesterol and a history of heart attacks.

Participants in the study were selected from medical centers located in Denmark, Finland, Iceland, Norway, and Sweden between 1988 and 1989. Participants ranged in age from 35 to 70, with an average of 59. Most were men. Participants were randomly assigned to receive a simvastatin daily dose or a placebo. The study was designed to follow participants for a minimum of three years. The study actually lasted a little over 5.4 years.

Terje Pedersen presented the study's results in 1994. He worked at the Oslo University Hospital, Centre for Preventive Medicine in Norway. The study found that simvastatin significantly reduced the number of deaths when compared with a placebo. The group that took simvastatin experienced a 35% decrease in LDL-cholesterol and 30% reduction in overall mortality compared to those who took a placebo. Heart attacks and strokes were reduced by 37% and 28%, respectively.

Interesting Numbers

According to the results of 4S, 30 people would need to be taking simvastatin in order to prevent one death from cardiovascular causes.

The study found that deaths due to other causes, such as cancer or accidents, were not increased. Women and older people, as well as those with diabetes, also benefited from the treatment.

Terje Pedersen's presentation received a standing ovation and a lot of attention. This was a significant breakthrough, as it settled the debate over whether or not elevated LDL-cholesterol was harmful.

Pedersen, it is interesting to note, had conducted another important trial prior to 4S. After his training, he worked in a small Norwegian hospital. Two of the consultants were interested in heart health. Pedersen was given the opportunity to run a small study on a heart drug, and he enjoyed it. Pedersen became the principal researcher for the Norwegian Timolol Trial. This trial showed that heart drugs could prolong life after a cardiac arrest. Pedersen's inspiration came from two Swedish doctors who assisted with the trial. He was not a lipidologist but learned quickly about lipids after deciding to do the 4S study.

(Continued)

(Continued)

A follow-up of the trial published in 2000 showed that simvastatin treatment continued for up to eight years reduced coronary heart disease deaths. Ten years after starting the study, those who continued to take simvastatin saw a 15% decrease in mortality.

This study had a big impact on medical opinion. Many other trials followed, leading to simvastatin being widely used. The 4S trial changed how doctors treat heart problems by providing strong evidence that lowering cholesterol with statins is beneficial. When the study results were published, Pedersen acknowledged that he knew it would be a big deal if the trial showed positive results, but he did not expect statins to become so popular.

Before this study, it was not clear if lowering cholesterol could help people with heart disease live longer. Many other studies have since confirmed these benefits, but the 4S trial was an important first step in understanding how statins can help people at risk of heart attacks.

Test Your Knowledge

1 You have total cholesterol of 202 mg/dl. Your HDL-cholesterol level is 50 mg/dl. And your triglycerides measure 105 mg/dl. Calculate your LDL-cholesterol.

   A 152;

   B 47;

   C 131.

Cholesterol, Lipoproteins, and Cardiovascular Health: Separating the Good (HDL), the Bad (LDL), and the Remnant, First Edition. Anatol Kontush.
© 2025 John Wiley & Sons, Inc. Published 2025 by John Wiley & Sons, Inc.

2 What lipoprotein is the main cause of atherosclerosis in humans?
   A VLDL;
   B LDL;
   C HDL.

3 What is the one of largest apolipoprotein known in mammals?
   A apolipoprotein A-I;
   B apolipoprotein B-100;
   C apolipoprotein B-48;
   D apolipoprotein E.

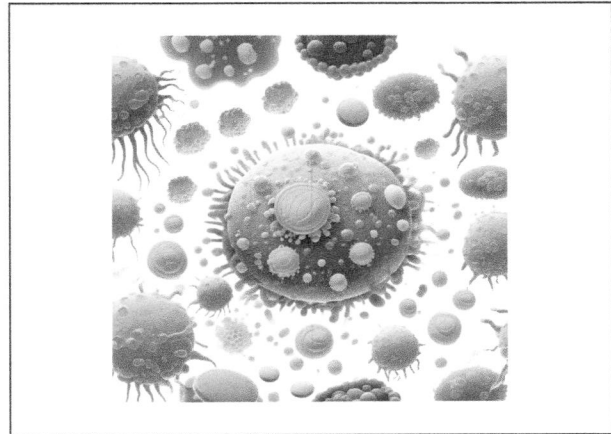

4   Akiro Endo tested how many fungi strains to determine if a mold possessed inhibitory activity
    against cholesterol synthesis?
    A more than 10;
    B more than 100;
    C more than 1,000;
    D more than 6,600.

Answers
1, C; 2, B; 3, B; 4, D.

Further Reading

Beheshti, S. O., C. M. Madsen, A. Varbo and B. G. Nordestgaard (2020). "Worldwide prevalence of familial hypercholesterolemia." Journal of the American College of Cardiology 75(20): 2553–2566.

Boren, J., M. J. Chapman, R. M. Krauss, C. J. Packard, J. F. Bentzon, C. J. Binder, M. J. Daemen, L. L. Demer, R. A. Hegele, S. J. Nicholls, B. G. Nordestgaard, G. F. Watts, E. Bruckert, S. Fazio, B. A. Ference, I. Graham, J. D. Horton, U. Landmesser, U. Laufs, L. Masana, G. Pasterkamp, F. J. Raal, K. K. Ray, H. Schunkert, M. R. Taskinen, B. van de Sluis, O. Wiklund, L. Tokgozoglu, A. L. Catapano and H. N. Ginsberg (2020). "Low-density lipoproteins cause atherosclerotic cardiovascular disease: Pathophysiological, genetic, and therapeutic insights: A consensus statement from the European Atherosclerosis Society Consensus Panel." European Heart Journal 41(24): 2313–2330.

Domanski, M. J., X. Tian, C. O. Wu, J. P. Reis, A. K. Dey, Y. Gu, L. Zhao, S. Bae, K. Liu, A. A. Hasan, D. Zimrin, M. E. Farkouh, C. C. Hong, D. M. Lloyd-Jones and V. Fuster (2020). "Time course of LDL cholesterol exposure and cardiovascular disease event risk." Journal of the American College of Cardiology 76(13): 1507–1516.

Falcone, G. J., E. Kirsch, J. N. Acosta, R. B. Noche, A. Leasure, S. Marini, J. Chung, M. Selim, J. F. Meschia, D. L. Brown, B. B. Worrall, D. L. Tirschwell, J. M. Jagiella, H. Schmidt, J. Jimenez-Conde, I. Fernandez-Cadenas, A. Lindgren, A. Slowik, D. Gill, M. Holmes, C. L. Phuah, N. H. Petersen, C. N. Matouk, L. Sansing, D. Bennett, Z. Chen, L. L. Sun, R. Clarke, R. G. Walters, T. M. Gill, A. Biffi, S. Kathiresan, C. D. Langefeld, D. Woo, J. Rosand, K. N. Sheth, C. D. Anderson and C. International Stroke Genetics (2020). "Genetically elevated LDL associates with lower risk of intracerebral hemorrhage." Annals of Neurology 88(1): 56–66.

Gencer, B., N. A. Marston, K. Im, C. P. Cannon, P. Sever, A. Keech, E. Braunwald, R. P. Giugliano and M. S. Sabatine (2020). "Efficacy and safety of lowering LDL-cholesterol in older patients: A systematic review and meta-analysis of randomised controlled trials." Lancet 396(10263): 1637–1643.

Hoogeveen, R. C. and C. M. Ballantyne (2021). "Residual cardiovascular risk at low LDL: Remnants, lipoprotein(a), and inflammation." Clinical Chemistry 67(1): 143–153.

Johannesen, C. D. L., A. Langsted, M. B. Mortensen and B. G. Nordestgaard (2020). "Association between low density lipoprotein and all cause and cause specific mortality in Denmark: Prospective cohort study." BMJ 371: m4266.

Kronenberg, F., S. Mora, E. S. G. Stroes, B. A. Ference, B. J. Arsenault, L. Berglund, M. R. Dweck, M. Koschinsky, G. Lambert, F. Mach, C. J. McNeal, P. M. Moriarty, P. Natarajan, B. G. Nordestgaard, K. G. Parhofer, S. S. Virani, A. von Eckardstein, G. F. Watts, J. K. Stock, K. K. Ray, L. S. Tokgözoğlu and A. L. Catapano (2022). "Lipoprotein(a) in atherosclerotic cardiovascular disease and aortic stenosis: A European atherosclerosis society consensus statement." European Heart Journal 43(39): 3925–3946.

Liou, L. and S. Kaptoge (2020). "Association of small, dense LDL-cholesterol concentration and lipoprotein particle characteristics with coronary heart disease: A systematic review and meta-analysis." PLoS One 15(11): e0241993.

Luo, J., H. Y. Yang and B. L. Song (2020). "Mechanisms and regulation of cholesterol homeostasis." Nature Reviews Molecular Cell Biology 21(4): 225–245.

Mortensen, M. B. and B. G. Nordestgaard (2020). "Elevated LDL-cholesterol and increased risk of myocardial infarction and atherosclerotic cardiovascular disease in individuals aged 70–100 years: A contemporary primary prevention cohort." Lancet 396(10263): 1644–1652.

Nissen, S. E., A. M. Lincoff, D. Brennan, K. K. Ray, D. Mason, J. J. P. Kastelein, P. Thompson, P. Libby, L. Cho, J. Plutzky, H. E. Bays, P. M. Moriarty, V. Menon, D. E. Grobbee, M. J. Louie, C. F. Chen, N. Li, L. Bloedon, P. Robinson, M. Horner, W. J. Sasiela, J. McCluskey, D. Davey, P. Fajardo-Campos, P. Petrovic, J. Fedacko, W. Zmuda, Y. Lukyanov, S. J. Nicholls and C. O. Investigators (2023). "Bempedoic acid and cardiovascular outcomes in statin-intolerant patients." New England Journal of Medicine 388(15): 1353–1364.

Norwitz, N. G., M. R. Mindrum, P. Giral, A. Kontush, A. Soto-Mota, T. R. Wood, D. P. D'Agostino, V. S. Manubolu, M. Budoff and R. M. Krauss (2022). "Elevated LDL-cholesterol levels among lean mass hyper-responders on low-carbohydrate ketogenic diets deserve urgent clinical attention and further research." Journal of Clinical Lipidology 16(6): 765–768.

Nurmohamed, N. S., A. M. Navar and J. J. P. Kastelein (2021). "New and emerging therapies for reduction of LDL-cholesterol and apolipoprotein B: JACC focus seminar 1/4." Journal of the American College of Cardiology 77(12): 1564–1575.

O'Donoghue, M. L., R. S. Rosenson, B. Gencer, J. A. G. López, N. E. Lepor, S. J. Baum, E. Stout, D. Gaudet, B. Knusel, J. F. Kuder, X. Ran, S. A. Murphy, H. Wang, Y. Wu, H. Kassahun and M. S. Sabatine (2022). "Small interfering RNA to reduce lipoprotein(a) in cardiovascular disease." New England Journal of Medicine 387(20): 1855–1864.

Patti, G., E. G. Spinoni, L. Grisafi, R. Mehran and M. Mennuni (2023). "Safety and efficacy of very low LDL-cholesterol intensive lowering: A meta-analysis and meta-regression of randomized trials." European Heart Journal – Cardiovascular Pharmacotherapy 9(2): 138–147.

Raschi, E., M. Casula, A. F. G. Cicero, A. Corsini, C. Borghi and A. Catapano (2023). "Beyond statins: New pharmacological targets to decrease LDL-cholesterol and cardiovascular events." Pharmacology & Therapeutics 250: 108507.

Ray, K. K., R. S. Wright, D. Kallend, W. Koenig, L. A. Leiter, F. J. Raal, J. A. Bisch, T. Richardson, M. Jaros, P. L. J. Wijngaard and J. J. P. Kastelein (2020). "Two phase 3 trials of inclisiran in patients with elevated LDL cholesterol." New England Journal of Medicine 382(16): 1507–1519.

Richardson, T. G., E. Sanderson, T. M. Palmer, M. Ala-Korpela, B. A. Ference, G. D. Smith and M. V. Holmes (2020). "Evaluating the relationship between circulating lipoprotein lipids and apolipoproteins with risk of coronary heart disease: A multivariable Mendelian randomisation analysis." PLoS Medicine 17(3): 22.

Rosenson, R. S. (2021). "Existing and emerging therapies for the treatment of familial hypercholesterolemia." Journal of Lipid Research 62: 100060.

Schoeneck, M. and D. Iggman (2021). "The effects of foods on LDL-cholesterol levels: A systematic review of the accumulated evidence from systematic reviews and meta-analyses of randomized controlled trials." Nutrition, Metabolism & Cardiovascular 31(5): 1325–1338.

Sniderman, A. D. (2021). "ApoB vs non-HDL-C vs LDL-C as markers of cardiovascular disease." Clinical Chemistry 67(11): 1440–1442.

Stefanutti, C. (2020). "Lomitapide – A microsomal triglyceride transfer protein inhibitor for homozygous familial hypercholesterolemia." Current Atherosclerosis Reports 22(8): 38.

Tokgözoglu, L. and P. Libby (2022). "The dawn of a new era of targeted lipid-lowering therapies." European Heart Journal 43(34): 3198–3208.

Wang, N., J. Fulcher, N. Abeysuriya, L. Park, S. Kumar, G. L. Di Tanna, I. Wilcox, A. Keech, A. Rodgers and S. Lal (2020). "Intensive LDL-cholesterol-lowering treatment beyond current recommendations for the prevention of major vascular events: A systematic review and meta-analysis of randomised trials including 327 037 participants." Lancet Diabetes & Endocrinology 8(1): 36–49.

Wolska, A. and A. T. Remaley (2020). "Measuring LDL-cholesterol: What is the best way to do it?" Current Opinion in Cardiology 35(4): 405–411.

Main Character Two: HDL, Carrier of "Good" Cholesterol

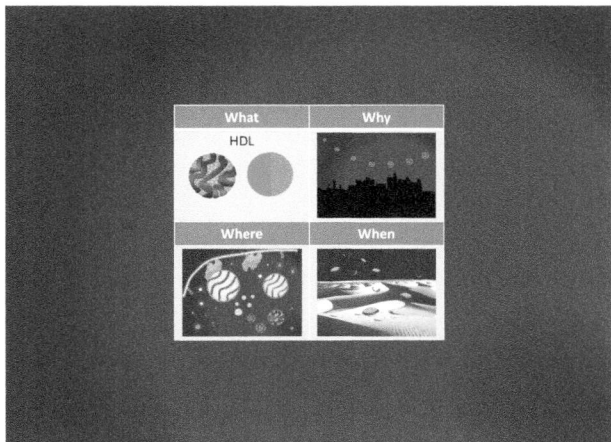

10

HDL – Why It Is Important

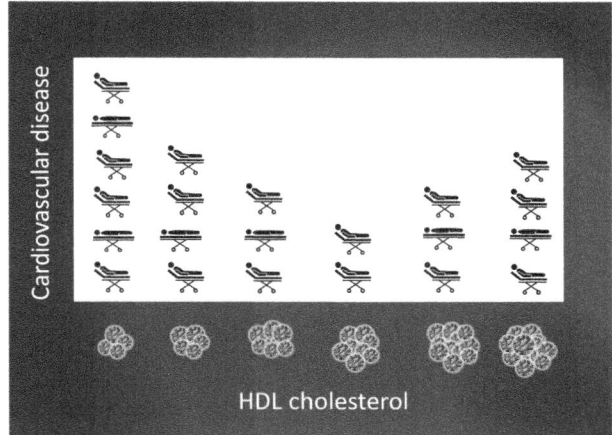

The U-shaped relationship between HDL-cholesterol and cardiovascular disease. The disease increases with both low HDL-cholesterol and extremely high HDL-cholesterol: "good" cholesterol may not be all that good! Note that higher HDL-cholesterol levels lead to larger HDL particle sizes.

The second main character of our story is The Good. This protagonist is also known as Good Cholesterol. The Good Cholesterol is the cholesterol that is measured in high-density lipoprotein (HDL). Some time ago and for a long time, this particle was thought to be only good for health. It was established that low levels of HDL-cholesterol in the blood are associated with more heart disease and stroke. High concentrations of HDL-cholesterol were therefore considered protective. When HDL-cholesterol concentrations are high, triglycerides or remnant cholesterol are often low. It was also considered a positive sign, as both are bad for health. High HDL-cholesterol and low triglycerides were thought to work together to prevent heart disease.

More recently, however, it was discovered that HDL is similar to LDL in having a bright and a dark side. Too much HDL in the blood turned out to be bad for health. Now, we know that as with many other things in the world, HDL is ambivalent and has its downsides.

Cholesterol, Lipoproteins, and Cardiovascular Health: Separating the Good (HDL), the Bad (LDL), and the Remnant, First Edition. Anatol Kontush.
© 2025 John Wiley & Sons, Inc. Published 2025 by John Wiley & Sons, Inc.

Low HDL-Cholesterol

Epidemiology

Our interest in LDL is a result of the epidemiological link between high levels of "bad" LDL-cholesterol and heart disease or stroke. In short, cardiovascular disease is more common when LDL-cholesterol levels are higher.

Another major finding in cardiovascular medicine is the relationship between HDL-cholesterol levels and heart disease. HDL particles from heart disease patients typically contain less cholesterol than healthy people, contrary to what is observed with LDL. The discovery was made in 1951 at Cornell Medical Center, New York. However, its significance took many decades to become apparent.

In later studies, it was found that low HDL-cholesterol concentrations are present not just in heart disease but also in other vascular diseases, such as myocardial infarction, stroke, peripheral arterial disease, and deep vein thrombosis. Reduced HDL-cholesterol is also a reflection of the severity and presence of coronary heart diseases. It is linked to heart disease markers such as coronary calcium and thickening in the carotid wall.

Doctors have found that people with diabetes and impaired glucose metabolism as well as various types of infections, among other conditions, also present with lower concentrations of HDL-cholesterol when compared to healthy individuals. Low HDL-cholesterol is a characteristic of many conditions, including heart disease.

However, the strongest argument in favor of the importance of HDL-cholesterol does not come from simple comparisons made between patients with heart disease and healthy controls. These data are sound, and they allow us to distinguish between people with and without heart disease. They provide initial support for studies on HDL. Most impressive, however, are the data that show how low HDL-cholesterol can predict heart disease many years before it occurs in large populations. People with low HDL-cholesterol levels at a certain time point are more likely to develop heart disease later in life compared to those who have normal levels. This is also true of strokes and other cardiovascular conditions. Normal concentrations of HDL-cholesterol are protective against heart disease and stroke. Even after a long-term monitoring of up to 30 years, low HDL-cholesterol levels are still reliable predictors for cardiovascular disease. The risk factor is also applicable to older individuals.

Framingham Heart Study, launched by Bethesda, Maryland's National Heart, Lung, and Blood Institute in 1948, was a landmark epidemiological research that identified low HDL-cholesterol levels as a risk factor for coronary artery disease. This study showed that low HDL-cholesterol was associated with the future development of myocardial infarction in men and women. Other large-scale prospective US and European studies made similar observations, including the Lipid Research Clinics Prevalence Mortality Follow-up (LRCF) Study, the Coronary Primary Prevention Trial (CPPT), and the Multiple Risk Factor Intervention Trial (MRFIT).

HDL-cholesterol is often called "good" cholesterol because of the association between HDL-cholesterol levels and heart disease. However, from a molecular perspective, HDL, LDL, and other lipoproteins carry identical cholesterol molecules. There is a general consensus that low concentrations of HDL-cholesterol in the bloodstream increase the risk of heart attacks and strokes.

Low HDL-cholesterol is associated with a substantial risk of cardiovascular disease. The simultaneous analysis of four large prospective studies conducted in the United States during the second half of the 20th century revealed that a 2% drop in HDL-cholesterol levels was associated with a similar 2–3% increase in cardiovascular disease risk. This may not seem like much, but when you multiply it by the 17.9 million deaths worldwide each year due to cardiovascular causes, the number is equivalent to a city the size of Nice or Miami being wiped off the face of Earth. This study shows that low HDL-cholesterol is a strong risk factor in cardiovascular deaths.

It is not surprising that the publication of these results caused a huge interest in HDL. This resulted in a rapid increase of research on the topic toward the end of the 20th and the beginning of 21st century.

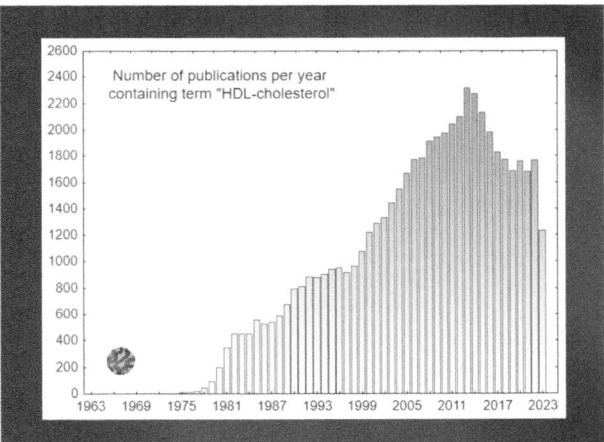

The number of articles published each year on HDL-cholesterol reached its peak in the 2010s.

All major calculators that estimate the overall cardiovascular risk for an individual now include HDL-cholesterol concentrations. These are the Framingham Risk Prediction Tool, the Prospective Cardiovascular Munster Study Score (PROCAM), and the Systemic Coronary Risk Estimation Approach (SCORE).

What does low HDL-cholesterol mean when we know that it is bad for our health? We must define this medical condition. Using the distribution of HDL-cholesterol concentrations in large populations, there are different cut-offs for defining low HDL-cholesterol among men and women. Women are known to have higher HDL-cholesterol than men. The most commonly used definition for low HDL-cholesterol is that it must be lower than 40 mg/dl in men or 50 mg/dl for women. However, this definition does not account for the well-known rise in HDL-cholesterol with age. Plasma HDL-cholesterol levels increase with age in both men and women. For example, in men, they rise by 7.9% between the ages of 50 and 65, while in women, they increase by 7.5%.

What It Is

For men, **low HDL-cholesterol** is defined as less than 40 mg/dl. Women's low levels are below 50 mg/dl. These limits are representative of the range in HDL-cholesterol for the general population. In surveys such as the National Health and Nutrition Examination Survey III (NHANES-III) and the Framingham Offspring Study, the average HDL-cholesterol levels for men and women respectively were 44 and 53 mg/dl.

Hypoalphalipoproteinemia is a condition where the circulating levels of HDL-cholesterol are lower than what is typically seen in a healthy population of the same age and sex. This is typically defined as being below the 5th, or 10th percentile.

Because of their medical importance, very low concentrations of HDL-cholesterol in humans are considered by the medical community to represent a distinct metabolic disorder termed "hypoalphalipoproteinemia." The term is derived from the words "hypo," meaning low, and "alpha-lipoprotein," which was an older name of HDL. This condition is characterized by lower HDL-cholesterol concentrations than in 95% of healthy individuals of the same gender and age from the same geographic area, termed the 5th percentile.

Meta-analyses of the results of multiple studies confirm that low HDL-cholesterol is a problem. HDL-cholesterol is inversely related to cardiovascular disease in large populations, including thousands of people. Low HDL-cholesterol is also associated with cardiovascular diseases in many ethnicities. The initial discovery that low HDL-cholesterol is a major cardiovascular risk factor in Caucasians was confirmed by other populations such as those of the Middle East and Asia where cardiovascular mortality is lower than in the United States. Low HDL-cholesterol is also a risk factor for heart disease, both in healthy people and patients with diabetes, high blood pressure, and other illnesses.

The conclusion that low HDL-cholesterol is responsible for heart disease can be theoretically biased. In fact, chronic disease patients are usually older than people in good health. HDL's lower cholesterol content could reflect the older age of patients. There are also more men with heart disease. The effect of ageing on HDL-cholesterol is, however, the opposite. It is actually increased in older people, not decreased. Men do have lower HDL-cholesterol than women. However, comparing patients with heart disease to healthy people of similar age and gender still shows lower HDL-cholesterol concentrations in their blood. These data demonstrate that low HDL-cholesterol is a strong predictor of heart diseases, regardless of gender and age.

It is interesting to note that most states with low HDL-cholesterol also have elevated plasma triglycerides. Low HDL-cholesterol levels are closely linked to obesity, insulin resistant, and inactivity, which all cause high triglycerides. This is another risk factor of heart disease as we will see later. This makes it difficult to assess the epidemiological significance of low HDL-cholesterol in isolation from other risk factors. Because of the connection between triglycerides and HDL-cholesterol, hypoalphalipoproteinemia is not included in the classification of dyslipidemias by the World Health Organisation. Low HDL-cholesterol can therefore just be a sign of high triglycerides.

According to epidemiologists, studies on HDL-cholesterol have been limited by the confounding of many variables. Specially designed mathematical methods can evaluate the contributions of all factors to mortality simultaneously. The "adjustment for other risk factors" method reveals that low HDL-cholesterol and increased cardiovascular risk are still associated even after this adjustment. Low HDL-cholesterol levels are an independent predictor for cardiovascular disease. This is because they do not depend on any other risk factors such as diabetes, age, or male gender.

Low HDL-cholesterol levels are linked not only to heart disease but also to many other serious disorders. A meta-analysis of multiple studies, for example, found that people who had low HDL-cholesterol levels at baseline were more likely to develop cancer. Low HDL-cholesterol levels are also linked to neurological disorders, such as cognitive decline and changes in the white matter of the brain. Some studies suggest that low HDL-cholesterol levels may even be linked to Alzheimer's. This long list of nonspecific conditions suggests that low HDL-cholesterol may be a condition commonly associated with poor health. Low HDL-cholesterol levels could be a sign that chronic inflammation is present in most diseases. In addition, low HDL-cholesterol has a negative association with longevity and a decreased chance of healthy aging.

History

Low HDL-cholesterol and Heart Disease

Who	When
Howard Eder	**1951**

Where	What
New York City NY, USA	Barr, D. P., E. M. Russ and H. A. Eder (1951) "Protein-lipid relationships in human plasma: II. In atherosclerosis and related conditions." American Journal of Medicine 11: 480.

Like the LDL story, HDL's history also began with biomarker discoveries and progressed to basic science before reaching therapy. HDL's story was more difficult to develop than LDL, but it did eventually evolve.

After World War II, the United States developed the most effective methods to isolate lipoproteins. Researchers used diluted ethanol solutions at a low temperature of 5°C to first separate alpha- and beta-lipoproteins, which were later referred to as HDL and LDL. Edwin Cohn developed this method at Harvard University in the 1940s.

David Barr's laboratory at Cornell Medical Center and New York Hospital used the fractionation technique extensively. Howard Eder, an assistant professor, and Ella Russ, a medical technician, worked together to apply this method to blood samples from healthy people

(Continued)

(Continued)

as well as patients suffering from a variety of diseases. In 1951, the first data obtained by this approach were published. Researchers measured the cholesterol concentrations in alpha- and beta-lipoproteins isolated from the serum of 20 young women and 24 young men. The researchers found that women had a higher concentration of cholesterol in alpha-lipoproteins than men. By contrast, men had more cholesterol in beta-lipoprotein than women. This result is presently well known as higher levels of HDL-cholesterol and lower levels of LDL-cholesterol in women than in men. Researchers correctly suggested that women's higher HDL-cholesterol could be linked to their protection against coronary heart disease.

Then, patients with various diseases were compared against healthy people. Researchers found that alpha-lipoproteins from 33 patients suffering from atherosclerosis had less cholesterol than those from healthy people. Alpha-lipoproteins from patients also contained less protein and phospholipid. Beta-lipoproteins from patients had higher levels of cholesterol, protein, and phospholipid than those of healthy people. The same observations were made on 35 patients with diabetes, and 12 patients with kidney diseases.

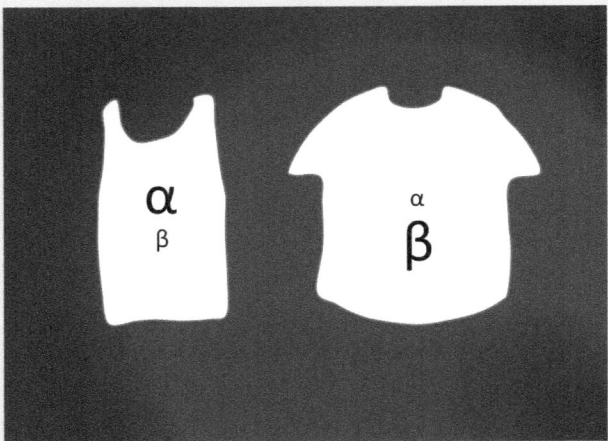

The statistical significance of the differences between groups was high. These differences were greater than those in the total cholesterol concentrations in plasma.

Researchers concluded that alpha-lipoprotein was deficient in atherosclerosis, diabetes, and kidney disease. This lipoprotein, in particular, contained less cholesterol. The concept of "good" cholesterol was born at this moment.

Genetics

So, it is evident that low HDL-cholesterol increases the risk of heart disease and stroke. Interestingly, however, genetics does not always agree.

It is possible to have genetically low HDL-cholesterol without an increased cardiovascular risk. The case of apolipoprotein A-I Milano, discovered in the 1970s, is probably the best known. Apolipoprotein A-I is the main protein present in HDL. As with many other proteins, the protein is subject to genetic variations, which are commonly referred to as mutations or variants. Apolipoprotein A-I Milano, a naturally occurring variant of apolipoprotein A-I, was discovered at the University of Milan in Italy. Researchers found, quite unexpectedly that carriers of the mutation had low HDL-cholesterol but were protected against heart disease. Since this surprising discovery, numerous studies have confirmed the paradoxical relationship between low HDL-cholesterol and reduced heart disease among carriers of apolipoprotein A-I Milano. A heterozygous variant involving the replacement of arginine by cysteine at position 151 of apolipoprotein A-I, which was later discovered in Paris, also showed cardioprotection despite low HDL-cholesterol in plasma. These results are consistent with those from Italy. However, no study has provided a clear mechanistic explanation up to now.

Other variants of apolipoprotein A-I have added more to the mystery surrounding apolipoprotein A-I Milano. Some heterozygous forms distinct from the Milano or Paris variants of apolipoprotein A-I deficiency also have plasma HDL-cholesterol levels below 50% normal, but with an elevated cardiovascular risk, not a reduced one. The Milano variant is not homozygous. People with homozygous forms of mutated apolipoprotein A-I reveal complete absence of HDL-cholesterol from plasma, and a reduction of HDL-cholesterol levels to undetectable. These subjects, unlike those who carry the Milano mutation, are at increased risk for premature coronary heart disease.

Open Question

Why is there no heart disease among some people who have low HDL-cholesterol, such as those with a mutated form of apolipoprotein A-I called apolipoprotein A-I Milano?

Tangier disease is another well-known example of genetically low HDL-cholesterol. The disease manifests as a deposition of cholesteryl ester in certain organs such as the tonsils which turn orange. This rare condition is caused by mutations of the ATP-binding transporter A1 (ABCA1), which is a key component in HDL particle formation. These patients have virtually no alpha-lipoproteins or apolipoprotein A-I in the blood. Parallel to this, triglyceride levels are slightly elevated while LDL-cholesterol is reduced by about half. In subjects with Tangier disease, a significant decrease in HDL-cholesterol can often, but not always lead to accelerated heart disease.

Fish-eye disease is a rare genetic condition that is linked to low HDL-cholesterol in the blood. This disease causes the front surface of the eye to become cloudy and develop small grey spots made of cholesterol. It usually starts in adolescence and can cause severe vision problems in adulthood. Fish-eye disease happens because of mutations in LCAT, an enzyme that plays a key role in lipoprotein transformations. When LCAT does not work well, it leads to lower HDL-cholesterol and delays the removal of cholesterol from tissue. Cholesterol buildup occurs in the eyes and sometimes the kidneys, which can lead to kidney failure. In fish-eye disease, some residual LCAT enzyme activity is present in lipoproteins containing apolipoprotein B. However, a complete loss of this enzymatic function leads to familial LCAT deficiency.

Both disorders can be inherited in different ways. In people with two copies of the gene for the disorder (homozygous), both familial LCAT deficiency and fish-eye disease are characterized by very low levels of HDL-cholesterol of less than 10 mg/dl. This represents classical forms of familial hypoalphalipoproteinemia. People who carry one copy of the gene (heterozygous) are usually healthy but have low levels of HDL-cholesterol. People with LCAT deficiency also have high levels of triglycerides and LDL-cholesterol. Studies have shown that this can sometimes lead to heart disease earlier than usual. However, the connection between LCAT deficiency and early heart disease is still not clear. In this disorder, having lower levels of LDL-cholesterol may ameliorate the risk for heart problems.

History

Apolipoprotein A-I Milano

Who	When
Guido Franceschini and Cesare Sirtori	**1980**
Where	**What**
Milan, Italy	Weisgraber, K. H., T. P. Bersot, R. W. Mahley, G. Franceschini and C. R. Sirtori (1980) "A-Imilano apoprotein. Isolation and characterization of a cysteine containing variant of the A-I apoprotein from human high density lipoproteins." Journal of Clinical Investigation 66: 901–917.

The history of apolipoprotein A-I Milano began in 1974 in Milan. It started with the hospitalization of Valerio Dagnoli, a railway employee from the small town of Limone sul Garda near Lake Garda. The doctors were amazed by the test results: he had very low levels of HDL-cholesterol and high levels of triglycerides, but his arteries and heart showed no significant damage. Intrigued, the doctors decided to investigate further and made a surprising discovery in his blood, as well as in his daughter's and father's blood. They found an anomalous protein that they named apolipoprotein A-I Milano after the city where it was discovered.

This groundbreaking apolipoprotein A-I Milano mutation was first identified by Guido Franceschini, Cesare Sirtori, and other researchers from the University of Milan in 1980. When they tested the entire population of Limone sul Garda, which consisted of around 1,000 people, they found that approximately 3.5% of them carried the mutation. Currently, there are about 40 certified carriers of the mutation residing in Limone sul Garda, along with seven to eight former residents who now live in Italy or abroad.

Researching in the municipal and historical archives of the parish church of San Benedetto, it was possible to identify a common factor among all individuals: the couple Giovanni Pomarelli and Rosa Giovanelli, who married in the late 1700s. The presence of the apolipoprotein could be traced back to 1644 or even earlier. Giovanni Pomarelli, born in the town in 1780, likely passed on the mutation to his children.

One of the main causes for the mutation and its transmission was the prevalence of marriages between close relatives. This was primarily due to limited access to the outside world, as there was no road connecting the town until the construction of the Gargnano – Riva road between 1929 and 1931. This road, with its impressive tunnels carved into the rock near the lake, opened up the county to tourism and global interaction.

Further studies conducted on the population of Limone revealed an unexpected cardioprotective effect associated with this mutation. The mutation alters one amino acid in the protein, replacing arginine with cysteine at position 173. As a result, apolipoprotein A-I Milano behaves abnormally but beneficially, effectively combating atherosclerosis and heart attacks. Cholesterol appears to be rapidly removed from the blood of the carriers of this mutation and transported to the liver for elimination.

In the years following this discovery, major laboratories in Northern Europe and the United States conducted studies on Limone. This research was made possible thanks to the collaboration of willing inhabitants, eager to contribute to science. Through this collaboration, scientists were able to synthesize apolipoprotein A-I Milano in vitro using bacteria. In the early 1990s, promising experiments were conducted on animals, showing that injecting a synthetic version of the protein into rabbits and mice could reverse plaque buildup in their blood vessels.

The studies went on with increasing intensity in order to explore potential medical applications. All residents of the town underwent blood tests and comprehensive examinations. Those who carry the mutation underwent a thorough cardiovascular investigation, including analysis of endothelial function, determination of peripheral vascular damage in carotid arteries and lower limbs as well as study of cardiac function and valvular systems. In November 2003, a group of researchers from the Cleveland Clinic Foundation, led by Steven Nissen, investigated an experimental drug containing apolipoprotein A-I Milano. This drug was administered to 47 subjects suffering from severe forms of atherosclerosis. Remarkably, within just six weeks, an average reduction of atherosclerotic lesions of 4.2% was achieved. However, later attempts to replicate this result using similar formulations of artificial HDL were largely unsuccessful. The therapy based on apolipoprotein A-I still awaits its development.

The town of Limone, proud of the recognition it received, decided to document the discovery of the apolipoprotein A-I Milano by establishing a Tourism Documentation Centre in the former municipal building, ensuring a lasting legacy.

Extremely High HDL-Cholesterol

These examples illustrate the wide divergence in epidemiology and genetics for low HDL-cholesterol, which is still puzzling. As data on the prevalence of extremely high HDL-cholesterol began to accumulate, the problem became even more serious. This rare metabolic condition termed "hyperalphalipoproteinemia" represents a dyslipidemic state in which circulating levels of HDL-cholesterol are higher than in 95% of age- and sex-matched healthy people from the same geographical area (the 95th percentile).

As hyperalphalipoproteinemia is rare, such studies require large numbers of subjects followed for a long time and have only been performed over the last decade. This research has shown that subjects with extremely high HDL-cholesterol levels have paradoxically higher mortality rates than those with normal or slightly elevated levels. These studies showed a U-shaped relationship between HDL-cholesterol and mortality, with elevated mortality both at low and extremely high HDL-cholesterol concentrations. This relationship was observed for not only cardiovascular and all-cause mortality but also mortality from infectious and inflammatory diseases.

What It Is

Hyperalphalipoproteinemia is a condition where the circulating levels of HDL-cholesterol are higher than what is typically seen in a healthy population of the same age and sex. This is typically defined as being above the 95th, or 90th percentile.

Many researchers are still puzzled by these paradoxical results. These data are very interesting because they clearly reflect the common sense knowledge that "too much good is always too bad." This can be expressed in other ways, such as "in every grain of good there is a piece or two of bad," and by a belief in a golden mean.

Studies have shown that high levels of HDL-cholesterol are associated with age-related macular deterioration, a condition affecting the eyes. Macular degeneration risk was increased when HDL-cholesterol levels were very high due to mutations in genes that code for CETP or hepatic lipase. This indicates that high HDL-cholesterol may also play a part in the disorder.

Similar to studies on low HDL-cholesterol, studies of extremely high HDL-cholesterol are also limited by confounding. For instance, those with extremely high levels often have a higher alcohol intake. Exercise and certain diets such as low-sugar and high-fat diets can also raise HDL-cholesterol. These associations make it difficult to draw firm conclusions about the causative role that extremely high HDL-cholesterol plays in vascular disease. We are eagerly anticipating further studies on the biological pathways that underlie this association.

Open Question

Is a low level of "good" cholesterol a general marker of poor health? Is it a more specific indicator of weakened cardiovascular health?

Genetics is less controversial in the case of extremely high HDL-cholesterol. Researchers have reported that accelerated vascular diseases can occur in patients with high HDL-cholesterol due to rare mutations, such as those found in SR-BI, CETP, or endothelial lipase. In animal studies, SR-BI deficiency can cause both highly elevated HDL-cholesterol and artery blockages. However, other studies of the same mutations failed to confirm a link between heart disease and extremely high HDL-cholesterol. These findings are still debatable but they at least show that extremely high HDL-cholesterol does not provide additional protection against heart disease.

It is best to maintain HDL-cholesterol levels within the optimal range for better health.

Incidence

Low HDL-cholesterol levels are often observed in the general population, especially in patients with cardiovascular diseases. Low HDL-cholesterol can be found in up to 60% of patients with coronary heart disease. In contrast, the prevalence of extremely high HDL-cholesterol is much lower, with carriers in general populations rarely exceeding 3%. This order-of-magnitude difference was the primary reason that only the link between low HDL-cholesterol and cardiovascular disease could be discovered in the early epidemiological studies.

Understanding the trends and prevalence of low HDL-cholesterol dyslipidemia is important to assessing cardiovascular risks and developing effective interventions. Low HDL-cholesterol dyslipidemia can be seen in both the general population and, in particular, those with coronary artery disease.

In the United States, about 37% of men and 39% of women have low HDL-cholesterol levels. Hong Kong has rates of 34% in men and women. In France, approximately 11% of men and 26% of women in the general population has low HDL-cholesterol. In Spain, 17.5% have low HDL-cholesterol, which is closely linked to cardiovascular diseases. In Turkey, 21% of the population has low HDL-cholesterol, defined as less than 40 mg/dl. This rate increases with increasing body mass index.

From a clinical perspective, it is important to note that low HDL-cholesterol often persists even after treatments are taken to normalize the lipid metabolism. Statins, the most commonly used class of lipid-lowering medications, do not or only mildly raise plasma HDL-cholesterol. As a result, the inverse relationship between HDL-cholesterol and coronary risk persists among coronary heart disease patients treated with statins. This relationship is, however, often weaker in these patients. This suggests that low HDL-cholesterol still remains a factor in heart disease, despite the LDL-cholesterol being lowered.

Even if the LDL-cholesterol level is low, patients who take lipid-lowering medications often have low HDL-cholesterol levels. It was observed in 66% of high-risk patients who had heart disease or other risks. This group of patients is still at a higher heart disease risk when their HDL-cholesterol levels are low. As HDL-cholesterol levels decrease, this risk increases. The Atherosclerosis Risk in Communities (ARIC) Study, which examined the relationship between LDL-cholesterol levels and cardiovascular risk, found this. Women with low levels of HDL-cholesterol and apolipoprotein A-I had a greater risk of cardiovascular disease, according to a study called Women's Health Study. Other studies, such as the Justification for the Use of Statins for Primary Prevention: An Interaction Trial Evaluating Rosuvastatin trial (JUPITER), however, found that there was no relationship between HDL-cholesterol and the risk of cardiovascular disease in participants who had low LDL-cholesterol.

Interesting Numbers

Around 30% of the general population have low HDL-cholesterol (defined as 40 mg/dl for men and 50 mg/dl for women). This percentage is strongly influenced by gender, ethnicity, and age. Only 2–3% of the general population have extremely high HDL-cholesterol (defined as >80 mg/dl for men and >100 mg/dl for women).

In Western societies, approximately 30% of the population has low levels of HDL-cholesterol. The occurrence of extremely high levels is less common, affecting only 2–3% of the population.

Open Question

Why does cardiovascular mortality increase in people who have extremely high HDL-cholesterol levels? Why is the mortality rate of infectious diseases also elevated in these people?

Practical Aspects: How to Measure HDL-Cholesterol

The central role of HDL-cholesterol in assessing the risk of heart disease and stroke emphasizes the importance of accurately measuring this parameter in individuals. Therefore, strict standards are in place for measuring HDL-cholesterol concentrations. A standardized method exists for measuring HDL-cholesterol in a small sample of human blood plasma. Initially, other lipoproteins are removed from the plasma through precipitation with a specific reagent and subsequent centrifugation for 10–20 minutes. The precipitated residue is discarded, and the remaining plasma, known as the "supernatant," is used to measure its cholesterol content using a photometer and a commercially available kit of reagents. The entire measurement process is typically completed within one hour.

How to measure HDL-C

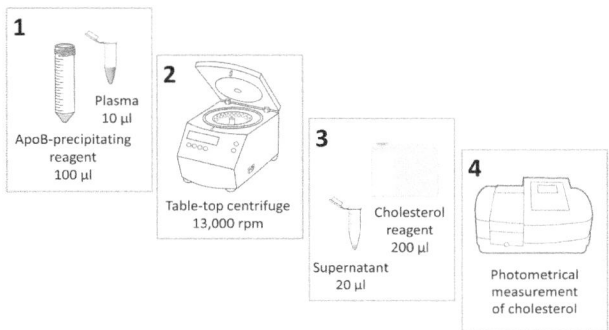

There are certain requirements to follow in order to accurately measure HDL-cholesterol levels as a factor of heart disease risk. HDL-cholesterol levels may vary from season to season, although the data are equivocal. LDL-cholesterol, on the other hand, is higher during winter. This may result in a more dangerous lipid profile in winter.

Second, it appears that HDL-cholesterol levels can be determined in samples taken both during fasting and without. Third, although the original method for measuring HDL-cholesterol after removing lipoproteins containing apolipoprotein B with a mixture of dextran sulfate and magnesium ions provides accurate results, more recent methods of directly measuring HDL-cholesterol levels may not always be good for those with abnormal lipid concentrations. High triglyceride can affect the results of these "direct" HDL-cholesterol tests.

As people age, HDL-cholesterol levels rise. Men's levels, for example, go up from 34 to 37 mg/dl when they reach age 65. Women's levels range from 41 to 44 mg/dl depending on their age group. The HDL-cholesterol of men is lower than that of women, regardless of where they are from (such as the United States, Sweden, or Iran).

Interesting Numbers

HDL-cholesterol was reported to drop by −1.8% with a two-day lag and by −5.6% with a two-week lag for every +5°C rise in outdoor temperature.

11

HDL – What It Is

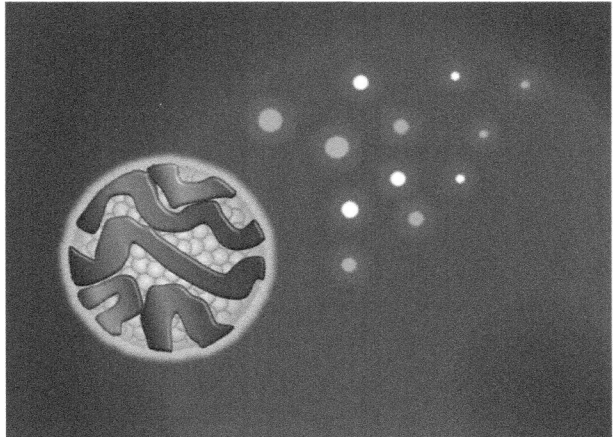

This image depicts an HDL particle moving through the bloodstream. The main proteins in HDL, apolipoprotein A-I and apolipoprotein A-II, are represented by dark-blue and light-blue ribbons, respectively. Phospholipids are depicted as yellow circles. In the background, red blood cells (erythrocytes) and white blood cells (leucocytes) are shown as red and white stars. The size of the HDL particle is magnified 1,000 times compared to the cells.

HDL, like LDL, is a lipoprotein composed of lipids and proteins. HDL is, however, different from LDL on many levels, including its physical properties, chemical makeup, biological activity, and metabolism. HDL typically has a smaller average diameter of around 10 nm, which is approximately 2.5 times smaller than the average diameter of LDL. The average HDL particle has a similar size to that of an antibody molecule. HDL has a lower molecular mass than LDL, ranging from 70 to 500 kDa.

HDL is also denser than LDL. The higher content of protein is responsible for this difference. HDL is made up of about half proteins and half lipids. The difference in density can be observed by comparing the denser appearance of meat (a source of proteins) to that of oil or butter (a source of lipids). While both HDL and LDL are denser than water and cannot float under normal physiological conditions (water has a density of 1.0 g/ml), HDL (density ranging from 1.063 to 1.21 g/ml) is considerably denser than LDL (density ranging from 1.016 to 1.063 g/ml).

Cholesterol, Lipoproteins, and Cardiovascular Health: Separating the Good (HDL), the Bad (LDL), and the Remnant, First Edition. Anatol Kontush.
© 2025 John Wiley & Sons, Inc. Published 2025 by John Wiley & Sons, Inc.

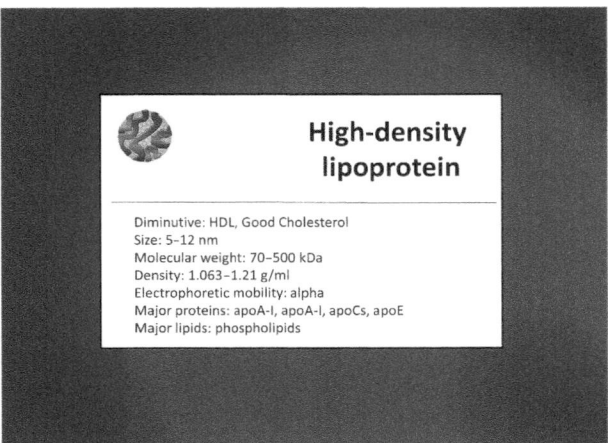

Proteins

The main building block of HDL particles is proteins. HDL contains more proteins than other lipoproteins and also has a greater variety of proteins.

Modern analytical techniques are able to detect hundreds of different proteins in HDL. This number is significantly lower for LDL or other lipoproteins. In the 20th century, antibodies were used to detect proteins in lipoprotein particles. However, these methods could cause cross-reactions with irrelevant molecules. Mass-spectrometric methods have greatly improved the detection of proteins and allowed identification of many of them within lipoprotein classes.

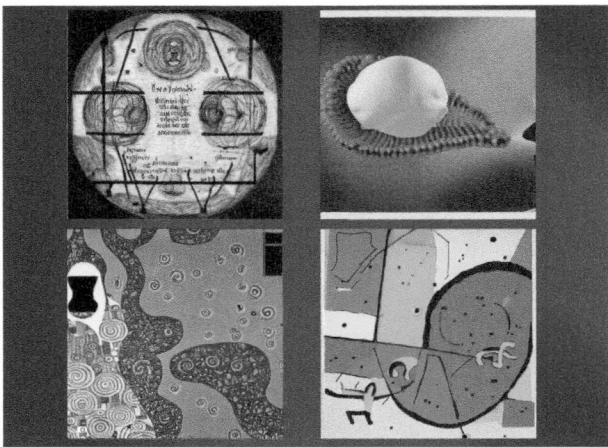

Here are images of HDL in the styles of Leonardo da Vinci, Salvador Dali, Juan Miro, and Gustav Klimt generated by Stable Diffusion AI. The images were generated based on prompts that included the name of the artist and the term "high-density lipoprotein." Source: Black Technology LTD/https://stablediffusionweb .com/ (accessed 9 February 2024).

Jay Heinecke, from the University of Washington, wittily presented a Lego model of HDL organization at a lipoprotein conference at the end of the 2000s. The model showed random figurines within a spherical lipoprotein particle, and it represented a change in how we understand lipoprotein composition. The success of this representation led to other models with dancing men, wine bottles, and others, all of which depicted HDL as a complex, living world populated by many personages.

A variation of the Lego model of HDL. The model was presented by Jay Heinecke in 2009 as a cartoon that showcased original findings from his laboratory at the University of Washington, Seattle.

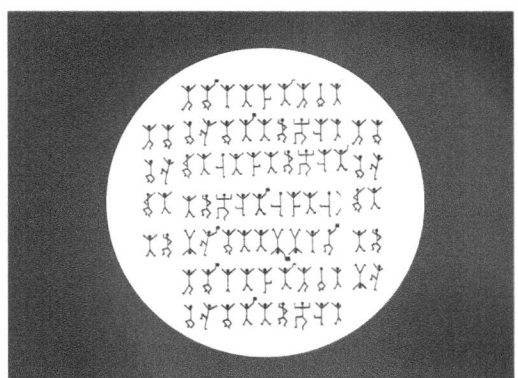 

The dancing men and the wine bottles models of HDL. Anatol Kontush introduced the dancing men model in 2010 during a Heart UK Meeting held in Edinburg. The city is the birthplace of Sir Arthur Conan Doyle, the author of The Adventure of the Dancing Men. This model was an ironical expansion of the Lego model proposed by Jay Heinecke. On the other hand, the wine bottles model was inspired by an upcoming seminar in Bordeaux, France – a region famous for its wine production.

HDL proteins can be divided into four groups: apolipoproteins, enzymes, lipid transfer proteins, and others. Apolipoproteins regulate the processing of lipoproteins in the body and maintain their physical structure. Enzymes are the catalysts responsible for chemically changing lipid molecules in lipoproteins. Lipid transfer proteins transport lipid molecules between lipoprotein particles.

It is important to keep in mind that some HDL proteins contain sugar residues. The glycosylation of these proteins can have a significant impact on their biological functions.

Apolipoproteins

The most important protein in HDL is apolipoprotein A-I. It gives HDL its circular shape and provides it with various functions through its interactions with cells and other biological entities. Apolipoprotein A-I makes up approximately 30% of the total mass of HDL. This corresponds to 70% total mass of protein, and the rest is made up by all the other proteins. Apolipoprotein A-I is crucial for HDL as it is found in almost all HDL particles. There are exceptions, however, where other apolipoproteins may replace apolipoprotein A-I as the backbone for HDL.

Apolipoprotein A-I is a small protein with a mass of 28 kDa and 243 residues of amino acids. The protein has unique physical properties despite its small size. It is amphipathic, which means it has both hydrophilic and lipophilic domains, interacting with aqueous and lipid environments, respectively. Apolipoprotein A-I can bind lipids, protecting them from undesirable interactions with water. In molecular detergents, such as soaps, that cover fat droplets, similar properties are observed. This increases fat solubility. Apolipoprotein A-I is a powerful biological detergent which effectively solubilizes lipids of all types. It forms micelles, which are stable and flexible enough to allow the apolipoprotein to move from one lipoprotein particle to another. Apolipoprotein A-I is also found in triglyceride-rich lipoproteins, such as chylomicrons and VLDL, but in lower quantities.

> **Amphiphilic (amphiphatic)** means having both hydrophobic (nonpolar) and hydrophilic (polar) properties.

Apolipoprotein A-II is the second most abundant HDL protein, accounting for about 15–20% of total HDL mass. The apolipoprotein has a molecular mass of 17 kDa and contains 77 amino acids. It is smaller than apolipoprotein A-I. Apolipoprotein A-II can also form HDL particles by itself, like apolipoprotein A-I. This makes it a structural HDL protein. Apolipoprotein A-II is present in approximately half of HDL particles. It often occurs alongside apolipoprotein A-I but is more hydrophobic. Apolipoprotein A-II can be viewed as an older version of the structural HDL proteins that evolved into apolipoprotein A-I. Apolipoprotein A-II circulates in dimers, which are made of two identical polypeptide chains. The chains are connected by a disulfide S-S link at a cysteine in the protein.

HDL contains several apolipoproteins in addition to A-I and A-II, including apolipoproteins A-IV, A-V, D, E, and others. These proteins, although present in small quantities, have a significant impact on lipoprotein processing. Apolipoprotein A-IV is one of the most notable examples. It is a 46-kDa large protein which is quite hydrophilic and is easily able to move between lipoproteins. Apolipoprotein A-V is another example. It circulates as a 39-kDa protein, primarily on triglyceride-rich particles, but it can also be found on HDL.

Apolipoproteins C are small apolipoproteins that regulate the lipolysis of triglycerides. HDL does it by exchanging apolipoproteins C with other lipoproteins.

Apolipoprotein D, originally called apolipoprotein A-III, is a 19-kDa protein that is associated with HDL. It has a hydrophobic pocket, which is used to carry small hydrophobic molecules. Apolipoprotein D transports fatty acids and other molecules efficiently.

Apolipoprotein E is a 34-kDa molecule which plays an important role in HDL structure, even though it is present in smaller amounts than apolipoprotein A-I. Apolipoprotein E has similar properties to detergents in terms of interacting with lipids as do apolipoproteins A-I and A-II. It can form large and spherical HDL particles on its own. These particles are processed by distinct pathways. Similar to A-II, apolipoprotein E is an evolutionary precursor to apolipoprotein A-I.

Apolipoprotein M is a 25-kDa protein primarily found not only in HDL but also in VLDL and LDL. It shares similarities with apolipoprotein D, including a hydrophobic pocket that allows it to bind and transport small hydrophobic molecules like sphingosine-1-phosphate, retinol, and retinoic acid.

Apolipoproteins in HDL are associated with cardiovascular disease, according to studies. Apolipoprotein A-I is the main protein in HDL. Its concentrations are closely linked to HDL-cholesterol. It can be used to determine how many HDL particles there are in the blood. Several studies have shown that HDL-cholesterol is not always a good predictor of heart disease. Instead, apolipoprotein A-I has been found to be a more accurate indicator. In a large INTERHEART study, which involved almost 30,000 participants in 52 countries, it was found that a high ratio between apolipoproteins B and A-I accounted for over 50% of heart attack risk. This ratio is more accurate in predicting heart attacks than the LDL- to HDL-cholesterol ratio. Although there is debate over whether HDL-cholesterol or apolipoprotein A-I levels are more accurate in predicting heart disease, HDL-cholesterol should be used to assess cardiovascular risk until this debate is resolved.

History

ABC Classification of Apolipoproteins

Who	When
Petar Alaupovic	**1971**

Where	What
Oklahoma City OK, USA	Alaupovic, P. (1971) "Apoliproproteins and lipoproteins. " Atherosclerosis 13: 141–146.

In the 1950s, the ultracentrifugation technique was developed to separate lipoproteins. Antigenicity was rapidly observed in lipoproteins at the start of these studies – the lipoproteins induced immune response, which involved the production of specific antibodies. The presence of proteins was thought to be responsible for this phenomenon. To study the proteins, delipidation was used – removing lipids from lipoproteins. It became apparent that LDL and LDL were two distinct entities. The proteins that HDL and LDL carried were different. These proteins were not able to exchange between the lipoproteins.

After electrophoretic separation, HDL was named alpha-lipoprotein and LDL was beta-lipoprotein. In order to distinguish between the two proteins in HDL and LDL, they were given the names proteins A and B. In contrast, VLDL antigenicity and LDL antigenicity appeared

---

(Continued)

(Continued)

similar in later studies, though not identical. These observations indicated that both VLDLs and LDLs carry the protein B. The heterogeneity in protein A has been observed since the 1950s, when studies were conducted on the terminal amino acids found in HDL.

Petar Alaupovic began working at Oklahoma Medical Research Foundation, Oklahoma City, in 1960 after completing his PhD from the University of Zagreb. His primary research focus was the lipoprotein protein fraction. Michel Macheboeuf had suggested in Paris, a decade before, that there might be a special affinity for proteins and lipids found within lipoproteins.

Alaupovic used an immunological approach when he studied the apolipoproteins. The apolipoproteins were delipidated by treating them with organic solvents such as heptane and then their ability to produce antibodies was assessed. This strategy required large-scale production of antibodies against apolipoproteins. Alaupovic collaborated with the City Zoo of Oklahoma, whose director was his personal friend. The researchers could use large animals, such as sheep and goats, which were kept in corrals. This method produced large quantities of antisera.

In fairness, this type of collaboration was very popular during these years. So, Gervase Mills' laboratory in London worked successfully with the London Zoo on the study of lipoproteins from different animals. John Chapman performed these studies, which resulted to the extensive characterizations of lipoproteins from many animal species.

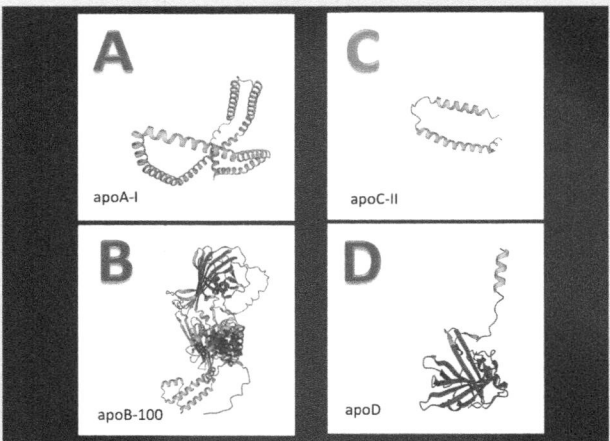

This method was used by Alaupovic's laboratory in 1964 to discover a new protein in VLDL. By gel electrophoresis, the protein was separated from the apolipoproteins A and B. Alaupovic suggested to call it apolipoprotein C, by using the letter following A and B. In 1966, he proposed for the first time a new classification of lipoproteins, based on the proteins they contain. This proposal was viewed as controversial, and it was not widely accepted for many years.

Apolipoprotein C's existence was only confirmed in 1969 by a second lab. The protein was also found to be heterogeneous and consists of two distinct forms. Alaupovic suggested using Roman numerals to differentiate between them. The proteins were designated C-I and C-II. It was found that the latter activated lipoprotein lipase. Very soon, Virgil Brown, along with his colleagues at the NIH, separated three distinct C apolipoproteins. These are now called apolipoproteins C-I, C-II, and C-III.

Two distinct proteins were found in HDL at the same time. Alaupovic assigned them the names A-I and A-II.

He proposed the ABC nomenclature for apolipoproteins in 1971 after a decade's worth of research. It was a well-thought-out system. His system defined three lipoprotein families characterized by apolipoproteins A, B, and C. The names of the respective family were LP-A, LP-B, and LP-C. Alaupovic suggested that each lipoprotein density class was made up of different families of lipoproteins, all of which had similar densities but a different protein composition. This view held that each lipoprotein was a polydisperse system of lipids and proteins characterized by a distinct apolipoprotein. Lipoproteins were defined by apolipoproteins rather than physical characteristics like size or density. Each apolipoprotein, like every other protein in the body, had a specific biological function. Each family covered the whole spectrum of density for plasma lipoproteins in a discontinuous way. This led to the postulate of the inherent heterogeneity in each lipoprotein class.

As a consequence, Alaupovic defined apolipoprotein in 1972 as a protein that binds neutral lipids and phospholipids into a polydisperse, soluble lipoprotein.

Alaupovic's immunological technique was used to develop other analytical methods. The first radioimmunoassay for determining concentrations of blood apolipoproteins has been developed soon. The first immunosorbent was also created to isolate apolipoproteins.

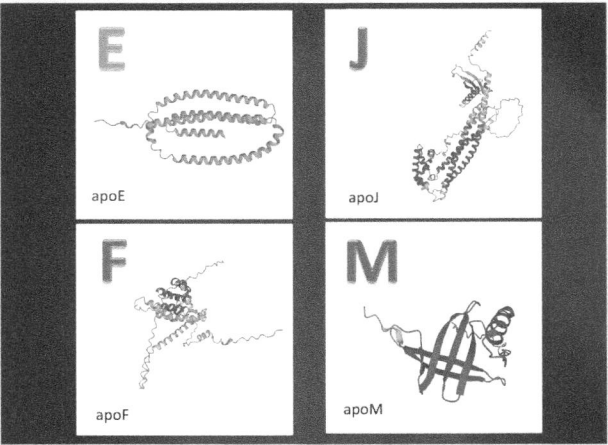

In 1973, two new apolipoproteins were discovered. Alaupovic, along with coworkers, identified apolipoprotein D in HDL. In VLDL, several laboratories found an arginine-rich protein that was later named apolipoprotein E. The protein was subjected under isoelectric focusing to reveal three bands, which according to ABC nomenclature were initially designated E-I, E-II, and E-III. Alaupovic, however, noted that the three bands were actually isoforms rather than distinct proteins. He therefore proposed using Arabic numbers from 1 to 3 which resulted in the apolipoproteins E1, E2, and E3.

(Continued)

(Continued)

> **What It Is**
>
> **Isoelectric focusing** is used to analyze and separate proteins and peptides from protein samples. It employs gel electrophoresis in order to separate proteins according to their isoelectric points, which is the pH (acidity) level at which the molecule does not have a net electrical charge.

This year was a milestone in the acceptance of the ABC classification. It was later included in books on biochemistry and medical literature.

In this respect, the concept of lipoprotein families was less fortunate. Alaupovic's discovery of apolipoprotein D led him to propose a fourth family, LP-D, to go along with the already proposed LP-A, LP-B, and LP-C. Apolipoprotein D, however, was found in particles containing apolipoprotein A-I. Other researchers have even suggested calling it apolipoprotein A-III. Due to this controversy, it was difficult to single out LP-D lipoproteins as a distinct family that is defined entirely by apolipoprotein D.

Later studies revealed VLDL to contain a combination of LP-B, LP-C, and LP-E, while LDL contained only LP-B. These observations complicated the concept that lipoproteins were divided into families. This was later replaced with the concept of individual proteins and lipids in lipoproteins.

Alaupovic's efforts finally paid off in 1978. Eight apolipoproteins were discovered, including A-II and B. C-II and C-III were also identified. D and E were added. The study of healthy individuals as well as patients with hyperlipidemia was extensive. This work was made possible by the largest antisera reserves in the world. Apolipoprotein A was found to be the most important component of HDL. Apolipoprotein A included both apolipoproteins A-I and A-II. Apolipoproteins A, B, C, and E were located in VLDL. It was hypothesized that normal plasma concentrations of the apolipoproteins are required for normal lipid transport in plasma.

The scientists logically moved on to ask the question: Are concentrations of apolipoproteins a better marker of atherosclerosis than concentrations of lipids? This question has remained unanswered for decades despite extensive research.

The molecular structure of apolipoproteins was characterized by later studies on purified apolipoproteins. Brian Brewer at the NIH sequenced apolipoprotein A-I in the 1970s. Its structure was then determined by others. Other apolipoproteins were also sequenced and structures determined. In the following decades, apolipoproteins F and H were discovered. Some letters, like G, K, and I, were strangely left out. Alaupovic's ABC world was finally mapped.

Enzymes

LCAT is an important HDL enzyme that helps convert free cholesterol into cholesteryl ester. Around 75% of LCAT in the blood is carried by HDL.

HDL enzymes also help in redox reactions, which eliminate harmful oxidized lipids. Paraoxonase, PAF-AH, and glutathione selenoperoxidase are three enzymes that participate in these reactions. The name "paraoxonase," or PON, comes from its ability to break down the toxic organophosphate paraoxon. The enzyme can also eliminate other toxic organophosphates to protect the nervous system. PON1 is a 43-kDa protein that is responsible for breaking down lactones such as homocysteine thiolactone. It is found mainly with HDL.

A second protein, PAF-AH, is also involved in the breakdown of pro-inflammatory oxidized phospholipids. It is carried by LDL and HDL. Glutathione peroxidases protect the body against damage caused by hydrogen peroxide and lipid hydroperoxides. Only HDL contains a type of glutathione peroxidase, called glutathione peroxidase 3.

What It Is

Paraoxonase (PON) is a major enzyme found in HDL. The enzyme is responsible for breaking down different molecules, including paraoxon, other toxic organophosphates chemicals, as well as lactones such homocysteine thiolactone.

The risk of heart attack can be evaluated through the activities and concentrations of the enzymes associated with HDL. Low activity of PON1 is linked with a greater chance of heart problems. There is however a disagreement over whether PON1 activity can really forecast cardiovascular risk. Low PON1 activity has been shown to not be specific enough to predict the risk.

PAF-AH can be a more accurate indicator of cardiovascular risks than PON1. This enzyme is mainly found in LDL, and its concentration increases along with LDL-cholesterol. Higher levels of PAF-AH have been associated with cardiovascular diseases. The relationship between PAF-AH and cardiovascular risks is however not strong. It can also be affected by other factors, such as inflammation.

Lipid Transfer Proteins

Transfer of lipids is essential for the fate of plasma lipoproteins. The transformation of HDL is governed by two key lipid transport proteins, CETP and PLTP. Both are mainly carried on HDL.

CETP is a 74-kDa protein that moves in the bloodstream between lipoproteins containing apolipoprotein A-I and those containing apolipoprotein B. The protein has a hydrophobic tunnel filled with two cholesteryl ester molecules, which is blocked at either end by an amphiphilic molecule of phosphatidylcholine. CETP can also transfer triglycerides through these interactions. The protein is able to change shape in order to accommodate lipoproteins of various sizes like HDL particles, LDL particles, and VLDL particles.

PLTP is a 78-kDa protein that belongs to the same family of proteins as CETP and lipopolysaccharide-binding protein. PLTP facilitates phospholipid exchange across plasma lipoproteins. This leads to HDL being converted into smaller and larger particles. The structural similarity between PLTP and lipopolysaccharide-binding protein allows PLTP to bind to lipopolysaccharide, a toxin produced by some bacteria. The synthesis of PLTP increases during inflammation and infection, indicating a protective role for this protein against infections.

HDL minor proteins are primarily those that regulate complement response and acute-phase reactions and protect against infection. They make up together less than 1% of HDL's total protein mass. The emergence of omics technology, which has greatly improved the sensitivity to detect proteins, led to their discovery in HDL. These studies showed that HDL contains a greater variety of proteins than previously believed.

Acute-Phase Proteins

Acute-phase response proteins are the body's reaction to an infection, tissue damage, trauma, or other insults. This response is usually seen as acute inflammation. These proteins significantly alter plasma concentrations when acute inflammation occurs. HDL, which is known primarily for its role as lipid-transporting particle, surprisingly contains a number of proteins that are involved in acute-phase response. This suggests that HDL could play a role in protecting against pathogens and in inflammation.

The serum amyloid A family is secreted in the acute phase during an inflammatory response. The main member of the family, SAA1, is a small 12-kDa protein that is carried by HDL. During acute-phase reactions, the production of SAA1 and SAA2 in the liver is increased, resulting in a significant increase of their circulating levels.

Another important protein involved in the acute-phase response and associated with HDL includes lipopolysaccharide-binding protein. This 60-kDa protein binds to the lipid A moiety in lipopolysaccharide. This removes lipopolysaccharide and protects the body from infection. Lipopolysaccharide-binding protein also binds and transfers phospholipids, similar to PLTP and CETP.

HDL carries fibrinogen as well, which is a protein that is synthesized in the liver and plays an important role in blood coagulation. SAA, lipopolysaccharide-binding protein, and fibrinogen are all elevated during the acute-phase response and are called positive acute-phase reactants. On the flip side, apolipoprotein A-I, apolipoprotein A-IV, and PON1 are all negatively affected by acute-phase response. They are called negative acute-phase reactants.

What It Is

Serum amyloid A (SAA) is a family of proteins that are secreted in the acute phase of inflammation. They are mostly transported by HDL.

Since their abundance is usually below one copy per HDL particle, there are questions as to how HDL-associated proteins are distributed throughout the HDL pool. These proteins could be localized in specific HDL subpopulations, with different origins or functions. Some proteins associated with HDL, such as complement C3, are similar in molecular mass to HDL particles on average, suggesting that they could only be associated with large HDL.

Other Proteins

Complement is an immune system component that helps eliminate pathogens from the body and remove damaged cells. The complement system is composed of 50 inactive proteins that circulate in the blood until they are activated. When activated, specific proteins are cleaved and stimulate cells to attack pathogens. To avoid damaging host tissues, the complement system must be controlled. HDL contains proteins that are involved in complement regulation such as C3, C4, and C9. The C3 and C4 proteins are important in activating complement, while C9 creates pores on targeted cells. HDL contains multiple complement components, which suggests that it could be used as a platform to host proteins involved in immune responses.

Other proteins in HDL contain serine proteinase inhibitor domains. Serine proteases are enzymes that cleave the peptide bonds of proteins. The catalytic triad of these enzymes consists of histidine, serine, and asparagine. This is what gives them their activity. Proteases are essential for protein degradation, and they require strict regulation via activation and inhibitory pathways.

Serpins, or serine protease inhibitors, block the activity of serine proteases and are important regulators for various biological processes, such as inflammation and coagulation. They also regulate angiogenesis and matrix degradation. Alpha-1-antitrypsin is a powerful inhibitor of serine proteases and one of the HDL proteins. It is not known what the exact function of serpins in HDL is, but it may prolong the time that they circulate in the blood. Serpins may also improve HDL's ability to remove cholesterol.

HDL transports a variety of proteins that have highly specialized functions. HDL is co-isolated with plasma retinol-binding protein, a 21-kDa transporter of retinol to peripheral tissues. Transthyretin is a 55 kDa thyroid hormone-binding protein that also binds apolipoprotein A-I on HDL. Transthyretin interacts with retinol-binding protein to prevent its loss through the kidney. Transferrin is another example, a 80-kDa glycoprotein that transports iron.

HDL also contains small peptides ranging in mass from 1 to 5 kDa. These peptides are fragments of larger proteins such as apolipoproteins, fibrinogen, and transthyretin. The presence of these peptides on HDL may serve as a pathway for peptide delivery or scavenging, slowing their renal clearance and proteolysis. Lipoproteins may act as versatile adsorptive surfaces for proteins and peptides because of their ability to bind various molecules.

The method used to isolate HDL has a strong influence on the composition of the HDL proteome. Proteomic analysis detects hundreds of proteins when human plasma HDL is obtained by ultracentrifugation in highly concentrated salt solution. The high ionic strengths used in the isolation method can alter the profile of HDL-associated proteins. It may be advantageous to use lipoprotein fractionation buffers that have a physiological pH and ionic strength for better protein characterization.

Interestingly, when HDL is isolated from human plasma using a non-denaturating approach called fast protein liquid chromatography (FPLC), proteomic analyses reveal the presence of numerous proteins that are not found in ultracentrifugally isolated HDL. This type of separation is, however, not possible without the co-elution of plasma proteins with HDL that are of the same size. Only if a protein is consistently detected in a variety of HDL fractions with different sizes, this suggests that it is truly associated with HDL. Notably, several abundant plasma proteins such as albumin, haptoglobin, transferrin, and alpha-2-macroglobulin are found in all apolipoprotein A-I-containing fractions. This indicates that they may be partially associated with HDL via nonspecific, low-affinity interactions.

Practical Aspects: How to Isolate Lipoproteins by FPLC

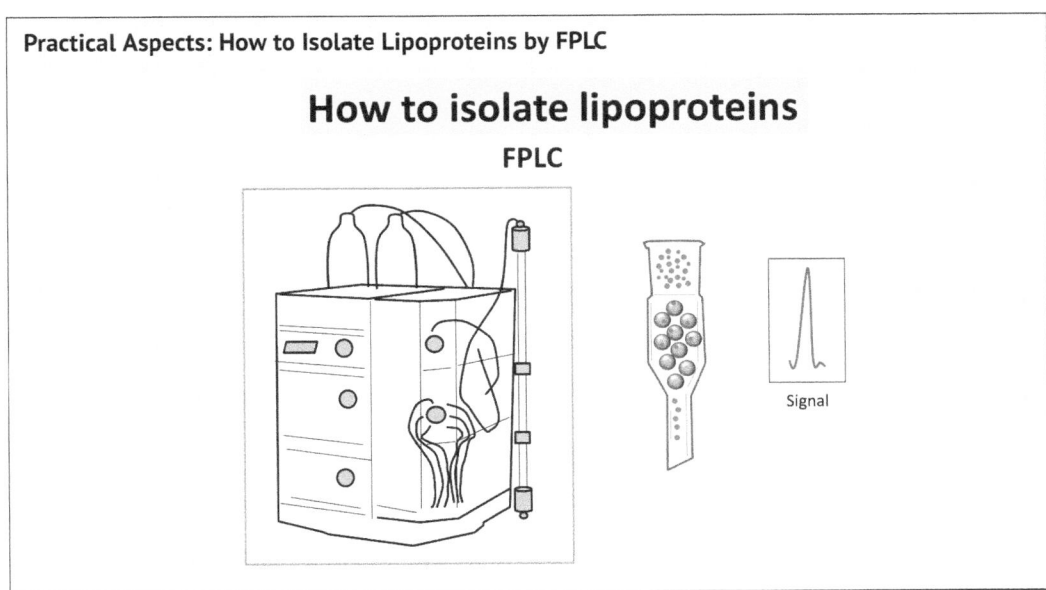

How to isolate lipoproteins
FPLC

Signal

(Continued)

(Continued)

FPLC is a quick and gentle method for separating lipoproteins, like HDL subclasses, from human plasma samples. It uses a special type of liquid chromatography called gel filtration. FPLC works by passing the mixture through a column filled with tiny beads made of a material called agarose. These beads have different affinities for the proteins in the mixture, allowing them to separate. The process is controlled by a pump that moves a buffer solution through the column at a constant rate. The buffer can be adjusted by drawing fluids from different reservoirs. FPLC resins come in different sizes and types depending on what needs to be separated. Lipoproteins eluting from the column are detected using spectrophotometry.

Lipids

Along with proteins, lipids are important in HDL. The HDL lipidome is mainly composed of phospholipids. In general, phospholipids account for 20–30% of total HDL mass. Cholesteryl esters make up 14–18%, free cholesterol accounts for 3–5%, and triglycerides make up 3–6%.

Phospholipids play an important role in the structure of HDL. By weight, they make up around half of the lipids found in HDL. HDL acts as a specific carrier for certain types of phospholipids. HDL is mainly composed of phosphatidylcholine and sphingomyelin. About 70% of HDL phospholipids are phosphatidylcholine, which is rich in polyunsaturated fats. Sphingomyelin is essential for maintaining HDL structure and function by affecting surface lipid rigidity. Lysophosphatidylcholines are a less abundant subclass of HDL phospholipids that have pro-inflammatory properties. Negatively charged phosphatidylinositol, phosphatidylserine, and phosphatidic acid are present in low amounts but can significantly affect the overall charge of HDL's surface, which can influence its interactions with enzymes, extracellular matrix components, and other proteins.

HDL and HDL-cholesterol are not the same. HDL is a type of lipoprotein composed of cholesterol (both esterified and nonesterified), phospholipids, proteins, and a small amount of triglycerides. HDL-cholesterol refers specifically to the cholesterol contained within HDL. Measuring the concentration of HDL-cholesterol in the blood can help determine the risk of developing heart disease. However, in clinical jargon, "HDL" is often used instead of "HDL-cholesterol," which is technically incorrect.

The surface lipid layer on HDL particles contains sterol molecules, which are mainly cholesterol. They help to regulate the fluidity of HDL particles. HDL contains small amounts of other sterols like oxysterols. The hydrophobic lipids are the HDL core. HDL's cholesteryl esters are formed by the trans-esterification, catalyzed by LCAT, of phospholipids with cholesterol. HDL triglycerides are formed by CETP-mediated exchange of cholesteryl ester from HDL with triglycerides from triglyceride-rich lipoproteins. Like cholesteryl ester, triglycerides also are hydrophobic. They are found in the lipid core.

HDL transports a variety of minor bioactive molecules, such as diacylglycerides and monoacylglicerides. It also carries small amounts of gangliosides, free fatty acids, sulfatides, lysosphingolipids, and glycosphingolipids. Sphingosine-1-phosphate is particularly interesting in this regard as it plays important roles in vascular biology. The molecule interacts with sphingosine-1-phosphate receptors found on endothelial and smooth muscle cells. These receptors regulate a variety of cellular processes, including proliferation, motility, and apoptosis. They also affect wound healing and immune response. HDL is the primary carrier of sphingosine-1-phosphate in the bloodstream.

HDL also contains small amounts of lipophilic antioxidants like tocopherols (vitamin E) or coenzyme Q10, which protect HDL's lipids against oxidation. Some HDL lipids are glycolipids and have sugar moieties. HDL particles have also been recognized as major carriers of microRNAs in the bloodstream.

Heterogeneity

HDL particles are estimated to number around 60 quintillions. They make up the majority of the lipoprotein particles in circulation, and they have the highest concentrations compared to the other types of lipoproteins. HDL particles are usually found in blood at a concentration of about 30 micromoles per liter.

Not all HDL particles have the same properties. This is known as heterogeneity in plasma HDL. The particles differ in terms of their chemical and physical properties, as well as how they are processed by the body and what they do on a biological level. It is not possible to understand this complexity by simply measuring HDL-cholesterol levels. A more detailed analysis of the HDL particle profile is needed.

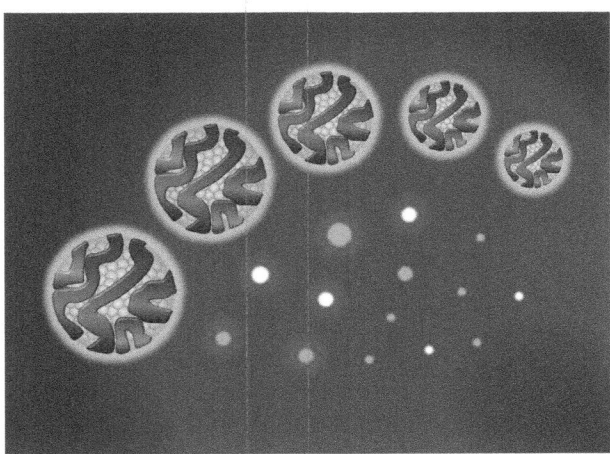

There are five main subtypes of HDL particles in the blood called subpopulations or subfractions of HDL. They are much smaller than red (erythrocytes) and white (leucocytes) blood cells. In this picture, the size of HDL is magnified by a factor of 1,000 as compared to that of the cells shown as red and white stars.

The size and the shape of HDL particles are the main differences among subpopulations. HDL particles can be either spherical or discoid and are typically 7–12 nm large. Discoid HDLs have a small diameter (less than 8 nm) and low lipid content (less than 30%). The majority of their composition is made of proteins (mainly apolipoproteins A-I), with a small amount of lipids such as phospholipids and free cholesterol. The levels of discoid HDLs in the bloodstream are low and the particles are not very stable. They can be separated using agarose electrophoresis. These particles are called pre-beta HDL because they have pre-beta-like mobility in the gel.

HDL is mainly composed of spherical particles. They are larger in size (more than 8 nm) and contain a core of cholesteryl ester and triglycerides. Different methods such as NMR, size exclusion chromatography, ion mobility, and ultrafiltration are available to measure HDL particle concentrations in blood. Ultrafiltration is useful for isolating protein-rich discoid pre-beta HDL particles with low molecular masses (40–70 kDa) that make up 5–10% of human plasma HDL.

HDL particles are also divided into subtypes according to their density: HDL1, HDL2, and HDL3. Early studies revealed that less dense HDL1 is only found in certain types of dyslipidemia. In the majority of people, only HDL2 and HDL3 are present. The density of HDL particles is usually between 1.063 and 1.21 g/ml. The method of sequential ultracentrifugation has been developed to separate HDL particles into the two subtypes. Large, light HDL2 has a density from 1.063 to 1.125 g/ml, while small, dense HDL3 is from 1.125 to 1.21 g/ml. These subtypes are dominated by round particles, as discoid particles are more dense than HDL3 and low concentrated in the blood. This method is not perfect because it can also co-isolate other lipoproteins such as Lp(a), which have similar densities to light HDL subtypes.

HDL is like a zoo. There is no such thing as an average HDL particle. They come in many shapes, sizes, densities, and other characteristics. HDL2 is a large, light particle that contains a lot of lipids. HDL3 is small and dense with fewer lipids. Ultracentrifugation is used to separate HDL2 and HDL3. Pre-beta HDL is a small and flat particle, while alpha HDL is large and round. Electrophoresis on agarose gels is used to separate pre-beta from alpha HDL. Small, dense HDL is a term used to describe small and dense HDL particles with a low molecular mass containing high amounts of protein and low amounts of lipid. The size of these particles is smaller than 10 nm, and their molecular mass is lower than 200 kDa. The term large HDL is used to describe big particles that are high in lipids, but low in proteins.

This method was further developed to allow the separation of HDL subtypes based on density and size. These included large, light HDL2b, 2a, and small dense HDL3a, 3b, and 3c. Scientists have previously used a technique called nondenaturing polyacrylamide gel electrophoresis to separate similar HDL subtypes based on its size and charge. Each HDL subtype has a range of sizes. HDL2b is the largest (between 9.7 and 12.9 nm), while HDL3c is the smallest (7.2–7.8 nm). HDL2b and 2a are the most common subtypes in individuals with normal cholesterol levels. The size of isolated HDL particles decreases with increasing density, and the protein content, as well as with the proportion of the surface lipids. This is similar to what was found for LDL subtypes. This indicates that the surface lipids and proteins are the main components of small, dense HDL3.

Two-dimensional electrophoresis is a technique that further separates different types of HDL particles based on size and charge. The method uses agarose gels for separating particles by charge, followed by another separation by size in polyacrylamide gels. To stain the gels, an antibody that detects specifically apolipoprotein A-I, the main HDL protein, is used. This staining can be employed to detect all particles that contain the apolipoprotein.

The initial separation of HDL particles by charge detects alpha, pre-alpha, and pre-beta HDL. The particles can be further separated based on size. Five major types of HDL particles are identified using this technique: very small discoid pre-beta-1, very small discoid alpha-4, small spherical alpha-3, medium spherical alpha-2 and large spherical alpha-1 HDLs. Each subtype has a different combination of apolipoprotein A-I, apolipoprotein A-II, phospholipids, free cholesterols, cholesteryl esters, and triglycerides.

The pre-beta HDL particle analysis reveals that it also contains LCAT, CETP, and other proteins. These particles do not, however, contain apolipoprotein E and apolipoprotein A-II. Pre-beta-1 HDLs are smaller than pre-beta-2 HDLs. The molecular mass of pre-beta-1 particles is 70–80 kDa, while pre-beta-2 particles are 325 kDa.

The differences between HDL particles are first of all due to the unique structure of apolipoprotein A-I. The HDL particle structure is defined by the properties of this protein. Apolipoprotein A-I binds lipids and changes shape depending on the amount of lipids bound. It is possible to create different sized lipid–protein particles. Apolipoprotein A-I is flexible and can collect lipids to form stable particles. It can be shaped into particles of various sizes and even shapes. The smallest HDL particle is flat, like a disc. Larger HDL particles have a round shape like a ball. Each HDL particle may contain between two and five apolipoprotein A-I molecules. The more apolipoprotein A-I molecules available for binding, the more lipid is bound to it. HDL disks are flat and do not have much lipid. They only contain two apolipoprotein A-I molecules. Three or more apolipoprotein A-I molecules may exist on round HDL particles. This allows the number of A-I molecules to increase as more lipids are added. Apolipoprotein A-I can accommodate varying amounts of lipids by undergoing specific movements that are caused by interactions between lipids and proteins. These movements include rotating, sliding, and changing shapes.

Apolipoprotein A-I is the protein responsible for HDL's structure. There are also other HDL proteins that vary from particle to particle. These particles can be separated based on their apolipoprotein A-I and A-II content. Two different subtypes of particles can be isolated, one that only contains apolipoprotein A-I (called LpA-I), and the other with both apolipoprotein A-I and apolipoprotein A-II (called LpA-I:A-II).

Apolipoprotein E is another important HDL protein. It helps increase the size of HDL by accumulating cholesteryl ester. These particles are excellent at binding to LDL receptors.

Different types of HDL particles vary in their protein and lipid content. Some proteins are found only in certain subtypes HDL particles. For example, only a small percentage of HDL particles contain apolipoprotein M, indicating the existence of subtypes that contain this protein. Similarly, a unique particle has been found that contains apolipoprotein L-I, apolipoprotein A-I, and haptoglobin-related protein.

As the HDL particle density increases, the amount of phospholipids, free cholesterols, cholesteryl esters and triglycerides, decreases. It means that larger HDL particles contain more lipids including cholesterol. When measuring HDL-cholesterol levels in blood, they reflect primarily the large HDL particle levels that are high in cholesterol. This measurement may not detect small particles that have low cholesterol content. In consequence, the concentration of cholesterol in large HDL2 particles can predict heart disease just as well as HDL-cholesterol.

When comparing the composition of lipids in terms of their percentage of total lipids, there are, however, very few differences between HDL particles. This indicates that the molecular makeup of these particles is in a constant balance between the subtypes. One notable difference is the decrease in sphingomyelin relative to total lipids as the density of HDL increases. Similarly, the proportion of free cholesterol also decreases with the density of HDL. Both sphingomyelin and free cholesterol reduce fluidity in lipid structures known as liquid-crystal lipids. Therefore, their lower abundance in small, dense HDL suggests higher fluidity on the surface lipids of this subpopulation.

Other types of lipids may also differ in HDL. For example, the amount of sphingosine-1-phosphate is not the same in all HDL particles. The HDL particles that are small and dense have a higher concentration of this lipid. This is because these particles are rich in apolipoprotein M, which carries sphingosine-1-phosphate.

Different subtypes HDL can have different effects on heart disease. The levels of HDL2 play a significant role in the difference in HDL-cholesterol between those with and without coronary heart disease. In studies, HDL2-cholesterol and HDL3-cholesterol concentrations were measured to estimate the level of each major subtype. There are mixed results, with some evidence suggesting that both HDL2-cholesterol and HDL3-cholesterol can be strong predictors of coronary heart disease.

We can better understand the situation when we use gel electrophoresis for HDL particle heterogeneity. Low levels of large alpha-1 HDL in the blood were found to be a strong indicator of cardiovascular events. This does not mean, however, that large HDL particles are directly protective against heart disease. This could be because a process which creates large alpha-1 HDL particle is the one that provides protection. This process is best illustrated by the production of large HDL when triglyceride-rich lipoproteins are lipolyzed by lipoproteins lipase.

By contrast, levels of small discoid HDL particles are often high in patients suffering from coronary heart disease. It is interesting that large alpha-1 HDL particles can better predict coronary heart disease than HDL-cholesterol. This test is however difficult to perform and is therefore not used as a standard in clinical practice for assessing cardiovascular risk.

When NMR is used to measure HDL particle concentrations, larger HDL (9.5–14 nm) are associated with lower risk of cardiovascular disease, whereas smaller HDL (7.3–8.2 nm) are associated with higher risk. This is in line with the results obtained when HDL is isolated by ultracentrifugation. The levels of large HDL particles in women are higher, and they increase with age. These particles can also be associated with longevity, and they can be increased by exercise. It is important to remember that many factors can affect the interpretation of how individual HDL subtypes protect against cardiovascular disease. There are strong inverse correlations, for example, between the large and small HDL particles, and between the large HDL particles and total LDL particles.

History

Heterogeneity of HDL

Who	When
Alex Nichols	**1951**

Where	What
Berkeley CA, USA	Jones, H. B., J. V. Gofman, F. T. Lindgren, T. P. Lyon, D. M. Graham, B. Strisower and A. V. Nichols (1951) American Journal of Medicine 11: 358–380.

In the 1940s, alpha- and beta-lipoproteins were separated using electrophoresis. Gunnar Bix, Arne Tizelius, and Harry Svensson of the University of Uppsala are credited with this breakthrough. In their first study, the researchers found that the beta-lipoprotein contained more cholesterol than the alpha-particles.

Lipoproteins were seen as broad bands, indicating that each class is composed of subtypes. In fact, ultracentrifugation experiments revealed that lipoproteins are a continuum of species. Frank Lindgren and his colleagues at the Donner Laboratory, San Francisco, used analytical ultracentrifugation to reveal the existence of many subtypes for the alpha- and beta-lipoproteins. These were renamed HDL and LDL. HDL particles were divided into three subtypes according to their density: HDL1, HDL2, and HDL3. Richard Havel's work using preparative sequential centrifugation led to the separation of LDL into VLDL and LDL.

Later development was ensured by coworkers of John Gofman who remained working at the Donner Laboratory. Alex Nichols, Frank Lindgren, and others developed methods to isolate plasma HDL subtypes. Ronald Krauss refined this technique at the Donner Laboratory and John Chapman at the French National Institute of Health and Medical Research in Paris. They used a gradient-density ultracentrifugation to isolate more than 10 LDL and HDL subtypes in a single run.

Electrophoretic techniques were also used to study the heterogeneity in lipoproteins. Alex Nichols used a gradient gel-electrophoresis method to separate HDL, and then LDL, into multiple subtypes. This method involved the migration of lipoproteins in their native, unaltered state through a polyacrylamide gel with increasing concentrations. Using this method, the particles were separated based on their size. The smaller the particles, the further they could travel in the gel.

(Continued)

(Continued)

Christopher Fielding, at the University of California in San Francisco, used classical electrophoresis on agarose gels and discovered a minor HDL type that was characterized as having pre-beta mobility similar to that of VLDL. The particle was different from the bulk of plasma HDL, which exhibited alpha mobility. The particle was discoid in shape and had a large amount of protein. This particle was called pre-beta HDL.

Bela Asztalos, at Louisiana State University in New Orleans and later at Tufts University in Boston together with Ernst Schaefer, combined the agarose gel and polyacrylamide gel electrophoresis into a two-dimensional technique. First, plasma samples were applied to agarose gel. After separation, samples were allowed to migrate in polyacrylamide-gradient gel. The HDL particles that were separated were then visualized with an antibody against the apolipoprotein A-I. This method detected more than 10 HDL subtypes, including pre-alpha and alpha-HDL.

Structure

The world of the HDL particle structure is full of strange things like double belts, belt buckles, picket fences, solar flares, and hairpins. All of these names are given to models proposed to explain the observations. These models, though they may appear to be very different, are all based on similar ideas. They demonstrate a direct link between particle structure and properties of HDL's main structural proteins. These proteins contain amphipathic parts that are attracted both to lipids and to water. They can interact with both water and lipids, keeping them separate. Apolipoprotein A-I is composed of eight helical domains that are both hydrophilic and lipophilic. Apolipoprotein A-II and apolipoprotein E can also form HDL when lipids are present. They also have helical domains that are attracted to both lipids and water.

The structure of HDL can be different depending on what it is made up of. There are three main types of HDL particles: ones with little lipid or without any lipid at all (called lipid-poor or lipid-free apolipoprotein A-I), ones that look like flat discs with little lipid (discoid lipid-poor HDL), and ones that are round with lots of lipid (spherical lipid-rich HDL).

Apolipoprotein A-I without lipids is completely exposed to water. The parts of the protein that prefer lipids are hidden inside, while the parts that prefer water stick out. The central and terminal parts of the proteins form a bundle, whereas another terminal part folds differently and is less organized. The termini are very close and interact, causing the lipid-free A-I to form clumps. Apolipoprotein A-I, which is lipid-free, can quickly change shape due to its instability. The dynamic nature of the apolipoprotein means that it is unrealistic to expect it to only have one structure. The protein is a mixture of different shapes that can be switched between.

Discoid HDL particles contain a small amount of lipids, and they are created by structural proteins such as apolipoproteins A-I and A-II. These proteins bind readily to lipids including phospholipids and the cholesterol in order to form stable particles. The particles can be created in the laboratory by mixing proteins with a small amount of phospholipids. Apolipoprotein A-I changes shape when lipids are present. This allows it to interact with lipids through its hydrophobic portions. Under an electron microscope, the particles have different sizes and a disc-like appearance. It is hard to isolate the discoid HDL particles from the blood, because they are not very stable and can quickly be modified by adding more lipids.

In a discoid particle, an "edge" of the particle causes apolipoprotein A-I to fold in a particular way. Two apolipoprotein A-I molecules in the form of rings are arranged into a "double-belt" structure with their helixes perpendicular to lipid chains and parallel to the flat surface of the HDL. The result is two rings stacked in opposite directions. Cryotomography can show the belt formed by apolipoprotein A-I molecules around HDL reconstituted in discoid form. This method can show how other molecules, such as apolipoprotein A-II and apolipoprotein E, fold similarly in small particles.

The terminal ends of the apolipoproteins A-I molecules are close together in this double-belt model. This concept is known as the belt buckle model. Computer analysis shows that in the double-belt arrangement, the central parts of both apolipoproteins A-I molecules are opposite to each other. It seems that the molecules can move in relation to one another between two stable configurations.

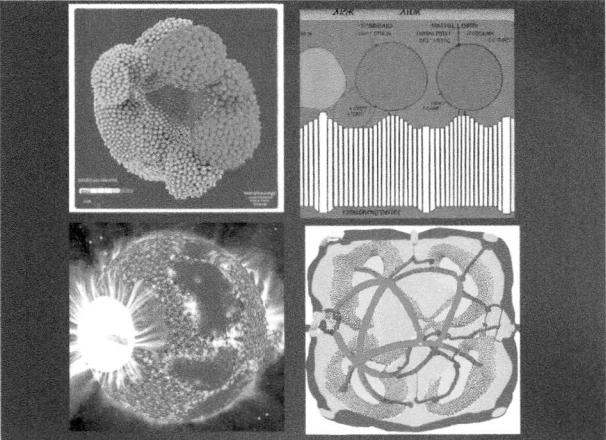

Artistic representations of different models of HDL such as (clockwise from the top left) double belt, picket fence, trefoil, and solar flares have been generated by Stable Diffusion AI based on a prompt involving the model name and "high-density lipoprotein." Source: Black Technology LTD/https://stablediffusionweb.com/ (accessed 9 February 2024).

Other models exist for the organization of apolipoprotein A-I in discoid particles. These include the picket-fence, double super-helix, and hairpin models. The hairpin model suggests that the two apolipoprotein molecules are not closed but have an open structure. This model is made up of a mixture of different apolipoproteins A-I orientations, according to data.

The picket-fence model suggests the apolipoprotein A-I wraps around the phospholipid bilayer with its helices paralleled to the phospholipid residues. Experimental data, however, has shown that the organization of apolipoprotein A-I is incorrect.

The double super-helix model, on the other hand, proposes that apolipoprotein A-I molecules form a super helix around an ellipsoidal phase of lipids. In this conformation, the apolipoprotein A-I appears as an open spiral with the lipids arranged in the middle of the helix. The double super-helix model and the double-belt model differ in that the ends of the apolipoprotein molecules are separated in the former while they are closer together in the latter.

A flexible part of the structure called hinge region can attach to and detach itself from the HDL surface depending on the size and amount of HDL lipid. In fact, some research has shown that the central part of apolipoprotein A-I can be arranged in a looped belt model. The solar flares model proposes two apolipoprotein A-I molecules arranged in a double-belt opposite structure with some parts sticking out of the water. This structure, however, may not be stable.

Another idea is that the apolipoprotein A-I molecules can trap lipids in their flexible pockets and create larger particles. The double-belt structure can be seen in saddle-shaped particles. These particles are twisted and have a non-flat shape. They get larger as more lipids are added. This model shows that the formation of discoid HDL begins with apolipoproteins A-I wrapped around a flat layer of lipid. Due to the apolipoprotein, discoid HDL's lipid portion has different regions, each with varying degrees of organization. Tightly packed lipids are found near the center, followed by moderately packed and finally disorganized lipids on the edges.

According to the data available, it appears that two molecules of the apolipoprotein A-I have a particular shape in discoid HDL. This is called the double-belt antiparallel model. The apolipoprotein can take on different shapes, particularly at the ends. The two chains of the apolipoprotein A-I do not wrap around the HDL particle smoothly but form a flexible hinge that can change size depending on the particle size or twist into saddle-shaped particles.

There are no edges in spherical HDL, so the structure of apolipoprotein A-I is less restricted. The overall structure of the apolipoprotein A-I is, however, similar for both HDL types. It takes on a double-belt shape. This shape is not affected by the number of A-I molecules per HDL particle. This indicates that all apolipoprotein A-I molecules are bent in order to fit into double-belt structures within spherical HDL. The trefoil model proposed by Sean Davidson at the University of Cincinnati shows that, in the case of three apolipoprotein A-I molecules per HDL particle, each apolipoprotein A-I molecule is bent 60° at specific points to form a trefoil shape. The surface of the lipid is divided into three equal 120° parts. The double belts may be twisted in different directions to accommodate the amount of lipids available if there are more than three apolipoprotein A-I molecules per HDL particle. This creates a cage-like structure. This twisting of the apolipoprotein A-I could explain why HDL's diameter decreases with increasing density, but the number of apolipoprotein molecules per HDL particle remains the same. This is seen in the small, dense HDL particle from human blood.

The trefoil works well for moving HDL parts, such as hinge areas, looped section, and "solar flares." Apolipoprotein A-I shows that it can change shape in spherical HDL. The specific shape of the protein depends on the size and shape of the HDL. The protein helps to keep the lipids arranged in spherical particle form, just as it does in discoid HDL.

The lipids of spherical HDL can be divided into two parts: the surface layer, which contains phospholipids and free cholesterol, and the core, which is composed of triglycerides and cholesteryl esters. The core lipids do not mix with the surface layer but still contact the proteins on HDL. Calculations reveal that the majority of surface area in spherical HDL particles is covered with protein. This ranges from 71% for large, light HDL2b, to 87% for small, dense HDL3c. The surface of HDL has a large amount of ordered lipids while the core of HDL is less ordered. At the same time, the surface of HDL contains many defects in the lipid arrangement. They are due to the surface being more curved than in other types of lipoproteins. The defects make it easier for proteins and lipids to attach to HDL.

History

Double-belt Model

Who	When
Jere Segrest	**1977**

Where	What
Birmingham AL, USA	Segrest, J. P. (1977) Amphipathi helixes and plasma lipoproteins: thermodynamic and geometric considerations. Chemistry and Physics of Lipids 18: 7-22.

After two decades of research, the major proteins and lipids in HDL were finally identified in the 1970s. However, their structure remained a mystery. There were no physical methods that allowed the study of nanometer-sized particles of HDL. The first models developed by researchers were based primarily on mathematical calculations, which included simple geometrical considerations that used the composition of HDL.

Researchers noticed that the sequence of amino acids in apolipoprotein A-I contained repeating units. This indicated the presence of the amphipathic helices. These helices are hydrophobic on the faces, which interact with lipids, and polar on the faces, which interact with water. This was the logical explanation for why these proteins could dissolve lipids to form lipoprotein particles that were stable, similar to detergents.

Jere Segrest, along with colleagues at the University of Alabama Birmingham, proposed in the 1970s that apolipoprotein A-I helices were arranged around discoid HDL in a circular pattern. The helices were perpendicular to acyl chains. The "belt model" was named after this model. Initially, the model was based on geometric considerations regarding lipid molecule shapes and sizes. According to the model, the smallest HDL particle is a disc made up of phospholipids and cholesterol. To protect the edges from water, apolipoprotein A-I, HDL's main protein, covers them. The LCAT enzyme reacts with the discs to produce cholesteryl ester from cholesterol that has not been esterified. Esters are hydrophobic and therefore are moved into

(Continued)

(Continued)

the core of the particle to avoid contact with water. The particles become spherical HDLs with a cholesteryl ester core. In the early publications, these small discs were referred to as "bicycle tire micelles."

Two belts are better than one!

Later studies refined the models. The authors pointed out that apolipoprotein A-I has two parts: a globular section at one end of the molecule and a binding end at the other. The model for the smallest HDL particles is made up of two apolipoprotein A-I molecules that are wrapped around a patch of lipids. Each molecule's lipid-binding portion is shaped as a ring with a curved side facing the lipids. Researchers found that when they looked at the different ways, two molecules can fit together, the rings aligned in an antiparallel configuration.

Over the years, this model has become a popular explanation of how apolipoprotein A-I is organized within discoid HDL. This model describes two ring-shaped apolipoprotein A-I molecules arranged in "double-belt" form. The helixes of the molecules are parallel to the disc surface and perpendicular to the fatty acid chains of lipids.

The model has gained wide acceptance in recent years. The original hypothesis by Segrest has been supported in many studies, using techniques such as crystal structure analysis, fluorescence spectroscopy, and infrared spectroscopy. Researchers were able, using cryotomography, to observe the belt of apolipoprotein A-I molecules surrounding discoid HDL that was reconstituted from apolipoprotein A-I and phospholipid in a test tube. Other apolipoproteins like apolipoprotein E, and apolipoprotein A-II, also have a similar belt-like structure in small HDL. These apolipoproteins are found on HDL in dimers.

The model's popularity was well deserved, but it could only explain the structure of discoid HDL. The problem was that most HDL particles in plasma are spherical, not disc-shaped. Spherical HDL particles have a round form with a cholesteryl ester and triglyceride core but lack the edges of discs. It was not clear if considerations for discoid HDL particles also applied to the spherical versions.

Sean Davidson's laboratory at the University of Cincinnati has successfully solved this problem. They carried out detailed studies of the molecular structure of HDL spheres and proposed the trefoil model for HDL. Researchers found that the spheres had three to four apolipoproteins A-I molecules, while the discs only had two. These molecules were arranged in a similar way to the discs, which was surprising. In order to explain this phenomenon, they bent certain parts of the apolipoprotein molecules A-I, which combined into a cage structure. The structure of the molecules allowed them to interact like they do in discs. The cage also held a neutral lipid core and supported other lipids at its surface. Sean Davidson elegantly converted Jere Segrest's double-belt structure from two dimensions into three dimensions.

12

HDL – Where It Comes From

HDL is made and broken down in different ways.

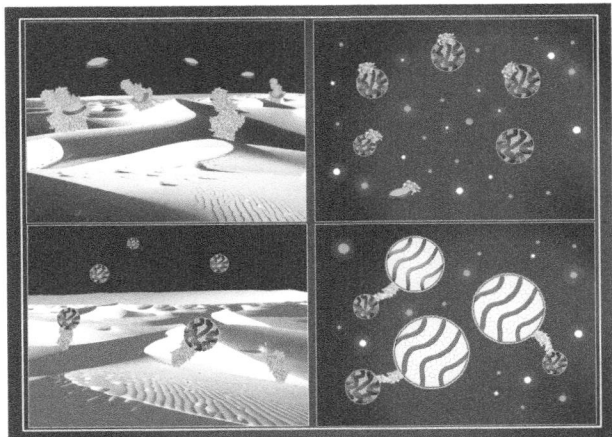

The processing of HDL involves several steps, illustrated in a clockwise direction starting from the top left. Initially, cells interact with discoid HDL, which is produced through the lipolysis of triglyceride-rich lipoproteins or secreted by cells. This interaction involves the ABCA1 transporter and increases the size of HDL by adding nonesterified cholesterol and phospholipids. LCAT converts cholesterol into cholesteryl esters, further expanding the size of HDL. CETP exchanges cholesteryl esters for triglycerides, creating a tunnel between HDL and triglyceride-rich lipoproteins. Large, mature HDL particles then bind to the SR-BI receptor on liver cells' surface to deliver cholesteryl esters to the liver.

First, the main components of HDL which are proteins and lipids come to HDL through different ways. The liver and intestines are the main producers of the proteins. Apolipoprotein A-I is mainly produced in the liver and the small intestine. The liver is responsible for most of the apolipoprotein A-I found in the blood. Apolipoprotein A-II is mostly produced in the liver but also in the intestine. The liver is the main place where many of the other proteins found in HDL are made. These include apolipoproteins C, SAA, apolipoprotein F, apolipoprotein M, LCAT, and CETP.

Apolipoprotein A-IV is produced in the intestine of most mammals including humans. It is then released into the bloodstream on newly formed chylomicrons. Apolipoprotein D and apolipoprotein E, on the other hand, are found in a variety of tissues, such as the liver and the brain. When we look at lipids, they come from widely different sources, such as cells and other lipoproteins.

Cholesterol, Lipoproteins, and Cardiovascular Health: Separating the Good (HDL), the Bad (LDL), and the Remnant, First Edition. Anatol Kontush.
© 2025 John Wiley & Sons, Inc. Published 2025 by John Wiley & Sons, Inc.

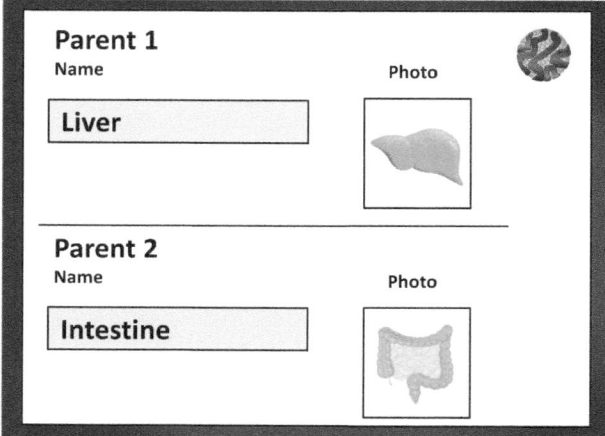

The liver and intestine are two organs that play a crucial role in determining the fate of HDL.

The majority of HDL particles in the blood are round particles formed by apolipoprotein A-I or small disks that have low lipid contents. These small HDLs are produced in the liver or intestines and secreted in the blood or lymph. They can also be derived by decomposing other lipoproteins such as VLDL and chylomicrons. These small HDLs can also be produced when spherical HDLs are transformed by PLTP.

Many receptors, enzymes, and lipid transfer proteins regulate the way HDL is processed by the body. Each of these factors can affect the amount of HDL-cholesterol in the bloodstream.

The newly formed HDLs, which are small and not stable, quickly accumulate lipids. The ABCA1 protein at the surface of the cells is involved in a process that allows them to get their first lipids. The protein helps the cells to release cholesterol and phospholipids that are then transported by small HDLs. In addition, cells secrete apolipoprotein A-I. It is unclear if apolipoprotein A-I is secreted with or without lipids.

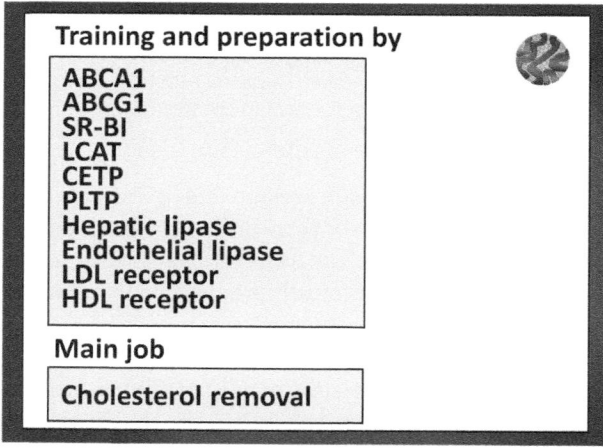

HDL processing in the blood is carried out by various proteins and enzymes, such as ABCA1, ABCG1, SR-BI, LCAT, CETP, PLTP, hepatic lipase, endothelial lipase, LDL receptor, and HDL receptor. The primary role of HDL is to remove cholesterol from cells and from the bloodstream.

ABCA1 produces small, spherical HDL particles that are lipid poor. They undergo many changes. These particles first get more lipids. These pathways are controlled at the surface of cells by proteins such as ABCA1, ATP-binding cassette transporter G1 (ABCG1), and SR-BI. The LCAT enzyme converts free cholesterol into cholesteryl ester. LCAT produces large, spherical HDLs because the free cholesterol is at the surface and the cholesteryl ester is at the core. The neutral lipids in the core of these particles are cholesteryl ester and triglyceride.

By removing the cholesteryl ester from their core, large HDL can be converted back into smaller HDL. HDL can be stripped of the molecules of cholesteryl ester in several ways. With the help of CETP, they can be transferred from HDL to lipoproteins that contain apolipoprotein B. SR-BI can help them be taken up by the liver and organs that produce steroids. The triglyceride in the HDL's core can be broken down by enzymes such as endothelial lipase and hepatic lipase. When CETP removes the cholesteryl ester molecules from HDL and replaces them with triglycerides, large HDL particle rich in triglycerides is formed. Hepatic lipase then breaks down the triglycerides to form small HDL particles. The combined action of CETP with hepatic lipase reduces HDL particle size, creates lipid-poor HDLs, and releases apolipoprotein A-I that interacts with ABCA1 during the next cycle adding lipids.

HDL particles can be removed from the blood in their whole form by the liver. This is done through LDL receptors, or unidentified HDL receptors.

Transporters

ABCA1

ABCA1 is a large membrane protein that has a molecular mass of 254 kDa. It helps form HDL, and by exporting cholesterol from the cell, it maintains a healthy cell cholesterol balance. ABCA1 is made up of two parts. Each part is divided into two sections: one that interacts with the cell membrane and another that is important for its transport function. Two sections of the protein pass through the membrane. ABCA1 can be found in a variety of cells. However, it is more common in macrophages or hepatocytes. ABCA1's structure is held together with disulfide S-S bonds between the different parts. The protein needs these bonds to bind the apolipoprotein A-I. In fact, ABCA1 is a channel that moves cholesterol and phospholipids out of the cell. This process creates HDL particles.

ABCA1 requires other proteins such as apolipoprotein A-I, which are present outside cells, to work properly. The transporter produces special cell membrane parts that are curved and stick out. These curved portions help apolipoprotein A-I bind to cell surfaces. Apolipoprotein A-I likes curved surfaces. The process of HDL production begins when apolipoprotein A-I inserts its tail in the curved areas of the membrane.

What It Is

The **ATP-binding cassette transporter A1 (ABCA1)** is a large membrane protein with a molecular mass of 254 kDa. It helps form HDL and maintains cell cholesterol balance. ABCA1 exports cholesterol to small discoid HDL and to apolipoprotein A-I.

ABCA1 is able to remove lipids from cells even when apolipoprotein A-I is not present. The result is large vesicles, called microparticles that lack apolipoproteins A-I and similar proteins. They are part of the process that removes cholesterol from the cells and can account for up to 30% of total cholesterol removed. HDL and microparticles are both produced simultaneously, but they do not come from one another.

The process of HDL creation is completed when apolipoprotein A-I separates from ABCA1. This process uses energy in the form of ATP to move lipids across the cell membrane and onto apolipoprotein A-I. ABCA1 helps make HDL particles on the cell surface, which have a diameter of 7.5 nm. These particles contain one apolipoprotein A-I molecule and three to four phospholipid molecules. They also contain one to two cholesterol molecules. The part of apolipoprotein A-I that sticks into the cell membrane is important for making these particles.

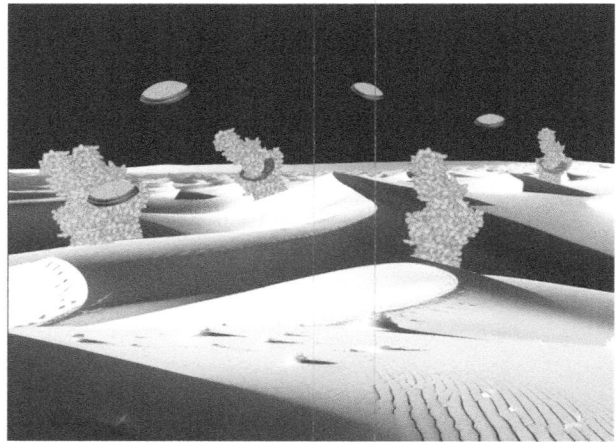

Lipid-free apolipoprotein A-I and small, discoid pre-beta HDL particles interact with the ABCA1 transporter to remove cholesterol and phospholipids from cells.

How HDLs form can be affected by the composition of a cell membrane. Phosphatidylserine is a negatively charged phospholipid that helps apolipoprotein A-I interact with the cell membrane. This makes HDLs more quickly form. It could be that phosphatidylserine alters the acidity at the membrane surface, causing changes to apolipoprotein A-I's shape.

ABCA1 also helps remove phospholipids and cholesterol from the cells to small discoid pre-beta-1 HDL. Interestingly, this particle can also be a product of the activity of ABCA1. Certain cells exposed to prebeta-1 HDL produce different types of HDL particles containing from two to four molecules of apolipoprotein A-I. These particles range in size from 9 to 12 nm and contain different amounts of phospholipids and free cholesterol. Other apolipoproteins such as apolipoproteins A-II, E, C-III, and M can also produce HDL particles when they interact with ABCA1.

ABCA1-mediated lipidation is a process that occurs mainly at the cell membrane. The pathway for HDL formation, involving ABCA1 and apolipoprotein A-I, may become activated when cells contain too much cholesterol. It is important because the formation of HDL can protect against atherosclerosis. Researchers have found that mice lacking ABCA1 who are fed a high-fat diet develop atherosclerosis more quickly. This could be because their macrophages cannot remove cholesterol as efficiently.

ABCG1

ABCG1 is a membrane transporter which helps in the remodeling and formation of HDL. It has a molecular mass of 76 kDa and is smaller than ABCA1. This protein is a half-type ABC transporter, which contains one transmembrane section and one transporter segment. ABCG1 can be found in macrophages and other cells.

Large spherical HDL particles interact with the ABCG1 transporter to remove cholesterol and phospholipids from cells.

ABCG1 prefers binding to spherical HDL. The protein does not readily interact with small lipid-poor HDL and apolipoproteins. These particles are formed when the lipid-free apolipoprotein A-I interacts with ABCA1 before acquiring more lipids.

What It Is

The **ATP-binding transporter cassette G1 (ABCG1)** is a membrane protein that aids in the remodeling and formation of HDL. ABCA1 exports cholesterol to large spherical HDL.

ABCG1 is a key player in making large HDL particles bigger. It does this by adding to them more cellular lipids. ABCG1 is also responsible for adding lipids to the small round particles created by apolipoprotein A-I and ABCA1. HDL's ability to stimulate ABCG1 to remove cholesterol is dependent on the phospholipid content. This suggests that ABCG1 is aided by phospholipids to drive the cholesterol removal process.

Enzymes

LCAT

LCAT transforms cholesterol in small HDL into cholesteryl esters. The HDL particles become large and spherical. LCAT also helps break down phosphatidylcholine whose fatty acid moiety moves to cholesterol, creating cholesteryl ester, and lysophosphatidylcholine. This process seems to be crucial for the removal and delivery of cholesterol to the liver. In fact, most cholesterol is transported to the liver in the form of cholesteryl esters.

HDL is responsible for the production of the majority of esterified cholesterol (approximately 75%) in the blood, but lipoproteins containing apolipoprotein B also play a role in this process. LCAT traps cholesterol within HDL particles by moving it to the center of the particle. This keeps the cholesterol there until the liver removes it. LCAT helps remove cholesterol from cells by increasing the gradient of free cholesterol between HDL and cell membranes. Free cholesterol, however, is less hydrophobic than cholesteryl esters and can be removed more easily from small HDL particles. Another way to deliver cholesterol to the liver is through the exchange of free cholesterol between HDL and cell membranes.

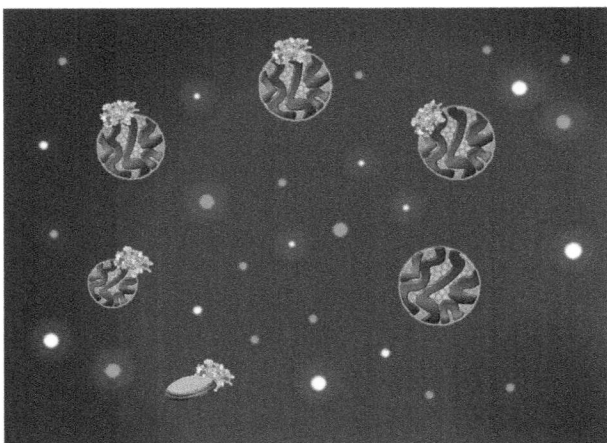

The LCAT enzyme (in pink) converts nonesterified cholesterol into cholesteryl esters, shifting cholesterol from the surface to the hydrophobic core of HDL particles. Consequently, the size of the particles expands.

LCAT begins cholesteryl esterification when the enzyme binds to HDL. Cryotomography allows us to observe that discoid HDL and LCAT form a complex. LCAT is 5 nm in length and smaller than discoid HDL. It can be seen at the surface of the particle as a small protrusion. LCAT can bind to the central part of the apolipoprotein A-I molecule because it does not move as much as the rest of the molecule. A portion of apolipoprotein A-I exposed to water interacts with and activates the enzyme. LCAT then transfers a fatty acid from phosphatidylcholine to the enzyme. The fatty acid then is combined with the cholesterol to form cholesteryl ester. LCAT can also break down other phospholipids, such as phosphatidylethanolamine, but sphingomyelin stops it from working.

Computer simulations can be used to study the process of cholesteryl esterification. These visualizations reveal that, after LCAT attaches to discoid HDL, certain domains of apolipoprotein A-I form tunnels which allow hydrophobic fatty acid chains and nonesterified cholesterol from the surface to reach the active site of LCAT. The cholesteryl esters, the products of this reaction, block these tunnels. It is necessary to transfer the newly synthesized molecules of cholesteryl esters to HDL. To achieve this, one apolipoprotein-A-I molecule forms a loop to allow the insertion of cholesteryl ester molecules in the middle of HDL particles.

Apolipoprotein A-I is therefore crucial in esterifying plasma cholesterol, and in forming mature spherical HDL particles by binding and activating LCAT within discoid particles. Other apolipoproteins such as apolipoprotein A-IV, apolipoprotein C-I, and apolipoprotein E, can also activate LCAT, but not as efficiently as apolipoprotein A-I. Apolipoprotein C-III, on the other hand, seems to reduce LCAT activity.

It is still a hotly debated issue whether LCAT is required for the efficient removal of cholesterol from tissues to the liver. Further research is needed to determine the relative contribution of esterified versus free cholesterol in this reverse cholesterol transport.

History

Cholesterol Esterification and Reverse Cholesterol Transport

Who	When
John Glomset	**1962**

Where	What
Seattle WA, USA	Glomset, J. A., F. Parker, M. Tjaden and R. H. Williams (1962) "The esterification in vitro of free cholesterol in human and rat plasma." Biochimica et Biophysica Acta 58: 398-406.

Esterification of cholesterol was first observed by Warren Sperry at Columbia University New York in 1935. It remained unclear until 1962 how this process occurred. John Glomset discovered LCAT in 1962, at the University of Washington Seattle. Glomset investigated changes in cholesterol concentrations within blood plasma when the temperature was 37°C. He found that phosphatidylcholine (called lecithin at that time) and cholesterol decreased simultaneously during the incubations. Esterification added fatty acids to cholesterol, and these fatty acids most likely came from phosphatidylcholine. Glomset hypothesized that the transfer took place under the influence of an unknown enzyme, which he initially called a fatty acid transferase. He then named the enzyme LCAT, or lecithin cholesterol acyltransferase. The name is derived from the fact that a fatty acid moiety (an "acyl group") of lecithin was transferred to cholesterol. Glomset hypothesized that this enzyme was the primary source of cholesteryl ester in humans.

The University of Washington studies were conducted on healthy individuals with normal plasma lipid concentrations. Kaare Norum and Egil Gjone at the University Hospital of Oslo in Norway found abnormal lipid concentrations in three sisters' plasma in 1967. They had high triglycerides and total cholesterol. The cholesterol esterification level in plasma was less than 10%. This is much lower than the esterification in healthy individuals, who esterify it by over 80%. Patients presented with symptoms such as anemia, proteinuria, and minor renal deficiencies. Researchers hypothesized that the sisters were genetically deficient in LCAT. Glomset's subsequent studies confirmed that the patients were completely devoid of LCAT. It was interesting to see that the HDL particles from the patients isolated at the Donner Laboratory, Berkeley, were discoid. Incubation with normal LCAT in a test tube converted the particles into their normal spherical shape. Plasma from the patients also contained large amounts of membrane-like structures mostly composed of phospholipids and cholesterol. These structures were termed lipoprotein-X.

(Continued)

(Continued)

Glomset, following these studies in 1968, suggested that the discoid particle was precursor to normal spherical HDL. LCAT's role was to convert cholesterol into cholesteryl esters within HDL. The liver cells then absorbed the cholesteryl ester formed. Nonesterified cholesterol is derived primarily from other tissues. In this way, LCAT is important in transporting cholesterol from tissues to the liver by converting the nonesterified form of cholesterol to an esterified one. This view suggests that HDL is the main transporter of cholesterol, as it plays a major role in the formation of cholesteryl ester. The process of the movement of cholesterol from peripheral tissues to the liver was termed reverse cholesterol transport.

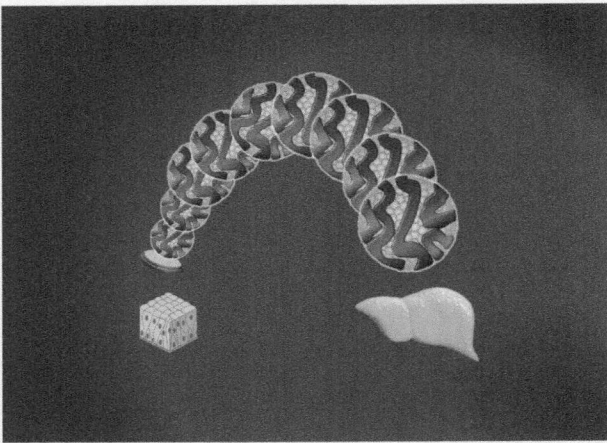

It was believed that the formation of normal HDL from discoid precursors is a rapid process difficult to observe in normal blood plasma. This conversion is blocked in patients who have LCAT deficiency. This allowed cholesterol to be removed from cells and accumulate, along with phospholipids, in the membrane-like structure of lipoprotein-X. These structures infiltrated kidneys, causing renal failure. They were also deposited on the cornea, causing impaired vision.

Only 100 people are known to have the disease caused by a partial or complete lack of LCAT. Albeit very rare, the disease provided important information on cholesterol turnover and transport in the blood. This condition, which can cause poor kidney function, has no treatment. Renal transplantation is the only effective solution. To address this issue, infusions of recombinant human LCAT are being developed.

Lipases

Lipases are enzymes that help the body process HDL. There are several types of lipases involved, such as lipoprotein lipase, hepatic lipase, endothelial lipase, and PAF-AH.

When triglycerides are broken down by lipoprotein lipase, the chylomicron and VLDL particles become less stable. This happens because triglycerides form the particle core which is destroyed. The particle surface is no longer needed and must be adjusted to the smaller core. The excess surface of chylomicrons or VLDL is removed, releasing phospholipids as well as free cholesterol and apolipoproteins. Existing HDL particles can pick up these molecules. The molecules can also be used to form new HDL. The concentration of HDL in the blood is then increased. Higher activity of lipoprotein lipase is therefore associated with higher HDL levels.

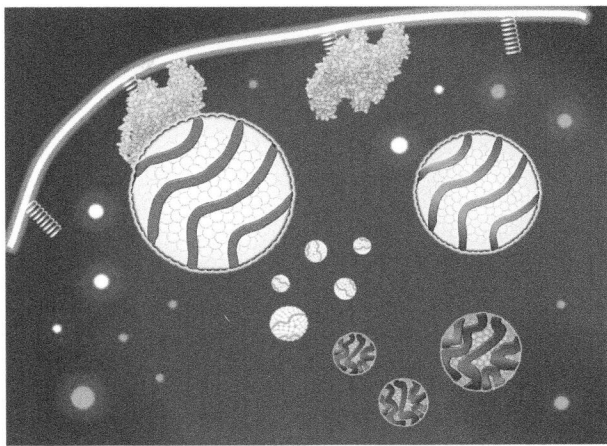

Breakdown of triglyceride-rich lipoproteins, such as chylomicrons or VLDL, by lipoprotein lipase results in the formation of core and surface remnants. The core remnants are composed of IDL particles, while the surface remnants consist of small apolipoproteins, phospholipids, and nonesterified cholesterol that merge with the pool of HDL.

It is important that moving cholesterol from VLDL and chylomicrons to HDL and then to the liver during lipolysis prevents it from staying in remnant lipoproteins which can induce buildup of cholesterol in the arterial walls. This pathway is termed reverse remnant-cholesterol transport (in short RRT). From a quantitative point of view, it can be more important for the protection from atherosclerosis than the classic reverse cholesterol transport from cells to the liver. In fact, the amount of cholesterol transported to HDL during lipolysis is greater than the amount removed from the cells. This process can prevent buildup of cholesterol in the arterial walls instead of removing it following its accumulation.

Different proteins can move between HDLs and VLDLs and affect the activity of lipoprotein lipase. Apolipoprotein C-III inhibits the lipase while apolipoprotein C-II activates it. Apolipoprotein A-II is essential for controlling lipoprotein lipase activity. It does this by dislodging apolipoproteins C from HDL. This is because apolipoprotein A-II has a higher affinity for the HDL surface than other proteins.

Hepatic lipase breaks down phospholipids that are on HDL's surface. This reduces the amount of HDL particles that circulate in the bloodstream because they are smaller and cleared by the kidneys faster. In contrast, a lack of hepatic lipase can lead to heightened levels of HDL-cholesterol.

Endothelial lipase, like lipoprotein lipase, is an enzyme that is active in dimers of two identical molecules. The lipase breaks down certain lipids present in lipoproteins. It primarily destroys phospholipids, but it can also break down triglycerides. This enzyme works better on HDL than other lipoproteins. Endothelial lipase lowers HDL-cholesterol levels in the body, just like hepatic lipase. Endothelial lipase also affects the HDL-cholesterol concentration in the blood by other functions, which do not include breaking down lipids. This may include binding to protein receptors on the cell surface.

HDL is also produced after lipolysis in brown adipose tissues. This is a type of tissue that is activated when the temperature drops. The main function of this tissue is to produce heat in order to keep the body warm. This is done by burning calories through a process known as thermogenesis. This can lead to elevated HDL-cholesterol levels. Adipose fat can be used as energy to lower total cholesterol levels in the blood and increase HDL-cholesterol. This can protect the heart. If carbohydrates are used instead of fat, the effect is the opposite though. Low levels of HDL-cholesterol could indicate that the body does not use fat very efficiently for energy.

It is interesting to note that the amount of cholesterol absorbed by the intestines directly correlates with the level of HDL in the blood. This indicates that the absorption of cholesterol in the intestines is another important factor in determining HDL level.

Lipid Transfer Proteins

CETP

Plasma CETP helps transport cholesteryl ester and triglyceride from one type of lipoprotein to another. This activity does not require energy and is driven by concentration differences between lipoproteins. CETP transfers triglyceride in exchange for cholesteryl ester from HDL to VLDL. Together with LCAT, this process helps eliminate cholesterol from the cells and return it to the liver.

CETP exchanges cholesteryl esters from HDL for triglycerides from lipoproteins rich in triglycerides through a tunnel between these particles.

HDL is a more preferred donor of cholesteryl ester than LDL for CETP. This is because plasma apolipoprotein F (also called lipid transfer inhibitor protein or LTIP) is primarily found in LDL. The apolipoprotein inhibits the transfer of lipids between LDL and VLDL. However, HDL to VLDL lipid transfers are not affected. Apolipoprotein C-I is another natural inhibitor of CETP in plasma.

CETP activity is affected by the level of triglycerides. It is increased when the triglyceride levels are high. A higher level of CETP activity results in a reduction of HDL-cholesterol and vice versa. CETP, along with hepatic lipase and SR-BI, helps recycle apolipoprotein A-I from large mature HDL. The production of apolipoprotein A-I with or without lipid is a critical step in HDL formation.

PLTP

Plasma PLTP helps in the transfer and exchange of phospholipids among lipoproteins. The protein may have a net effect on HDL, such as a reduction in particle size. PLTP can also cause the separation of apolipoprotein A-I and apolipoprotein A-II from HDL. The protein is essential for maintaining HDL levels, but its elevated activity appears to promote atherosclerosis in animals. Remember that results from animal studies on HDL should not be directly applied to humans because of fundamental differences in lipoprotein and cholesterol processing between species.

Receptors

SR-BI

SR-BI, a receptor in the body, plays a key role in the movement and transport of cholesterol. SR-BI binds a variety of partners, including different types of lipoproteins. The receptor is found mainly in the liver and in certain tissues that are involved in cholesterol metabolism. Unlike ABC transporters, SR-BI allows cholesterol moving between lipoproteins and cell membranes without requiring energy. According to the differences in cholesterol concentrations, cholesterol can be moved from lipoproteins into cells or from cells into lipoproteins through SR-BI.

SR-BI's main function is to take in cholesterol from HDL particles and release them into the bloodstream without destroying. SR-BI also plays a role in moving cholesterol from cells to HDL particles. This receptor is selective for cholesteryl esters, but it can also transport nonesterified cholesterol from HDL into cells. SR-BI is found in specific areas of cell membranes where HDL particles are bound.

SR-BI, at a molecular level, forms a channel that allows cholesterol to move through the membrane. SR-BI's outer part is required for binding HDL particles and taking in their cholesterol. The outer part also aids in interactions between the receptor and cell membranes.

SR-BI is a cholesteryl ester transporter that helps cholesterol move between HDL and cell membranes. HDL particles which interact with SR-BI are larger and contain more lipids compared with those HDL particles which interact with ABCA1 or ABCG1. The liver can remove cholesteryl ester from HDL faster when HDL particles have more lipids and are bigger. When HDL interacts with SR-BI, the receptor releases particles with less lipids.

The level and the activity of SR-BI in the liver have a great impact on how HDL-cholesterol is processed by the body. If SR-BI is present in large amounts, it will remove more cholesteryl esters from the blood. This reduces the level of HDL-cholesterol. If there is not enough SR-BI, the amount of HDL-cholesterol in the blood increases. This can paradoxically worsen atherosclerosis because the cholesterol is not transported properly from the arteries to the liver. However, in healthy people with normal levels of lipids in their blood, SR-BI does not play a major role in removing cholesterol from the blood. Instead, cholesterol gets transported to the liver via the LDL receptors by other lipoproteins that contain apolipoprotein B and are helped by CETP.

Removal

The protein parts of HDL are removed from the blood at different speeds. Apolipoprotein A-I and apolipoprotein A-II remain in the bloodstream for approximately four to five days. Apolipoprotein A-II breaks down slightly slower than apolipoprotein A-I. Other proteins, such as apolipoprotein A-IV and apolipoproteins C and E, found in HDL and triglyceride-rich lipoproteins are removed faster from the blood than apolipoprotein A-I.

The main sites where HDL proteins are broken down are the liver and the kidney. The liver is in charge of removing the majority of cholesterol carried by HDL. The liver excretes the cholesterol into the bile. The rate of reverse cholesterol transport seems to be dependent on the sending of cholesterol into the bile. The hepatic SR-BI controls this process. The small intestine is another place where cholesterol can be removed. This occurs via trans-intestinal transport, allowing the elimination of plasma cholesterol directly into the intestine through the enterocytes.

Special receptors can remove HDL particles in their entirety from the body. These HDL particles are quickly taken up by liver cells and stored in the cell's compartments. Some HDL particles are then transferred to another part of the cell where they can be broken down. Cells can release again a small proportion of HDL particles.

Certain receptors like cubilin may be involved in HDL absorption in the kidneys. This process may also be influenced by a receptor called ectopic beta-chain membrane-bound ATP synthase. The protein can also be involved in HDL release from the cell. The LDL receptor can be used to remove large HDL with apolipoproteins E from the blood.

Apolipoproteins such as apolipoprotein A-I, which carry no lipids, or very little lipids, are removed quickly from the bloodstream by the kidneys. Low levels of HDL-cholesterol and apolipoprotein A-I can be linked to the ability of the kidneys to filter blood.

History

Cellular Cholesterol Transporters

Who	When
Monty Krieger	**1996**

Where	What
Cambridge MA, USA	Acton, S., A. Rigotti, K. T. Landschulz, S. Xu, H. H. Hobbs and M. Krieger (1996) "Identification of scavenger receptor SR-BI as a high density lipoprotein receptor." Science 271: 518-520.

According to John Glomset, reverse cholesterol transport happens when HDL takes cholesterol from cells in different parts of the body and brings it to the liver. In the 1980s, this process was observed experimentally. Before then, it was thought that other lipoproteins such as LDL were responsible for the first step in this process.

In 1982, Christopher Fielding and Phoebe Fielding demonstrated that HDL could take up cellular cholesterol. They discovered that cholesterol was removed from cultured cells using a minor and unidentified type of HDL. Later studies enabled the isolation of this particle, which was named discoid pre-beta HDL. This type of HDL accounts for about 5% HDL in plasma. This discovery proved that HDL is important in the first stage of transporting cholesterol from cells using LCAT. This latter produces cholesteryl ester from nonesterified cholesterol and increases HDL-cholesterol.

Scientists have argued for many years about how exactly the transport of cholesterol between HDL and cells occurs. Some believed that cholesterol simply moves on its own through diffusion. Others thought that active transport is the main way cholesterol gets from cells to HDL particles. In the early 1990s, it was found that HDL can remove cholesterol from cells using both processes. For the spontaneous diffusion, phospholipids present in HDL are necessary. For the active transport, apolipoproteins are required.

In 1996, Monty Krieger's lab at the Massachusetts Institute of Technology discovered the first HDL receptor. It is called scavenger receptor class B type I (SR-BI). Researchers found that the receptor is involved in moving cholesterol from cells into HDL and vice versa. HDL phospholipids are crucial for this process. SR-BI is a docking receptor for HDL that takes in only cholesterol, not protein. It is found mainly in the liver and other tissues that produce steroid hormones where it interacts with HDL, LDL, viruses and other partners. It is a receptor that helps the liver cells pick up cholesterol from HDL. However, its mechanism differs from the LDL receptor.

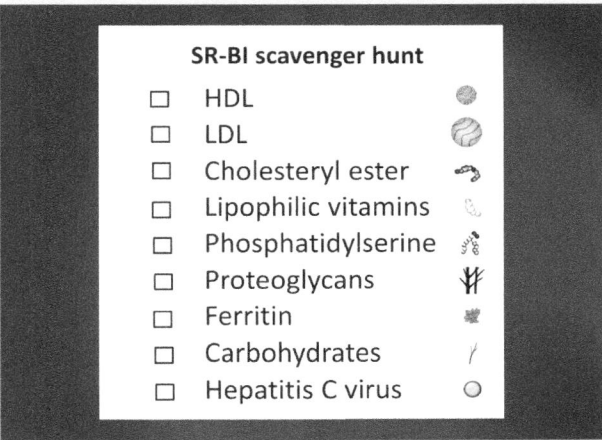

A little later, it was found that patients with Tangier disease have a problem with a particular pathway of cholesterol transport. This involves apolipoproteins. The condition is marked by low HDL-cholesterol, accumulation of cholesterol in some cells, and accelerated atherosclerosis. This discovery led researchers to search for the gene that causes Tangier disease. In 1999, the laboratories of Michael Hayden at the University of British Columbia, Vancouver, Gerd Assmann at the University of Munster, and Gerd Schmitz at the University of Regensburg in Germany, independently identified the gene as ATP-binding cassette transporter A1 (ABCA1). The existence of this transporter was predicted by Jack Oram and Shinji Yokoyama at University of Washington, Seattle three years earlier.

ABCA1 was quickly found to be involved in the movement of cholesterol, phospholipids, and other lipid molecules through the outer cell membrane. This process is active and requires energy from the cells. These molecules are then taken up by discoid HDL particles, or apolipoproteins with little or no lipid. The activity of ABCA1 can affect the level of HDL-cholesterol in the blood and make someone more or less susceptible to heart disease.

(Continued)

(Continued)

Scientists observed that ABCA1 is crucial in removing excessive cholesterol from cells. It is also important in the creation of HDL particles.

The pathway of exporting cholesterol from cells to small discoid HDL was discovered. These particles, however, make up only a small portion of plasma HDL. It was still unclear how the cells transferred cholesterol to the normal HDL particles, which make up most of the plasma HDL. The laboratory of Alan Tall, at Columbia University in New York, discovered a protein named ATP-binding cassette transporter G1 (ABCG1). The study showed that this protein helped move cholesterol from the cells to HDL. This protein is important for immune cells known as macrophages that can be filled with cholesterol, contributing to heart disease. ABCG1 moved cholesterol to large HDL, but the ABCA1 did not. The SR-BI receptor was also found to be involved in the transfer of cholesterol between cells and large HDL. This pathway does not, however, require energy in the form of ATP.

The study of mechanisms for cholesterol removal in cells was largely completed by this discovery.

Open Question

Can removing cholesterol from the blood through HDL and the SR-BI receptor protect against heart disease in a similar way as removing cholesterol through LDL and the LDL receptor?

13

HDL – What It Does

Low levels of HDL-cholesterol in the blood are a major factor that increases the risk for heart disease. However, even if someone's HDL-cholesterol is high, they could still be at heart disease risk. The amount of HDL-cholesterol in the blood, also known as "quantity of HDL," does not necessarily reflect the effectiveness of HDL to protect against heart disease. In addition, the ability of HDL to positively influence biological processes within cells and tissues is also important to determine its capacity to prevent heart disease. These properties are referred to as "quality" or "functionality" of HDL.

HDL particles are beneficial to the body. They can reduce inflammation, fight oxidation, and protect cells from dying. They also widen blood vessels, fight infection, prevent blood clots, and remove excess cholesterol from the cells. HDL is able to perform so many functions because of the wide range of proteins it contains. HDL is responsible for a variety of protective activities. There is therefore a great deal of diversity in the functions of HDL particles. The small, dense, and protein-rich HDL particles are particularly powerful at carrying out these activities. It is easy to understand, since this particular subtype of HDL contains a lot of different proteins. HDL functions are carried out by certain groups of proteins in the particles. In addition, the HDL lipid composition can influence how proteins are attached to the particle surface. HDL subsets can act to assemble specific proteins that perform certain functions.

> **Open Question**
>
> What biological functions of HDL are involved in the protection of arterial walls from atherosclerosis?

Removal of Cholesterol

HDL's ability to remove cholesterol from the cells of the body and send it to the liver to be eliminated has been long regarded as a primary way how HDL protects against heart disease. Low levels of HDL-cholesterol are thought to be a cause of cardiovascular disease. They are believed to reflect insufficient cholesterol removal from tissues, including arterial walls. The process is known as reverse cholesterol transport, or RCT, and is in part ensured by HDL. According to the mainstream hypothesis, HDL is central in removing cholesterol and maintaining cholesterol levels in cells, including those of the arterial wall.

Cholesterol, Lipoproteins, and Cardiovascular Health: Separating the Good (HDL), the Bad (LDL), and the Remnant, First Edition. Anatol Kontush.
© 2025 John Wiley & Sons, Inc. Published 2025 by John Wiley & Sons, Inc.

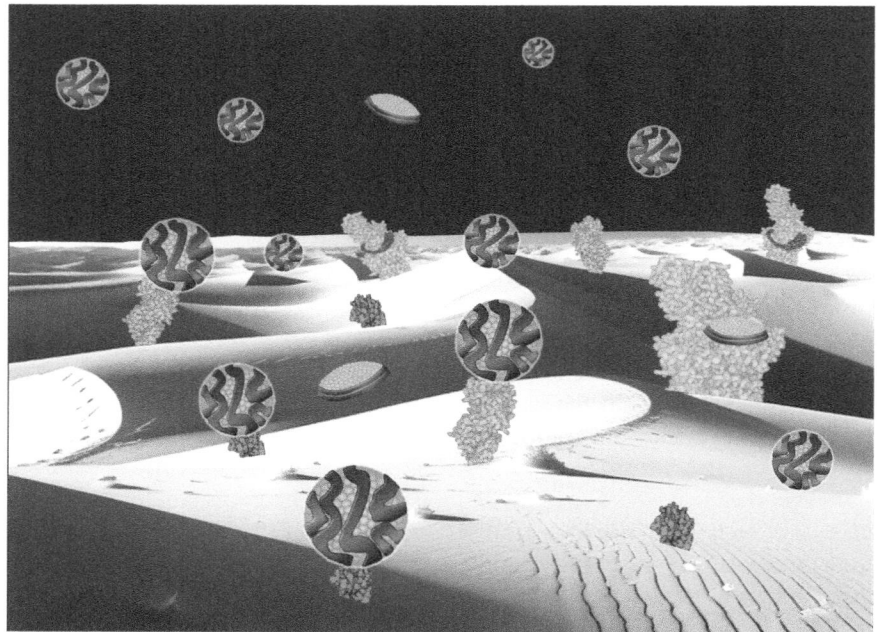

Cholesterol is removed from cells through multiple pathways. First, lipid-free apolipoprotein A-I and small discoid HDL can achieve this by utilizing the ABCA1 transporter (in light blue and green). Second, large spherical HDLs remove cellular cholesterol by utilizing the ABCG1 transporter (in violet). Third, cholesterol can be exchanged between cells and HDL through the SR-BI receptor (in dark green), which specifically binds spherical HDL. Lastly, cholesterol can directly move from the cell membrane to HDL particles through diffusion.

Reverse cholesterol transport is based on the observation that while all cells can produce cholesterol, only the liver has the ability to eliminate it in large quantities. This occurs through the excretion in the bile of cholesterol, either in its native form or as bile acids. The cells of the arterial wall are moderately able to remove cholesterol. The reverse cholesterol transport pathway is important in protecting against cholesterol buildup within arterial cells. When there is an imbalance between picking up cholesterol from lipoproteins and removing it through HDL particles, macrophages in the arterial walls become foam cells.

The ability of HDL particles to remove cholesterol from cells is called cholesterol efflux capacity. The reverse cholesterol transport begins with the cholesterol efflux. Recent research suggests that the process of removing cholesterol from the cells and transporting it out of the body is a better indicator of heart disease risk than simply measuring the levels of HDL-cholesterol in the blood. HDL-cholesterol levels are not a good indicator of how effectively the body can eliminate cholesterol from its cells. It is possible to have low HDL-cholesterol and still be at a lower risk of heart disease, if the body is able to effectively eliminate cholesterol. It has been observed that individuals with the genetic variant of apolipoprotein A-I Milano have lower levels of HDL-cholesterol but still a reduced risk of heart disease.

HDL crosses the endothelium in order to reach the cells of the arterial wall. The endothelium controls the movement between blood vessels and tissues. Aortic endothelial cells can move HDL and apolipoprotein A-I from one side to the other of the cells in lab experiments. Cell surface receptors facilitate this movement.

Both apolipoprotein A-I and HDL, which are lipid-loaded, interact with surface receptors after passing through the endothelium to remove cholesterol from the cells. ABCA1 is the main receptor for apolipoprotein A-I, while ABCG1 and SR-BI are important partners for spherical HDL. ABCA1 and ABCG1 are involved in the removal of cholesterol, which requires energy in the form of ATP. The movement of cholesterol via SR-BI, on the other hand, is not energy dependent.

ABCA1 removes cholesterol and phospholipids from cells. Apolipoproteins with low lipid content are involved in this process. They remove lipids from specialized membrane areas called caveolae. These lipid-free apolipoproteins are then lipidated to form small HDL particles. These particles are able to further remove cholesterol and phospholipids efficiently from some cells.

Tangier disease is characterized by a defect in the ABCA1 transporter, indicating that it is crucial for lipid removal. In fact, blocking ABCA1 reduces the removal rate of cholesterol from cells to apolipoprotein A-I. ABCA1 accounts for a significant portion of the cholesterol removal (at least 80%) in macrophages that are loaded with cholesterol. In this process, small pre-beta HDL particles play an important role. ABCA1 plays a key role in removing cellular cholesterol and forming HDL particles.

ABCA1 removes lipids by interacting with cellular cholesterol and phospholipids. This involves ABCA1 forming small areas where lipids gather, followed by binding apolipoprotein A-I before removing the cholesterol. This results in the formation of small HDL particles with low lipid content that can then grow to larger HDL particles richer in lipids.

ABCA1 is involved in the transport of cholesterol from the plasma membrane, endoplasmic reticulum, Golgi network, and lysosomes to apolipoprotein A-I. The cellular cholesteryl ester must first be converted to nonesterified cholesterol to leave the cells.

Although ABCA1 plays a vital role in the removal of cholesterol from cells, macrophages still release some cholesterol via other pathways in the absence of ABCA1. The ABCG1 transporter helps to move cholesterol from cells to large HDL particles rich in cholesteryl esters. ABCG1 is responsible for moving cholesterol within the cell membrane to interact with large mature HDL, but not with apolipoproteins with little lipid content. ABCG1 is less important than ABCA1 at removing cholesterol from macrophages and has a subtle impact on atherosclerosis.

HDL particles, in addition to the specific cholesterol efflux via the specialized transporters, can also cause a nonspecific cholesterol efflux. This type of efflux is slow, not limited, and can go in both directions via passive diffusion. The SR-BI is responsible for nonspecific cholesterol movement between HDL particles and the plasma membrane. It works by binding HDL particle to the receptor and rearranging the lipids within the cholesterol-rich regions of the membrane. How much cholesterol moves through SR-BI is determined by the amount of phospholipids found in HDL. The receptor is important in moving cholesterol from macrophages, without requiring energy.

Other pathways are also available for the efflux of cholesterol. The first involves cholesterol moving from the plasma membranes directly to HDL particles, without a receptor. This receptor-independent passive diffusion pathway may be more important if cells are not loaded up with cholesterol. It can remove cholesterol even better than SR-BI. This pathway is less effective when the cells are loaded with cholesterol, because it cannot be accelerated. This type of diffusion does not involve hydrophobic molecules of cholesterol passing through water. Instead, they move directly between HDL particles and the hydrophobic portions of the plasma membrane.

A process known as retroendocytosis is another way by which HDL can remove cholesterol from cells. This pathway involves HDL particle being taken into small compartments within the cell where it is enriched with lipids, and then released. However, only a small amount of absorbed HDL is released this way, which is not enough to account for all the cholesterol removed during the efflux.

HDL particles are also really good at taking in cholesterol and other lipids which are released from lipoproteins rich in triglycerides during lipolysis. When lipoprotein lipase destroys the triglyceride core of these lipoproteins, large amounts of surface lipids like nonesterified cholesterol and phospholipids are released. Together with released apolipoproteins, these surface remnants join the existing pool of HDL in the blood. This pathway increases the size of HDL and expands the pool. Small HDL particles which are poor in lipid are better at taking in cholesterol and phospholipids than larger ones. The size of small HDL increases and they become good donors of cholesterol to the liver through the SR-BI receptor in the pathway of reverse remnant-cholesterol transport.

The HDL protein and lipid components both play a part in the removal of lipids from cell membranes and triglyceride-rich lipoproteins. The HDL particles have more proteins at their surface than other lipoproteins, which may explain why they appear to be more effective in this process. The proteins help HDL interact with molecules outside of the cell such as receptors. Apolipoprotein A-I is the most important HDL protein for removing lipids. Its primary job is to transport and dissolve phospholipids. Apolipoprotein A-I has a unique shape which allows it to bind tightly to the cell membranes in small HDL particles with less cholesterol. This facilitates the removal of cell cholesterol. When apolipoprotein A-I forms more cholesterol-rich particle, however, it changes shape and does not bind to cell membranes as tightly. The HDL particle can then be released. HDL enzymes are also important in removing cholesterol from the organs and transferring it to the liver. LCAT is an important enzyme in this process. However, even without LCAT, cholesterol can be transported to the liver.

HDL's ability to remove cholesterol is influenced by many factors, including proteins. Lipids play an important role as well. The type of lipids in HDL affects how well it removes cholesterol. HDL can acquire cholesterol in HDL and remove it from the cells by using lipids. How well HDL can remove cholesterol depends on the physical state of lipids at the surface. The lipids can incorporate cholesterol more easily when they are fluid. Liquid crystal unsaturated phospholipids have a high fluidity and are very effective in removing cholesterol. Sphingomyelin, on the other hand, reduces the fluidity of lipids and can lower the removal rate of cholesterol. Sphingomyelin, however, has a strong affinity for cholesterol and can prevent it from returning to the cells from HDL. This direct interaction counteracts the effects of reduced fluidity.

HDL particles are divided into different subtypes. Each subtype has a different ability to remove cholesterol from the cells. Different HDL subtypes interact with different receptors in order to remove cholesterol from cells or to deliver it to them. HDL particles with less lipid remove cholesterol more effectively from the cells than HDL particles with more lipids. HDL particles that are smaller, denser, and have less lipid interact with ABCA1. The structure of small HDL particles with less lipids is very effective at removing cholesterol. These small particles are often disc shaped and have defects in the lipid arrangement. This allows them to absorb a large amount of cholesterol and other lipids. The smaller the HDL particles, the more effective they are at removing cholesterol. The smallest HDL type, called lipid-free apolipoprotein A-I, is best for accepting cholesterol from the cells. The larger HDL particles with more lipids, on the other hand, are better at removing cellular cholesterol through SR-BI and ABCG1.

History

The HDL Hypothesis
The discovery of Tangier disease by Donald Fredrickson in 1960 was the first report of a familial HDL deficiency. This disorder is characterized by low HDL-cholesterol and an undetectable alpha-lipoprotein, as well as the accumulation of cholesterol in tissues such as tonsils. A defect

in the production of HDL was proposed to explain these observations. Alternative hypotheses included rapid removal of HDL from the blood. The absence of HDL caused the accumulation of cholesterol in tissues, and this could be explained by HDL's role in removing the cholesterol from cells. The knowledge that many tissues are unable to break down cholesterol led to these considerations. This excess cholesterol is toxic to cells and must be removed.

Who	When
George Miller and Norman Miller	**1975**
Where	**What**
UK	Miller, G. J. and N. E. Miller (1975) "Plasma-high-density-lipoprotein concentration and developement of ischaemic heart disease." Lancet 1: 16-19.

George and Norman Miller formulated the HDL hypothesis in 1975. They studied data that showed the amount of cholesterol increases in the body when HDL-cholesterol levels in the blood drop. They proposed that HDL transports cholesterol to the liver, where it is broken down and removed from the body. The researchers also suggested that when HDL-cholesterol levels in the blood drop, this can cause a buildup of cholesterol within the arteries. This could accelerate the development of atherosclerosis. A number of conditions can reduce HDL-cholesterol levels, including high plasma cholesterol and high triglycerides. Other factors, like being overweight, male, and diabetic, may also increase the risk. They concluded that low HDL-cholesterol levels could contribute to the formation of arterial blockages because they prevented cholesterol from being removed from the arterial wall. It is important to emphasize that Millers' hypothesis did not mention any potential benefits from raising HDL-cholesterol. Other investigators made this leap of faith later.

Later studies supported the concept of Miller and Miller. In the United States, large-scale epidemiological research, like the Framingham Heart Study, showed that the risk of coronary artery disease increases by 3% for women and 2% for men with every 1 mg/dl drop in HDL-cholesterol. David Gordon and Basil Rifkind proposed in 1989 that HDL played an important role in the protection of the heart.

In the 1990s, these strong findings were followed up by solid data which clearly demonstrated links between low HDL-cholesterol levels and heart disease. In the 2000s, the concept of HDL function was developed after thorough biochemical and biological studies. The concept stated that low HDL-cholesterol and heart disease are linked because HDL can remove cholesterol from arterial walls. Other beneficial effects of HDL on the arterial wall could be a

(Continued)

(Continued)

further benefit to this lipoprotein. The conclusion was virtually unanimous: HDL-cholesterol has to be elevated in order to combat heart disease. Studies done in experimental animals largely supported this strategy. Indeed, raising HDL-cholesterol delayed the development of atherosclerosis in mice and rabbits. The mainstream ignored the rare voices that spoke out against this approach.

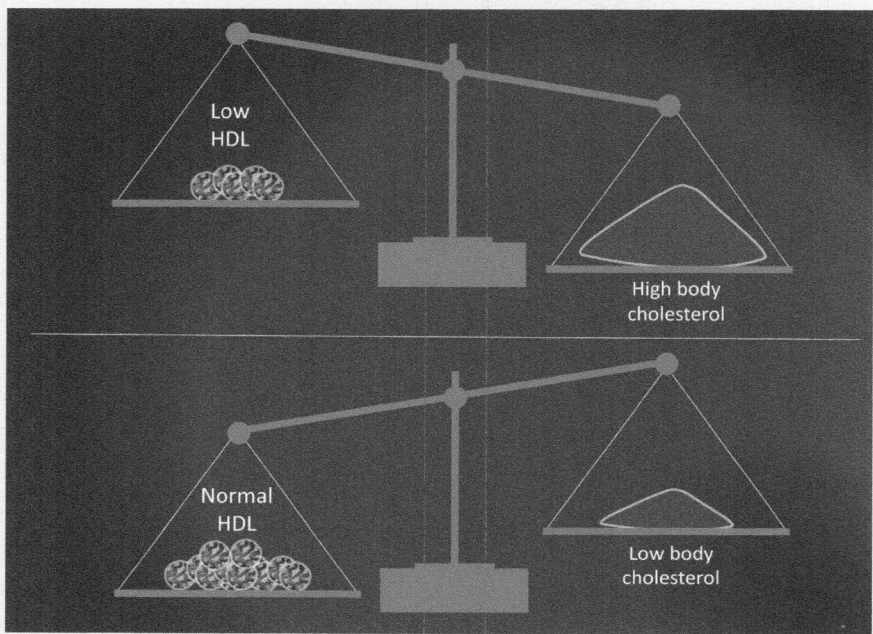

The HDL hypothesis reached its peak at this time. General public became aware of the notion of "good" cholesterol. Costly therapeutic developments by the biggest pharmaceutical companies like Pfizer, Merck, Eli Lilly, and Hoffmann-La Roche were underway. There were big hopes that together with statins that decrease LDL-cholesterol, drugs which increase HDL-cholesterol would finally eradicate heart disease.

Inhibition of Oxidation

Antioxidative activity of HDL refers to the ability to prevent damage caused by oxidation in the body. LDL is believed to be the main target of protection against oxidative damage by HDL. However, HDL can also inhibit the production of harmful reactive oxygen species and reduce oxidative stress in cells. This has been observed in endothelial cells and macrophages.

It is unclear how important HDL's antioxidative activity is compared to its other effects on vessel health. It appears unlikely though that HDL's ability to prevent LDL oxidation can actually delay the development of atherosclerosis.

HDL can protect LDL and other lipoproteins from damage by oxidative stress. This damage is caused by different types of oxidants, including free radicals. The latter are also called one-electron oxidants because they carry at least one unpaired electron which makes them unstable and extremely reactive. Free radicals can cause oxidation of both the lipids and proteins in LDL through self-accelerating chain reactions. HDL particles are very effective at protecting both the lipids and proteins in LDL from oxidation by free radicals, leading to the formation of stable termination products which are harmless.

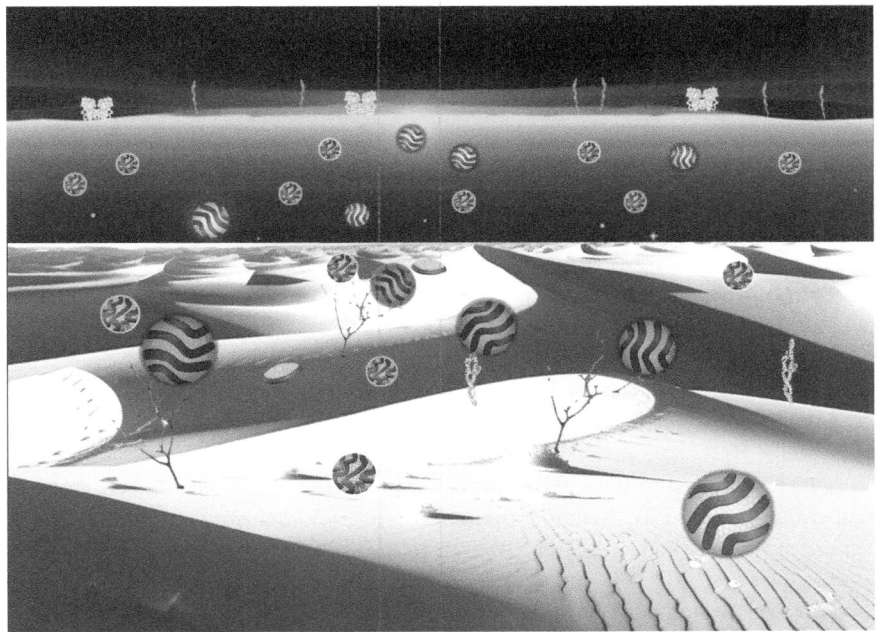

HDL has the ability to protect LDL particles from oxidation in the arterial intima. These LDL particles are held onto the surface of arterial wall cells by proteoglycans. HDL achieves this protection by removing oxidized lipids from LDL, breaking them down, and preventing further oxidation. As oxidized lipids have the potential to induce inflammation, HDL serves as an anti-inflammatory agent.

The first step in the protection against oxidative damage conveyed by HDL is removing oxidized lipids from LDL or cells. When oxidized LDL and HDL are incubated together, oxidized lipids, such as lipid hydroperoxides, are quickly transferred from the LDL to the HDL. Adding a hydroperoxyl OOH group to a phospholipid or cholesteryl ester molecule makes it more water soluble, allowing it to escape from LDL more easily than non-oxidized molecules. The second step in how HDL protects LDL from oxidation is by deactivating oxidized lipids. HDL can neutralize the harmful effects of certain types of oxidized lipids by breaking them down into harmless substances.

However, the specific ways in which HDL protects cells from oxidative damage are not well understood. It is known that HDL needs to interact with certain receptors on cell surfaces in order to have its antioxidant effects. The SR-BI receptor seems to be responsible for the antioxidant effects seen in endothelial cells, while both SR-BI and ABCG1 are involved in HDL's ability to reduce the production of free radicals in macrophages.

The role of HDL components in protecting against oxidative damage is important. Apolipoprotein A-I is a key component that helps deactivate harmful substances, like lipid hydroperoxides, which can cause damage to cells. Apolipoprotein A-I does this by reducing the hydroperoxides into harmless lipid hydroxides, stopping the chain reaction of damage. Another component, apolipoprotein A-II, also helps inactivate lipid hydroperoxides but to a lesser extent. The amount of apolipoprotein A-I in HDL and its oxidative status are important factors in how well HDL can protect LDL from being damaged by free radicals. During this process, oxidized forms of apolipoprotein A-I and apolipoprotein A-II are formed which can be converted back to non-oxidized proteins by specialized enzymes called methionine sulfoxide reductases, which are found in many tissues and help safeguard from oxidative harm. Other apolipoproteins found in HDL, such as apolipoproteins A-IV, A-V, E, J, and M, also contribute to protecting LDL from oxidative damage in laboratory settings.

Enzymes in HDL, such as PON1, PAF-AH, and LCAT, can help safeguard from oxidative harm. They break down detrimental substances like some types of oxidized phospholipids. However, these enzymes are not very effective against other oxidized molecules like lipid hydroperoxides. PON1 has been shown to have strong anti-atherosclerotic effects in living organisms. It is found exclusively in HDL and is attached to lipids through its hydrophobic part. Another enzyme called PON3, also found in HDL, has similar properties to PON1. It is more likely, however, that PAF-AH, rather than PON1, is responsible for breaking down oxidized phospholipids in HDL. LCAT may also break down certain types of oxidized phospholipids produced during lipoprotein oxidation.

The lipid composition can also affect the antioxidative activity of HDL. The fluidity of the phospholipid layer in HDL particles plays a role in how efficiently harmful substances are transferred from LDL to HDL. Lastly, HDL particles contain small amounts of antioxidants like tocopherols, which may contribute slightly to their antioxidant properties.

Different types of HDL have varying levels of substances that help protect against oxidative damage. PON1 is usually found in large HDL but can be found in small particles when separated by ultracentrifugation. LCAT, PAF-AH, and apolipoprotein A-IV are more abundant in small, dense HDL. Because of these differences, not all HDL particles have the same ability to protect LDL from oxidative damage caused by free radicals. The smaller and denser HDL subfractions typically show stronger antioxidative activity against LDL oxidation. This is due to their ability to deactivate lipid hydroperoxides through a two-step process involving the transfer of these oxidized lipids to HDL and their subsequent inactivation by apolipoprotein A-I. The hydrolysis of oxidized lipids by enzymes associated with HDL also seems to be mainly linked to small, dense HDL3. The preferential localization of enzymatic activities with this type of HDL may explain this association.

Reduction of Inflammation

HDL's anti-inflammatory activity is the ability to reduce inflammation in the arterial wall cells. HDL inhibits certain processes within cells involved in inflammation. This requires the interaction with specific proteins on cell surfaces. HDL is an important component of the body's defense system because it has these properties. Both laboratory and clinical research are pointing to a link between HDL and its ability to reduce inflammation, and its role in the prevention of atherosclerosis.

HDL can reduce the expression of proteins that attract monocytes to the surface of the cells lining blood vessels. HDL also reduces the release of proteins by cells that are responsible for promoting inflammation. These effects limit inflammation by reducing white blood cell adhesion to the blood vessel wall.

HDL can also affect directly certain types of blood cells involved in inflammation. Monocytes can get activated to produce inflammatory substances when they come into contact with lymphocytes.

HDL particles prevent lymphocytes from activating monocytes. HDL also reduces the amount of monocytes and the growth of their precursors. In addition, HDL and apolipoprotein A-I both reduce the movement and activation of neutrophils, another type of immune cell. As a result, HDL can have a significant impact on the immune system by stopping cell activation.

HDL can reduce inflammation in several ways. By removing cholesterol from the cells, HDL can help reduce the production of molecules causing inflammation. HDL can also remove oxidized lipids from cells, which may otherwise cause inflammation.

Different HDL components play a part in preventing inflammation. Apolipoproteins A-I, A-II, and A-IV are important in HDL's anti-inflammatory effects. It is no surprise that HDL3 with its dense structure and high protein content has a better ability to stop inflammation than HDL2 which has fewer proteins. HDL's phospholipid component helps it to have anti-inflammatory properties. The type of fatty acid in this lipid is a factor that can influence how effectively it reduces inflammation. In this respect, polyunsaturated fats found in fish oil have a particularly positive effect.

Protection from Cell Death

The act of protecting cells can be described as cytoprotective. The cells in the arterial walls, such as macrophages, endothelial cells, smooth muscle cells, and lymphocytes, can die by apoptosis, which is a cell death that is programmed, or necrosis, which is unregulated. Dead cells can cause atherosclerotic plaques in the arteries. The buildup of modified LDL can cause macrophage death.

HDL protects macrophages as well as endothelial cells from death through apoptosis. This is done by removing harmful forms of cholesterol from the cells, such as oxidized cholesterols. HDL has antioxidant properties which reduce the production of harmful substances in cells such as superoxide anion or hydrogen peroxide. HDL interacts directly with toxic oxidized lipids to neutralize and remove them. HDL not only protects against cell death but also encourages the growth and migration of certain types of cells that can prevent cholesterol buildup in arteries.

HDL is a powerful antiapoptotic agent, which means it can help prevent programmed cell deaths. This is achieved by maintaining mitochondrial integrity, preventing DNA fragmentation, inhibiting intracellular pathways that lead to apoptosis, and activating survival mechanisms. HDL increases the production of nitric oxide in endothelial cells, which improves blood vessel function and cell viability. The interaction between HDL and SR-BI is crucial for protecting endothelial cells from apoptosis. The ABCG1 transporter also plays a significant role in the antiapoptotic effects of HDL on macrophages.

HDL contains two groups of molecules that are important in protecting cells: apolipoproteins and minor lipids. Apolipoproteins A-I and E appear to be the two proteins that play a major role in protecting cells against death. These proteins help cells survive through their interaction with receptors like SR-BI and ABCG1 that activate pathways to promote cell survival and eliminate harmful lipids. Certain types of sphingolipids found in HDL, such as sphingosine-1-phosphate, also contribute to protecting cells from apoptosis. Sphingosine-1-phosphate is especially effective at preventing cell death and improving the survival of endothelial cells by increasing the production of nitric oxide.

The HDL3 particles, which are small and dense, can shield cells against damage caused by oxidized LDL. This finding is quite logical, considering that these particles are rich in apolipoprotein A-I and sphingosine-1-phosphate, both known for their exceptional cell-protective properties.

HDL helps cells stay alive and makes blood vessels relax better. This happens when HDL attaches to the SR-BI receptor on cell surfaces.

Defense Against Infection

Infection can make atherosclerotic plaques worse and unstable, which can lead to heart attack and stroke. HDL is a natural defense system that has multiple ways of fighting infections. HDL is the first to remove lipopolysaccharide and clear it out through the liver. This stops it from causing problems within the cells. Apolipoprotein A-I binds to lipopolysaccharide, neutralizing its harmful effects. These effects are attributed to the terminal region of apolipoprotein A-I. Lipopolysaccharide-binding protein is another component of HDL that helps in fighting infections. The protein can also quickly bind and neutralize lipopolysaccharide.

What It Is

Lipopolysaccharides are large molecules found in the outer part of some bacteria. These molecules are toxic and made of a lipid and a polysaccharide. These are also known as endotoxins, and they are mainly produced by certain bacteria such as *Escherichia coli* or *Salmonella*.

HDL contains a factor which can kill the parasite trypanosoma, which is responsible for sleeping sickness. This parasite, which can cause harm to animals in Africa, is naturally destroyed by our plasma. This parasite can only be killed by a subtype of HDL. Apolipoprotein L-I, a minor HDL protein, is a key player in the killing of trypanosomes. It maximizes the pathogen's killing ability by working with hemoglobin-binding protein, haptoglobin-related protein, and apolipoprotein A-I. The trypanosome lytic factor is formed by these components, which allows it to bind specifically to the pathogen. The factor is a subpopulation within dense HDL particles. It has a unique set of properties, including a high molecular mass of 490 kDa and a large particle size of 15–21 nm.

Relaxation of Vessels

HDL particles improve the function of blood vessels by releasing nitric oxide and reducing the production of reactive oxygen species. HDL particles can also help stimulate the production of another substance, prostacyclin. This helps relax blood vessels. HDL binds to the SR-BI receptor located on the surface cells of the blood vessels. This causes a series of events to occur inside the endothelial cells, which result in an increase in nitric oxide production.

HDL is also important in the removal of cholesterol from the endothelial cells. It helps activate signals within the cell which improve its function. This pathway uses the ABCG1 transporter to maintain healthy blood vessel function.

Apolipoprotein A-I is an important HDL component that aids in the widening of blood vessels. This is done by interacting directly with SR-BI on the surface of endothelial cells and activating specific signals within cells. PON1 may also be involved in the widening of blood vessels.

One of the important substances in HDL is sphingosine-1-phosphate, which can affect the lining of blood vessels. Sphingosine-1-phosphate interacts with specific receptors for this molecule and activates signals inside cells. When HDL interacts with the receptors for sphingosine-1-phosphate, it helps produce nitric oxide which widens blood vessels. Small, dense HDL3 particles, which have a lot of sphingosine-1-phosphate, are especially important for this process.

Attenuation in Clotting

HDL particles can have multiple effects on blood clots, and they may also help slow the progression of arterial blockages. In laboratory studies, HDL was found to inhibit platelet adhesion, fibrinogen binding, and granule release, as well as the production of substances that promote blood clotting. HDL inhibits blood coagulation factors. HDL also has antithrombotic properties on endothelial cells.

Just like how HDL helps relax blood vessels, it can also prevent blood clotting by boosting the production of nitric oxide. This molecule prevents platelets from adhering together and forming blood clots. HDL does this by interacting with the SR-BI receptor found on the surface of platelets. HDL also helps reduce the cholesterol in platelets, further preventing clotting.

HDL also inhibits the activity of certain blood coagulation proteins. These proteins form complexes at the surface of some lipids in the blood. Apolipoprotein A-I, the main protein in HDL, prevents these lipids from activating the proteins. HDL's phospholipid component also plays a part in its ability to prevent clotting. It removes cholesterol from platelets and reduces the formation of certain lipids that encourage clotting. Some of the negatively charged phospholipids in HDL such as cardiolipin have anticoagulant properties that contribute to its effect on blood coagulation.

Improvement of Sugar Processing

Low levels of HDL-cholesterol are common in diabetes. It is due to the high plasma triglycerides and activity of CETP in diabetics. Hence, it was believed that there was no causal relationship between HDL-cholesterol levels and glucose metabolism. Recent data however suggest that low HDL-cholesterol levels may precede the development of diabetes and contribute to this disease.

Low HDL-cholesterol may be linked to diabetes because HDL improves the function of beta-cells which produce insulin. HDL can affect sugar metabolism in a number of ways. For example, it can reduce blood glucose levels and increase insulin in diabetics. In lab tests, it also stimulates insulin secretion by beta-cells. The ABCA1 transporter can activate insulin secretion by beta-cells when apolipoprotein A-I or A-II are present in HDL. HDL's ability to affect insulin secretion may be due to the removal of excess cholesterol from beta-cells through ABCA1. These findings suggest HDL's ability to remove cholesterol could be responsible for the many benefits it provides, such as reducing inflammation, protecting cells from dying, preventing blood clots, and helping with diabetes.

Open Question

Can changes in lipoprotein metabolism help prevent diabetes?

HDL can also improve insulin sensitivity by protecting beta-cells that produce insulin in the pancreas from dying. HDL can keep these cells alive, preventing their death. HDL can also improve insulin sensitivity by reducing inflammation in muscles caused by harmful lipids. HDL and apolipoprotein A-I can have a positive effect on glucose processing in cells of adipose tissues.HDL can also help cells take in glucose more efficiently, speed up the process of turning glucose into energy, and promote the burning of fatty acids for fuel. This highlights the role that HDL plays in the body's energy production.

14

HDL – How It Stops Working

HDL production in the blood and its regeneration can be delayed by certain diseases. These changes can lower HDL-cholesterol levels.

Causes

The concentration of HDL-cholesterol is controlled by genes, which can be passed down from parents. Heritability normally ranges from 40% to 60% but can be as high as 80%. Genes that influence HDL processing in the body are those that produce HDL components, as well as lipases and lipid transfer proteins. However, only a small percentage (less than 10%) of HDL-cholesterol variation can be explained by the genes that have been identified. The effects of each gene on HDL-cholesterol levels vary between populations and are typically small. However, certain genes have a strong influence.

Remember that the size of a gene's effect on HDL metabolism does not always reflect its importance. The *HMGCR* gene, for example, plays a major role in LDL metabolism yet only affects LDL levels by a very small amount (1 mg/dl). It is unclear how these small genetic variations (each with an effect of less than 5% and the majority of them less than 1–2%) can explain such a high percentage of up to 80% of HDL-cholesterol heritability. There are probably other genetic factors not yet discovered.

Genetic studies help us understand if there is a relationship between HDL-cholesterol and cardiovascular risk. Mendelian randomization is a way to achieve this. It is similar to a natural randomized trial. According to this concept, people with variations in a gene, such as *CETP*, have a lower cardiovascular disease risk when their HDL-cholesterol levels are higher. By contrast, there is no association between heart disease and variations in the genes coding for ABCA1 or hepatic lipase. One study reported that only 6 of the 15 genetic variations which affect HDL-cholesterol levels are associated with heart attacks. All of these genetic variations affect LDL-cholesterol and triglyceride concentrations as well, making it difficult to draw any clear conclusions. On the other hand, nine of the ten genetic variations that are related to LDL-cholesterol levels have been linked with heart attacks. This controversy shows we need to do more research in order to better understand the role that different genes play in HDL processing.

Transporters

There are many mutations of the *ABCA1* gene. Over 100 of them have been identified. Tangier disease is caused by mutations of this gene. This condition is characterized with very low HDL-cholesterol levels. Homozygous Tangier patients have HDL-cholesterol levels less than 5 mg/dl, and their apolipoprotein A-I levels are below 4 mg/dl. It is because the ABCA1 transporter, which shuttles cholesterol from cells into HDL in healthy people, does not function properly.

Heterozygous carriers of the mutant gene have lower HDL-cholesterol levels, but they are not as low as in people with homozygous Tangier disease. Low HDL-cholesterol puts people at a higher risk of heart problems, such as heart attacks and strokes. However, lowering LDL-cholesterol levels can reduce this risk for people with Tangier disease.

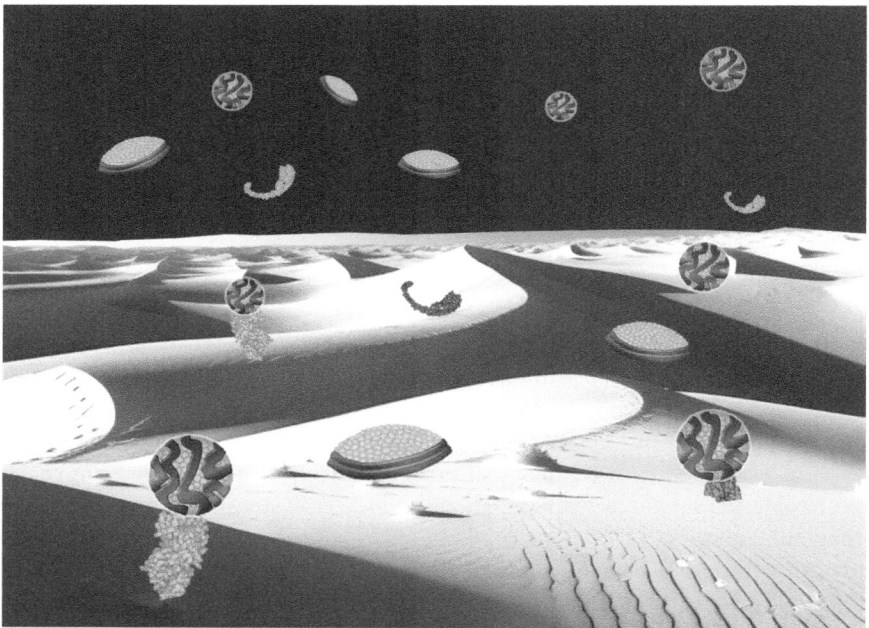

In the absence of the ABCA1 transporter, lipid-free apolipoprotein A-I and small discoid HDL cannot remove cholesterol from cells.

Apolipoproteins

A homozygous deficiency in apolipoprotein A-I is a rare disorder where HDL-cholesterol levels are low and there is no apolipoprotein A-I in the blood. It leads to a slower rate of removal of cholesterol from tissues, and a buildup of cholesterol. The carriers of this disorder have low HDL-cholesterol due to small HDL particles that contain apolipoproteins A-II, E, and A-IV, instead of large lipid-rich HDLs. This condition may cause xanthomas and loss of transparency in the cornea. It can also increase the risk of heart disease. Heterozygous apolipoprotein A-I deficiency has about half of the normal HDL-cholesterol and apolipoprotein A-I levels in blood. However, they usually do not have any symptoms.

Mutations in the *APOA1* gene can lead to hypoalphalipoproteinemia, increasing the risk of heart disease. However, there are exceptions such as the Milano and Paris variants of apolipoprotein A-I, which have lower HDL-cholesterol but paradoxically reduced the risk of heart disease. These variants alter the structure of HDL.

A mutation in the *APOC3* gene can result in higher HDL-cholesterol and a reduced risk of atherosclerosis. Apolipoprotein C-III inhibits lipoprotein lipase. Its loss of function can accelerate the lipolysis and production of HDL.

In large population studies, the *APOE* gene was studied to determine its relationship with HDL-cholesterol levels. In the majority of these studies, individuals with E2 alleles had higher HDL-cholesterol levels. Those with E4 alleles had lower HDL-cholesterol.

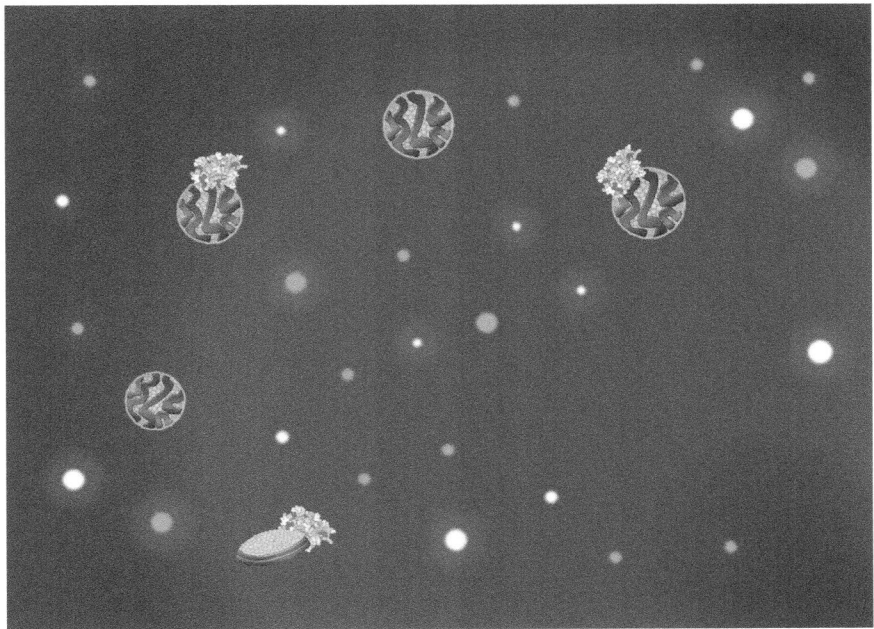

When the LCAT enzyme is not functioning properly, it disrupts the formation of large HDL. The size of HDL is reduced in this condition.

Enzymes

LCAT deficiency is a condition that can cause familial LCAT deficiency and fish-eye disease. A lack of LCAT can lead to problems such as cloudy corneas in fish-eye disease, anemia, and kidney failure caused by lipid accumulation. A faulty LCAT that is not working properly can cause this deficiency. In people with homozygous LCAT deficiency, both familial LCAT deficiency and fish-eye disease are characterized by low levels of HDL-cholesterol and apolipoprotein A-I, which are features of familial hypoalphalipoproteinemia. LCAT deficiency can also cause lipoprotein-X to appear. Heterozygous people with one copy of a faulty gene that causes LCAT deficiency usually have low HDL-cholesterol levels but show no symptoms.

PON1 has two variants that are well known, Gln192Arg and Met55Leu. They are related and have an effect on PON1 activity. The Gln192Arg is the most important factor in determining PON1 activity and its relationship to coronary heart disease. The variant, in particular, is associated with higher PON1 activities, lower markers of oxidative stress, and lower risks of death. A review of numerous studies indicates that there is however not a strong link between genetic changes in PON1 and cardiovascular risk.

Familial lipoprotein lipase deficiency also called Type I hyperlipoproteinemia and familial chylomicronemia is a rare genetic disorder which leads to a massive buildup of chylomicrons in the blood. People with this disorder can have very high levels of triglycerides and low levels of HDL-cholesterol. Lipoprotein lipase mutants can affect HDL-cholesterol levels by preventing lipids being transferred from triglyceride-rich lipoproteins to HDL during lipolysis. The disorder does not however increase the risk of heart attack because the chylomicrons are too big to penetrate the arterial walls.

On the other hand, when lipoprotein lipase activity is moderately decreased, cholesterol does not move as well to HDL during triglyceride breakdown by lipoprotein lipase. This causes HDL-cholesterol to drop and small remnant lipoproteins to build up in the arteries, leading to cholesterol buildup in the arterial walls.

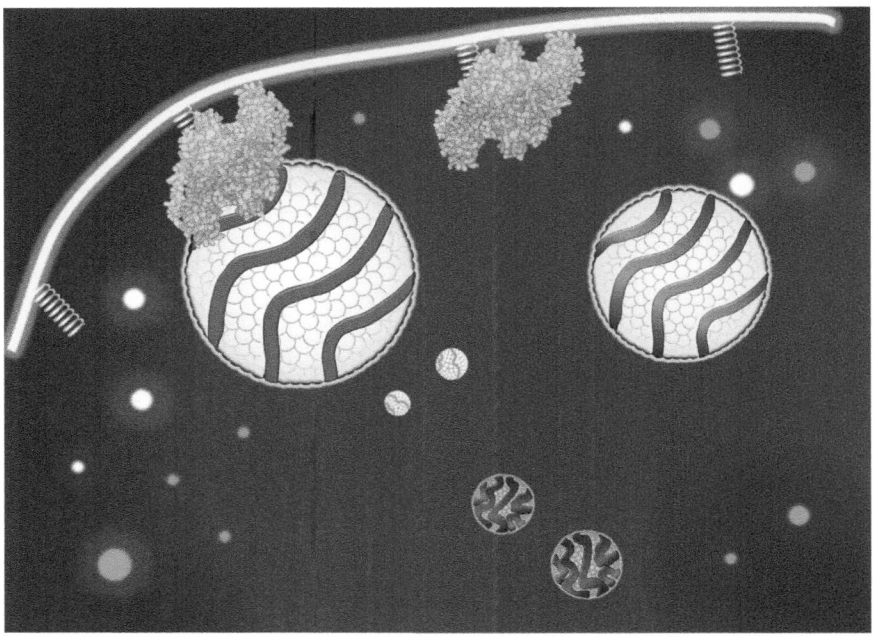

When the breakdown of triglyceride-rich lipoproteins by lipoprotein lipase is not functioning properly, it causes a delay in the formation of core and surface remnants. This, in turn, leads to a reduction in the flow of apolipoproteins, phospholipids, and nonesterified cholesterol to the pool of HDL. As a result, levels of HDL-cholesterol are reduced.

HDL-cholesterol levels can be reduced by increasing the activity of hepatic lipase, which breaks down HDL triglycerides. This will also cause HDL to be eliminated from the blood faster. Endothelial lipase also breaks down triglycerides and phospholipids in a similar manner. This lipase can cause inflammation and reduce HDL concentrations when it is active.

By contrast, complete hepatic lipase deficiency is a rare genetic condition that causes high levels of cholesterol, triglyceride, HDL-cholesterol, and apolipoprotein A-I in the blood. This happens because the body cannot break down triglycerides properly. This leads to HDL and LDL filled with triglycerides. This condition can cause high HDL-cholesterol, which is generally considered to be good for the heart. Paradoxically, these variations in the hepatic lipase gene may however increase cardiovascular risk. Too little or too much activity of hepatic lipase can therefore harm the heart.

Some changes in the *LIPG* gene, which codes for endothelial lipase, can also affect the way the body processes HDL-cholesterol. These changes may cause issues with the enzyme's ability to break down HDL. HDL-cholesterol can rise above normal levels, which is not necessarily beneficial though.

Lipid Transfer Proteins

Genetic variations influence CETP, one of the most well-known lipid transfer proteins. Some of these mutations are common while others are rare. A deficiency of CETP usually results in higher levels of HDL-cholesterol and normal or lower levels of LDL-cholesterol. In severe cases, where CETP activity is completely lost, HDL-cholesterol levels can rise up to fivefold the normal level. This condition occurs more often in Asians and results in an accumulation of very large HDL and diverse LDL particles. The increase in HDL-cholesterol levels can be as low as 10–30% when there is a partial CETP deficiency. Mutations in the *CETP* gene often cause familial hyperalphalipoproteinemia. In addition, the inability to secrete CETP can contribute to elevated HDL-cholesterol.

Increased activity of CETP leads to a greater transfer of cholesteryl esters from HDL to lipoproteins that are rich in triglycerides. This ultimately results in fewer cholesteryl esters remaining in HDL and lower levels of HDL-cholesterol.

The majority of variations in the *CETP* gene are linked to a lower CETP activity and higher levels of HDL-cholesterol. This is associated with a reduced risk of heart disease. These specific variations of the *CETP* gene are also more common in people who live longer, and they are linked to higher HDL-cholesterol levels. These variations in the *CETP* gene associated with higher HDL-cholesterol are also related to lower triglycerides, LDL-cholesterol, and remnant cholesterol levels. However, it is difficult to determine if the cardioprotective effects are due to an increase in HDL-cholesterol, or a decrease in LDL-cholesterol or remnant cholesterol. These results support the notion that low CETP activity is causally linked to a lower risk of heart disease.

Others

Different versions of the *SCARB1* gene coding for SR-BI are associated with elevated levels of HDL-cholesterol. The effects of these variants on health outcomes in humans are inconsistent.

The *LAG3* gene is also associated with elevated heart disease risk and high HDL-cholesterol. The *PLTP* gene is associated with higher levels of HDL-cholesterol and is more commonly found to have variations in individuals with hyperalphalipoproteinemia.

Consequences

Various factors such as diabetes, infections, and inflammation can cause changes in HDL particles, affecting their structure, composition, and function. These modifications may be important for cardiovascular diseases because of their relationship with dyslipidemia and inflammation. These modified HDL particles have been referred to as functionally deficient HDL, or dysfunctional HDL. They are similar to chameleons because they can change their properties just like chameleons do.

Functionally deficient HDL refers to HDL particles with reduced levels of their normal biological functions. HDL dysfunction, on the other hand, means that HDL has completely lost its ability to perform its normal functions in protecting from atherosclerosis, such as reducing inflammation and oxidative stress.

Although laboratory tests can detect dysfunctional HDL, its presence in the body has not been definitively proven. The specific test used to measure HDL function can determine the extent of its loss. When tested for anti-inflammatory activity or the ability to relax blood vessels, HDL may be completely dysfunctional. Tests measuring antioxidative capacity or cholesterol efflux show that HDL can be functionally deficient with a partial loss of function.

Abnormal Composition

The protein and lipid components of HDL may be altered in conditions that are associated with an increased risk of cardiovascular diseases such as dyslipidemias. The modifications to the proteins carried by HDL may promote atherosclerosis, even if there is no reduction in HDL-cholesterol levels in the bloodstream.

The amount of apolipoprotein A-I in HDL can be reduced by several diseases which increase the risk of heart problems. This is especially true for people with high LDL-cholesterol and low HDL-cholesterol. In addition, HDL content of other proteins, such as apolipoprotein E, apolipoprotein C-I, apolipoprotein C-III, and SAA, are also elevated for people with high LDL-cholesterol. This indicates chronic inflammation.

The proteins of HDL can change during the acute phase. Some proteins like apolipoprotein A-I, decrease, while others like lipopolysaccharide-binding protein increase. Inflammatory conditions can cause a decrease in the HDL apolipoprotein A-I because the liver produces less and SAA replaces it. SAA can facilitate inflammation and artery blockage. It can cause HDL to stick to the proteoglycans in the arterial walls and be quickly removed from circulation. Moreover, SAA increases the production by immune cells of cytokines, which keep inflammation going.

Apolipoprotein A-I can also undergo other changes in HDL, apart from being replaced with SAA. The apolipoprotein can undergo modifications when it is exposed to substances which cause oxidation, secreted from cells of the arterial wall. It is no surprise that levels of oxidized apolipoprotein A-I in the blood are higher among patients with heart diseases and smokers. Myeloperoxidase is an enzyme produced by activated neutrophils and is a significant source of reactive nitrogen and chlorine species in arterial walls. It binds apolipoprotein A-I and causes oxidation of specific amino acids. This modified HDL is less effective at removing cholesterol from the cells.

Apolipoprotein A-I can also be glycosylated in addition to oxidative modifications when glucose levels are high, as they are with diabetic patients or older individuals. The apolipoprotein can also be modified by the proteolytic degradation induced by arterial cells.

When there is inflammation, certain enzymes associated with HDL such as PON1, PAF-AH, and LCAT can become depleted or stop working. This can occur in conditions such as diabetes, metabolic syndrome, and premature coronary artery disease.

HDL functions can be affected by changes in the amount of lipids. HDL can have too much triglyceride, and not enough cholesteryl ester. This occurs when the HDL-cholesterol is low and triglycerides are high in the blood. It is common in diabetics and people with metabolic syndrome. Chronic inflammation is also present in this condition.

Simple terms, low HDL-cholesterol means there are less circulating HDL particles and less cholesterol in HDL. The result is a decrease in the size of HDL particles, which have less cholesteryl ester. These changes can be seen often in acute-phase responses and inflammation.

Deficient Processing

The changes in HDL proteins and lipids are closely related to the way HDL is processed by the body when diseases occur. HDL processing is affected by conditions such as high cholesterol, diabetes, obesity, inflammation, and infection. All of these conditions are often linked and it is difficult to separate their pathways.

Dyslipidemias

HDL processing is affected in atherogenic dyslipidemia. The liver produces more apolipoprotein B-100, which leads to higher blood triglyceride levels. Due to CETP's overactive actions, high triglycerides lead to lower HDL-cholesterol levels and higher triglycerides within HDL. HDL particles become smaller and denser when there is too much triglyceride in the blood. These HDL particles, because they are unstable and have lost some of their apolipoprotein A-I, are quickly broken down and removed from the blood by the kidneys. The number of HDLs in the blood drops as a result.

Decreased lipoprotein lipase activity can also lower HDL levels by reducing the production of surface components of triglyceride-rich lipoproteins, which are needed for HDL particle formation. This deficiency reduces the transfer of nonesterified cholesterol from the triglyceride-rich lipoproteins to HDL. As a result, the removal of remnant cholesterol to the liver for excretion can be delayed. This can lead to an accumulation of remnant cholesterol in the blood and potential buildup in the arterial wall.

In atherogenic dyslipidemia, levels of HDL2 particles, which are large and cholesterol rich, decrease with the overall HDL-cholesterol level. The levels of the small, lipid-poor particles do not change much. This leads to an increase in the proportion of small HDL. This shift from larger to smaller HDL particles is a sign that HDL formation has been delayed.

Low HDL-cholesterol levels, known as hypoalphalipoproteinemia, can also be caused by a lack of apolipoprotein-A-I, reduced LCAT activity, increased hepatic lipase activity, or a dysfunctional ABCA1 protein. High insulin levels can accelerate the breakdown of HDL. HDL-cholesterol levels can also be negatively affected by excess body fat. More fat around the abdominal area is particularly harmful.

Smoking can also be linked to low HDL-cholesterol. Smoking can negatively affect large HDL2 and small HDL3 as well as HDL particle concentration. Smoking just one cigarette may temporarily lower HDL-cholesterol by as much as 10%. Inhaling secondhand smoke can also lower HDL-cholesterol. Smoke particles may play a significant role in lowering HDL. Smokers' low HDL-cholesterol levels are probably due to LCAT problems, as it is less active.

On the other hand, a deficiency in CETP, SR-BI, or hepatic lipase can cause high levels of HDL-cholesterol, which is known as hyperalphalipoproteinemia.

In the absence of cellular cholesterol transporters, levels of HDL particles in the blood are very low.

Impairment of Function

Conditions that cause cardiovascular disease can weaken HDL's protective functions, including its ability to remove cholesterol from cells and fight inflammation. High levels of lipids, diabetes, and inflammation are all conditions that can lead to cardiovascular disease. However, it is unclear to what extent these changes in HDL functions actually affect the development of heart diseases. HDL dysfunction is a concept that has been studied mainly in lab tests.

Defective Removal of Cholesterol

The ability to remove cholesterol from cells could be affected if there are less HDL particles. This can cause problems in removing cholesterol from the arteries and accelerate atherosclerosis. Even if HDL particles are present in sufficient numbers, their ability to remove cholesterol may be reduced because they are modified. Low levels of HDL-cholesterol can also cause problems in the removal of cholesterol from cells. These individuals have been shown to excrete less cholesterol than those with normal levels of lipids. It is debated whether low HDL particle levels are the problem or if the particles themselves have intrinsic issues.

Some forms of low HDL-cholesterol may be characterized by HDL that is not effective at removing cholesterol. There are, for example, several mutant forms of apolipoprotein-A-I which are linked to low HDL-cholesterol. These HDL types are less effective than normal HDL at removing cholesterol. However, not all mutations of apolipoprotein A-I cause a reduction in cholesterol removal. Apolipoprotein A-I Milano is one example that is good at removing cholesterol. If changes in the cellular HDL receptors like ABCA1 are responsible for low HDL-cholesterol, HDL's ability to remove cholesterol may also be compromised.

There are contradictory findings regarding how HDL removes cholesterol from the cells of people who have unhealthy lipid levels. In 2011, Daniel Rader at University of Pennsylvania showed that removal of cholesterol from macrophages by HDL in a test tube can predict both the existence and the future appearance of heart disease. Many other researchers reported similar observations for patients with different forms of dyslipidemia and other disorders. However, some other studies have shown that HDL is able to normally remove cholesterol from cells in these people. In people with diabetes, the ability of HDLs to remove cholesterol can also be reduced. It is likely that they have lower levels of a protein known as SERPINA1, which helps apolipoproteins bind phospholipids and makes ABCA1 function better. If a small HDL does not have enough SERPINA1, this could increase heart disease risk in diabetics.

Changes in the body's lipoprotein processing can lead to abnormalities in HDL lipid structure, making it less effective at removing cholesterol. HDL from people with a genetic CETP deficiency may struggle to accept cholesterol from macrophages, leading to particles that are overloaded with cholesterol and other lipids. This makes it difficult for these very large particles to acquire more cholesterol. As a result, people with very high HDL-cholesterol levels may also struggle to acquire nonesterified cholesterol when triglyceride-rich lipoproteins are broken down. Additionally, HDL may have difficulty removing cellular cholesterol when it contains too much triglyceride.

Laboratory studies have shown that glycation can affect cholesterol removal. Oxidative stress is another factor which can affect HDL's ability to remove cholesterol. Myeloperoxidase can bind HDL, causing oxidative changes that prevent cholesterol removal.

Inflammation can reduce HDL's ability to remove cholesterol from the cells. Inflammation affects cholesterol removal primarily through ABCA1. In an environment where inflammation is promoted, HDL can be affected by oxidation. Inflammatory HDL has less phospholipid which makes it difficult to transport cholesterol.

All these changes can cause HDL to transport cholesterol differently during inflammation, shifting its focus toward macrophages and other cells along the arterial walls instead of liver cells. Inflammation reduces the removal of cholesterol from artery cells, possibly because macrophages require more cholesterol to fight infection.

Low levels of HDL-cholesterol can occur when people are prone to inflammation. It can cause problems in the removal of cholesterol from the arterial intima. Inflammation is common in people with heart disease, which can lead to other problems such as high LDL-cholesterol and cell damage. How well HDL removes cholesterol from cells is a good indicator for heart disease.

Deficient Protection from Oxidation

HDL particles from people with low HDL-cholesterol and certain medical conditions, such as diabetes, have a reduced ability to protect lipids against oxidation-induced damage. HDL may be less effective at preventing artery damage when it lacks the protective properties against oxidative stress. The reduced ability of HDL to protect from oxidation can be due to both abnormal glucose and lipid processing. Oxidative modification also plays a part. While studies in the laboratory have shown a reduced ability to protect LDL from oxidation, there are however no direct indications that this occurs in real-life situations.

The deficiency of apolipoprotein A-I can cause an accumulation of harmful lipid peroxidation products in HDL particles.

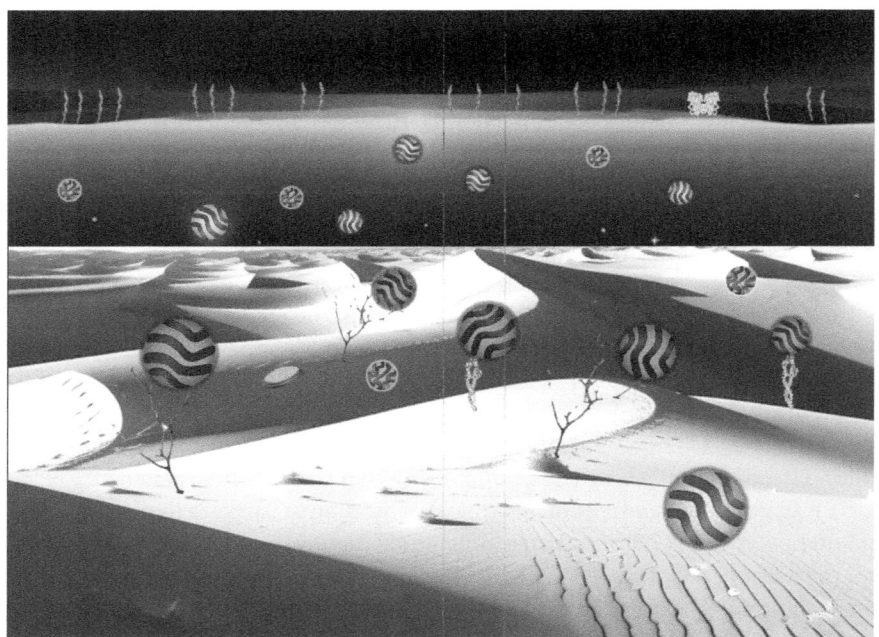

Reduced levels of HDL particles or issues with their composition can diminish HDL's ability to protect LDL particles from oxidation. Consequently, this can lead to increased uptake of LDL by cells, resulting in the formation of foam cells and heightened inflammation within the arterial wall.

Apolipoprotein A-I's ineffectiveness at deactivating oxidized lipids could be why HDL particles high in triglycerides do not protect against damage. The shape of apolipoprotein A-I is altered when triglycerides replace cholesteryl ester in the core. This may affect the ability of certain parts of A-I to interact with oxidized lipids. In addition, certain parts of apolipoprotein A-I may become oxidized when there is an oxidative stress in the arteries. This can prevent HDL from deactivating oxidized lipids. When sugar levels in the blood become too high, apolipoprotein A-I can undergo glycation. This may also cause HDL to lose its protective properties.

There may be changes in enzyme activity that makes HDL less efficient at breaking down oxidized lipids. The activities of PAF-AH and PON1 enzymes are actually lower in HDL of people with diabetes than healthy people. PON1 also undergoes greater glycation when people have diabetes.

Studies have shown that HDLs lose their ability to protect LDLs from damage by cells of the arterial walls during acute inflammation. This deficiency is primarily due to the replacement of apolipoprotein A-I with acute-phase proteins such as SAA. This protein is found mainly in small, dense HDL particles. They might not function well under inflammatory conditions. In addition, the loss of HDL antioxidative function during acute inflammation can be linked to a decrease in the activities of certain enzymes.

HDL's ability to resist damage from free radicals can also be affected by changes in its lipids. These changes can include an increase in saturated fatty acid and sphingomyelin, which both make the surface of HDL rigid. This can reduce HDL's capacity to remove from LDL harmful oxidized lipids.

Insufficient Reduction of Inflammation

HDL is a powerful anti-inflammatory agent. HDL's anti-inflammatory properties may be impaired in dyslipidemic people. In diabetes, the glycation of apolipoprotein A-I can lead to a loss of HDL function. Apolipoprotein A-I may also be oxidized to a large extent, which can contribute to the loss of HDL's anti-inflammatory properties. Apolipoprotein C-III may contribute to this deficiency in diabetes.

HDL's anti-inflammatory properties can become weaker and even pro-inflammatory during an acute-phase response. It can cause inflammation rather than reduce it. In this condition, HDL cannot stop the production of molecules which cause inflammation in arteries, such as oxidized lipids. Additionally, HDL can undergo changes in its function by acquiring important components. Through mass spectrometry analysis, researchers discovered the presence of hemoglobin beta and alpha chains in pro-inflammatory HDL. This particular type of HDL contains a significant amount of hemoglobin in its oxyhemoglobin state, along with distinct chemical and physical properties.

HDL's capacity to inhibit oxidation is reduced during acute inflammation. This may be a part of a defensive mechanism that activates pathways to kill bacteria by promoting oxidation. This immune response can involve a decrease in HDL-cholesterol, a rise in HDL triglycerides, and changes to the HDL protein components. These changes can affect HDL's ability to remove cholesterol from the cells, and its ability to fight oxidative stress and inflammation. These HDL modifications may be intended to move cholesterol from the liver into immune cells such as macrophages to enhance their function during infection.

The body's reaction to a disease or injury can be beneficial in the short term but harmful over the long term. When the body's reaction is unable to repair the injury (like when an atherosclerotic plaque develops in a blood vessel), it can lead to long-term changes in the cholesterol levels. This can accelerate the formation of plaque, which in turn leads to heart disease. HDL can cause cardiovascular disease when it is unable to reduce inflammation as it should. It may be why older people who had problems with their blood vessels and inflammation when they were young are more likely than younger people to develop heart disease. Changes in cholesterol levels seen in metabolic syndrome patients may be an evolutionary response to repair the damage.

HDL is not working properly in those at risk of heart disease, or who have already been diagnosed with it. For example, HDL is no longer able to counteract oxidized LDL's negative effect on aortic cell relaxation in patients with diabetes and obesity. HDL in patients with stable coronary disease or acute coronary syndrome cannot stimulate the production of nitric oxide that relaxes blood vessels. This could be because of changes in HDL's composition and the way it binds to cells. Modifications in apolipoprotein A-I, due to factors such as high blood sugar and oxidation, can cause HDL to malfunction.

15

HDL – What To Do To Correct It

HDL is an attractive target in the treatment of heart disease. Increasing HDL-cholesterol levels has been thought to reduce the risk of heart disease. Research also indicates that improving HDL function can be achieved by raising HDL-cholesterol and apolipoprotein-A-I levels, as well as by normalizing HDL processing in the body.

This is a particularly interesting idea because even though people may be treated with statins or other drugs that lower LDL-cholesterol, they still face a high risk of cardiovascular issues. Even if people meet the LDL-cholesterol goals, their blood pressure and blood sugar can still indicate a high risk of heart events.

This idea, however, does not take into account the fact that low HDL-cholesterol levels are often associated with other conditions which increase the risk for heart disease – for example, high levels of lipoproteins rich in triglycerides. It is remarkable that some genetic conditions that are associated with low HDL-cholesterol levels do not increase cardiovascular risk. The data available on the cardiovascular risks of people with low HDL-cholesterol, including genetic forms, do not always support that increasing HDL-cholesterol is protective against heart problems.

Low HDL-cholesterol levels do not necessarily indicate a problem in the HDL particle's function. Theoretically, it is therefore possible to reduce the risk of heart disease by improving HDL function without increasing HDL-cholesterol levels. When evaluating the effectiveness of HDL therapies, it is important to consider both HDL-cholesterol and HDL function. It is unclear whether HDL-cholesterol should be used as a therapy target when there is an increase of it but no improvement in the removal of cholesterol from blood.

HDL-focused strategies should help reduce plaque formation and remove excess cholesterol. This can result in a reduction of inflammation and the stabilization or regression of plaques. It is important to increase HDL's ability to remove cholesterol, as well as its ability to combat inflammation and oxidative stress. A treatment that targets HDL should speed up the transfer of cholesterol from arterial walls into the bile. This treatment should also reduce inflammation in the arterial wall and oxidative stress.

> **Open Question**
>
> Can the risk of heart disease be decreased by selectively enhancing specific HDL functions that protect against atherosclerosis? Can this be achieved by accelerating the removal of cholesterol from arterial wall cells?

Researchers have carried out several large-scale studies to determine whether raising HDL-cholesterol levels can reduce heart disease and mortality. These trials were conducted

Cholesterol, Lipoproteins, and Cardiovascular Health: Separating the Good (HDL), the Bad (LDL), and the Remnant, First Edition. Anatol Kontush.
© 2025 John Wiley & Sons, Inc. Published 2025 by John Wiley & Sons, Inc.

at the end of the 20th century. The studies such as the Armed Forces Regression Study (AFREGS), the Bezafibrate Infarction Prevention (BIP) trial, the Veterans Affairs High-density lipoprotein cholesterol Intervention Trial (VA-HIT), the HDL Atherosclerosis Treatment Study (HATS), and the Arterial Biology for the Investigation for the Treatment Effects of Reducing Cholesterol (ARBITER) trials have shown that raising HDL-cholesterol levels may slow the progression of artery blocking and reduce the risk of cardiovascular problems. These studies were conducted on patients who had not been treated with statins or other lipid-lowering medications. In these studies, however, increasing HDL-cholesterol led to a decrease in other types of lipids in the blood that are associated with heart disease, such as triglycerides and LDL-cholesterol. It was unclear whether raising HDL-cholesterol alone could protect against heart diseases without impacting other blood lipids. Interestingly, the inverse relationship between plasma triglycerides and HDL-cholesterol indicates that raising HDL-cholesterol lowers triglycerides. This, in turn, may lower remnant cholesterol, which is harmful to the heart. However, later studies in patients taking statins largely yielded negative results and did not confirm that increasing HDL-cholesterol is an effective way to combat heart disease.

Many studies indirectly supported the notion that increasing HDL-cholesterol can be beneficial for cardiovascular health. However, a large meta-analysis of 299,310 participants in 108 trials involving lipid-modifying medication found no correlation between HDL-cholesterol changes and heart disease or mortality. Changes in LDL-cholesterol, on the other hand, had an impact on heart disease. A second study showed that HDL-cholesterol changes did not significantly affect stroke risk.

Scientists have looked at how lower cardiovascular risk and higher HDL-cholesterol levels are connected in different groups of people. This was done by analyzing data from different studies. Research shows that heart disease risk in the United States has decreased significantly from 1976–1980 until 1988–1994. Then, it further improved slightly from 1988–1994 till 1999–2004. At the same time, HDL-cholesterol levels increased from 50 mg/dl to 51 and 52 mg/dl. Over a 10-year period, the Framingham Heart Study showed that dyslipidemia decreased when HDL-cholesterol levels increased and triglyceride levels dropped. In this study, the average HDL-cholesterol levels of both men and women were increased by 2 and 3 mg/dl, respectively. These changes predicted future events.

Another study used data from the Framingham Offspring Study between 1975 and 2003 to determine if raising HDL-cholesterol levels with lipid therapy can reduce the risk of cardiovascular problems. The study showed that a 1% increase in HDL-cholesterol was associated with a 2% drop in cardiovascular risks, similar to what other studies found. The study was limited by a few factors, such as the small number of cardiovascular incidents and other factors which could have affected the results. However, it provided evidence that increasing HDL-cholesterol reduces the risk of heart disease.

Researchers examined the relationship between the changes in HDL-cholesterol levels and the risk for heart disease among another group consisting of 4,501 male physicians. The researchers measured the HDL-cholesterol level in 1982 and then again 14 years later. Results showed that those with an increased HDL of +2.5 to +12.5 mg/dl were at a lower risk of developing heart disease by 44%, and those with >= +12.5 mg/dl were at a lower risk by 57%. Another study, which looked at stable heart patients who received different medications or placebos over a period of 30 months, confirmed these findings. Even after taking into account changes in LDL-cholesterol levels, the analysis revealed that higher HDL-cholesterol levels were predictive of a lower risk for cardiovascular events. For every 1% rise in HDL-cholesterol, the number of events decreased by 2%.

Understanding why HDL-cholesterol levels increase is important. Genetic studies indicate that some genetic factors that raise HDL-cholesterol do not actually reduce the risk of cardiovascular disease. The best ways to raise HDL-cholesterol levels are by improving the way HDL is produced and processed in the body. One way to do this is by accelerating the rate at which apolipoprotein A-I acquires lipids. Fibrates, statins, and niacin are some of the drugs that can influence this step. Another method would be to reduce lipid exchanges between HDL and lipoproteins rich in triglycerides, as well as to prevent the kidney or the liver from removing HDL. CETP inhibitors can be used as an example of this kind of treatment.

The types of HDL particles that are produced by treatments that accelerate HDL formation or improve its processing can vary. As an example, apolipoprotein A-I mimetics and reconstituted HDL primarily increase smaller HDL particles that have less cholesterol. CETP inhibitors, statins, and niacin however increase larger HDL particles with more cholesterol. It may be more important to have small HDL with more function than large HDL with lots of cholesterol.

Lifestyle

There are several lifestyle changes that can effectively increase HDL-cholesterol. Diet plays a central role in achieving this goal. However, exercise has a more significant impact on HDL-cholesterol compared to dieting alone. Quitting smoking also contributes to higher HDL-cholesterol in the bloodstream. Additionally, losing weight and consuming moderate amounts of alcohol are also key components of these lifestyle changes.

Exercise

A regular aerobic exercise program can increase HDL-cholesterol levels from 6% to 25%, with an average increase of 4.6%. Acute aerobic exercises can lead to an increase as high as 43%. It is common among doctors to recommend regular aerobic exercise for raising HDL-cholesterol levels. The effect of the exercise increases with the number of exercise hours. A sedentary lifestyle and lack of exercise are linked to low HDL-cholesterol levels. Physical training can increase HDL-cholesterol levels in people with normal cholesterol as well as in those with cardiovascular disease. Exercise can have a different effect on each individual. For example, men and people with heart disease may respond better to exercise.

It is interesting that physical activity has a greater effect on reducing heart disease risk than what would be expected based on the studies on the relationship between heart disease and HDL-cholesterol. According to epidemiological studies, for every increase of 1 mg/dl in HDL-cholesterol, cardiovascular risk is reduced by 2–3%. Exercise only increases HDL-cholesterol by 1–2 mg/dl in large-scale studies. This would lead to only a 3–6% risk reduction, much less than up to 40% observed in the studies. This indicates that other factors, besides HDL-cholesterol, contribute to the heart-protective effects of exercise. For example, exercise can lower triglyceride, which is also beneficial to the heart.

Exercise reduces the activity CETP, which is likely why it increases HDL-cholesterol levels in the blood. Exercise can also influence other proteins which participate in HDL processing, like hydrolytic lipases. Regular and acute exercise both increase the activity of lipoprotein lipase which is responsible for producing more HDL-cholesterol. Exercise also increases LCAT and PON1, both of which are linked to HDL-cholesterol. Exercise can help improve HDL-cholesterol function and remove cholesterol from the cells. Exercise can protect LDL from damage due to oxidative stress. Exercise may also have an anti-inflammatory effect on HDL.

Alcohol

Numerous studies have shown that alcohol consumption can raise HDL-cholesterol levels in the blood. Men who have two drinks per day (about 20–30 grams of alcohol) or women who have one drink per day (about 10–15 grams of alcohol) can increase HDL-cholesterol levels by 10%.

HDL-cholesterol may increase as people drink more alcohol. The HDL-cholesterol level can be increased by up to 15% if they drink 30–40 grams alcohol per day. However, alcohol can also have a negative effect on the triglyceride level in the blood. In many studies, moderate alcohol consumption was linked with a lower risk of cardiovascular diseases. This makes the effect on HDL-cholesterol levels potentially important for public health. However, there are studies that indicate the positive impact of moderate alcohol consumption on heart disease is mainly attributed to the favorable social and economic factors associated with this lifestyle.

Moderation in alcohol consumption is associated with a minimal risk for the heart. Alcohol consumption in excess, which is equivalent to four drinks a day or more, can increase the risk of heart problems. It is still true, even if it raises HDL-cholesterol. This increased risk is due to the negative effects of drinking heavily on other risk factors such as blood pressure and inflammation.

Moderate alcohol consumption has a greater effect on small HDL3 than large HDL2 levels. Heavy drinking, on the other hand, is linked to a preference of elevating HDL2 level.

Alcohol can alter the composition of HDL. It tends to raise the content of polyunsaturated fats and phospholipids in HDL. Alcohol can also affect HDL's ability to remove cholesterol from macrophages and other cells. It can also influence other steps in the removal of cholesterol from the body, including increasing ABCA1 expression and increasing the activity of certain enzymes.

Nutrition

Many nutrients modify the levels of HDL-cholesterol. Different fats and carbohydrates strongly influence HDL-cholesterol. Other nutrients like vitamins and polyphenols can also affect how HDL works and help increase its levels.

There are ways to improve HDL-cholesterol, such as exercising, eating fish, drinking red wine in moderation, consuming fruits, and being female.

Fats in Food

Different types of fats may increase cholesterol in the blood and therefore the risk of cardiovascular disease. Diets high in fat tend to raise HDL- and LDL-cholesterol. The ketogenic diet is high in fat and low to moderate in protein. It contains very little carbohydrate. This diet shows how strongly HDL- and LDL-cholesterol levels can be raised in certain individuals, especially in lean healthy subjects.

Low-fat diets can lower HDL- and LDL-cholesterol. However, this is not always true, since losing weight and reducing fat intake may keep HDL-cholesterol stable. Low-fat diets can also speed up the breakdown and slow down the production of apolipoprotein A-I. A low-calorie, high-fat diet may also affect cholesterol levels and reduce the body's ability to remove cholesterol from the blood. It can also decrease the activity of PON1, which helps to protect against heart disease. The diet may also alter the profile of HDL in the blood, by decreasing the production of pre-beta HDL.

Cholesterol

HDL-cholesterol increases when we consume more cholesterol. This is due primarily to higher HDL2-cholesterol. Increasing dietary cholesterol can raise LDL-cholesterol higher than HDL-cholesterol, leading to an unhealthy level of lipids in the blood and a faster development of cardiovascular disease. Sources of dietary cholesterol can also influence the balance between HDL- and LDL-cholesterol. For example, adding one egg per day to a normal diet with regular fat content can raise HDL-cholesterol without affecting LDL-cholesterol.

Phospholipids

Different types of phospholipids, such as those from soybeans, safflowers, eggs, and fish roe, can have a positive effect on blood lipid levels, especially on HDL-cholesterol. These phospholipids mainly work in the intestines by interfering with the absorption of cholesterol. They can also increase the secretion of bile acids and cholesterol.

Saturated Fat

When it comes to regulating lipid concentrations in the blood, dietary fatty acids are even more effective than dietary cholesterol. Consuming higher amounts of most fatty acids can raise HDL-cholesterol, with saturated fatty acids having the strongest impact. Increasing saturated fat intake compared to monounsaturated fats and polyunsaturated fats or carbohydrates can raise HDL-cholesterol in both animals and humans, as long as dietary cholesterol intake remains constant. This effect may be due to a slower removal of apolipoprotein A-I from the blood.

However, reducing overall fat intake, especially saturated fat and cholesterol, can lower HDL-cholesterol by decreasing the production and secretion of apolipoprotein A-I by the liver. Saturated fat can also negatively affect the anti-inflammatory properties of HDL and impair blood vessel function in humans.

Monounsaturated Fat

When saturated fats are replaced with monounsaturated fats in the diet, it can improve the blood lipid profile by lowering the ratio of LDL-cholesterol to HDL-cholesterol. When monounsaturated fats are used instead of carbohydrates, they can increase HDL-cholesterol and decrease the ratio of total cholesterol to HDL-cholesterol. Adding monounsaturated fats to the diet can also raise HDL-cholesterol. On the other hand, trans isomers of monounsaturated fats, which are created during hydrogenation, have negative effects on blood lipids compared to their cis counterparts. When consumed, trans-monounsaturated fatty acids typically increase LDL-cholesterol and decrease HDL-cholesterol.

Polyunsaturated Fat

It is possible to increase HDL-cholesterol and lower the ratio of total cholesterol to HDL-cholesterol by replacing carbohydrates with polyunsaturated fats. However, if saturated fats are replaced by polyunsaturated ones, total cholesterol, LDL-cholesterol, and HDL-cholesterol can be decreased.

This is because all fatty acids increase HDL-cholesterol when substituted for carbohydrates, but the effect becomes weaker as the fatty acids become more unsaturated.

The addition of omega-3 fatty acids, such as eicosapentaenoic, to the diet can increase HDL-cholesterol. This increase is small and usually does not exceed 5%. The increase in HDL-cholesterol caused by polyunsaturated fats is in line with studies showing that higher levels of n-6 and n-3 polyunsaturated fatty acids in the blood correlate with higher HDL-cholesterol. Polyunsaturated fats increase HDL-cholesterol mainly by increasing the cholesterol content of large HDL2 particles. The addition of n-3 polyunsaturated fats shifts cholesterol from smaller HDL3 to larger HDL2.

Polyunsaturated fats can raise HDL-cholesterol by reducing the production of triglyceride-rich particles and increasing the production of spherical HDL particles. They also increase the activity of LCAT. Polyunsaturated fats can also help normalize HDL processing in the blood by affecting hepatic lipase and lipoprotein lipase. These effects could be beneficial to heart health.

A diet rich in polyunsaturated fatty acids can also help reverse cholesterol transport. This process can be accelerated in mice using fish oil rich in polyunsaturated fats. Fish oil helps improve the ability of HDL to remove cholesterol from macrophages. ABCA1 is responsible for this effect. Fish oil is thought to make HDL particles' surfaces more fluid. This allows them to remove cholesterol better from cells. Consuming polyunsaturated fatty acids has also been shown to increase HDL's anti-inflammatory properties in humans.

These fats also have more positive effects on heart health. They can lower blood pressure, improve immunity, and decrease platelet aggregation. These effects are independent of their impact on HDL or triglyceride concentrations in the blood. These actions are what might explain why polyunsaturated fats can help reduce cardiovascular disease risk.

In general, clinical trials have shown that substituting saturated fats for polyunsaturated ones leads to a reduction in cardiovascular events. Some studies, however, have found no significant effects. It is therefore recommended that polyunsaturated fats should account for around 10% of your total energy intake.

A recent systematic review of controlled trials found that omega-3 dietary supplements did not have any effect on cardiovascular events, or mortality from all causes. However, they significantly reduced cardiovascular deaths by 7%. Eicosapentaenoic acid ethyl ester by itself significantly reduced cardiovascular mortality and events by 18%.

Carbohydrates

Too much carbohydrate consumption can cause unhealthy cholesterol levels. A high carbohydrate intake has been linked to lower levels of HDL-cholesterol and higher levels of blood triglycerides. Diets that have a low glycemic index and do not spike blood sugar levels are good for the heart because they increase HDL-cholesterol. The added sugars found in processed food are an important part of our daily diet, and they can have a negative impact on blood cholesterol. Sugar can also cause chronic inflammation of the arterial walls if consumed in excess.

In studies, it was found that too much sugar added to food is linked with lower HDL-cholesterol. Sugary drinks have also been associated with lower HDL-cholesterol. Too much fructose (a sugar bonded to glucose to form the disaccharide sucrose) can cause health issues and increase the production of LDL-cholesterol. It is recommended that sugar added to foods be reduced in order to improve blood cholesterol and treat hyperlipidemia. Carbohydrate-restricted diets, where only 10–15% of energy comes from carbs, have been shown to help with weight loss and improve cholesterol concentrations.

By contrast, the Mediterranean diet, which is low in carbohydrate, has been shown to raise HDL-cholesterol. In 2003, the World Health Organization advised that no more than 10% of calories should come from sugar. In this respect, fructose is particularly harmful as it is processed differently by the body than glucose.

Polyphenols

Polyphenols, which are found naturally in plants, can increase HDL-cholesterol and improve HDL function. Polyphenols are found in berries like blueberries and strawberries. In a study, moderate consumption of berries over eight weeks raised HDL-cholesterol by 5.2% among middle-aged adults without changing total cholesterol or triglyceride levels. Grapefruits and grapefruit juice, which are rich in polyphenols, can also increase HDL-cholesterol. Anthocyanins, a polyphenol type found in berries, play a role in raising HDL-cholesterol.

Dark chocolate contains polyphenols, also known as flavonoids. After a week of dark chocolate consumption, healthy volunteers saw their HDL-cholesterol increase by 9% while LDL-cholesterol decreased by 6%.

Polyphenols are also found in vegetables like garlic. Taking garlic powder tablets daily increased HDL-cholesterol levels by 11.5% and decreased LDL-cholesterol by up to 13.8% in patients with mildly high levels of cholesterol.

Vitamins

If there is an effect, it is not very strong. Taking vitamin A, C, D, or E supplements does not change HDL-cholesterol levels in blood consistently for different groups of people.

Therapy

Apolipoproteins

HDL's apolipoproteins are important for its biological actions on cells and tissues. Apolipoprotein A-I, among the various types of apolipoproteins found in HDL, has a number of beneficial effects and can help slow the progression of atherosclerosis. This leads to the idea that increasing apolipoprotein A-I levels may be more effective than increasing HDL-cholesterol.

When apolipoprotein A-I is combined with lipids, it forms small particles of pre-beta HDL. These particles start the process of removing cholesterol from the body. Animal experiments show that injecting apolipoprotein A-I directly into the bloodstream reduces inflammation and makes atherosclerotic lesions more stable. However, this method is not without its challenges. These include rapid kidney filtration and the difficulty of manufacturing apolipoprotein A-I free from impurities.

A possible alternative is to increase the levels of the apolipoprotein through gene therapy. This can help remove cholesterol from cells and improve the structure of the lesions.

Using viral vectors for the delivery of the *APOA1* gene is a promising method to increase the levels of apolipoprotein A-I and HDL and reduce atherosclerosis. The studies on animals that carry this gene provide solid evidence that HDL is directly involved in the slowing of atherosclerosis. Mice possessing human apolipoprotein A-I are protected against atherosclerosis as shown in 1991 by Edward Rubin and Ronald Krauss at University of California, Berkeley. These data, however, are not available for humans.

Reconstituted HDL

Reconstituted HDL is an artificial version of HDL that is made in the lab from purified human apolipoprotein A-I, phospholipids, and possibly some other proteins and lipids. Reconstituted HDL can be injected into the bloodstream to quickly increase HDL-cholesterol and reduce atherosclerotic lipid plaque. This method is similar to using apolipoprotein A-I infusions, except that in the latter case, the protein is infused without lipids. This therapy is considered promising as it gives patients HDL externally to stabilize their lesions, especially if they have heart problems.

Reconstituted HDL, when administered to humans leads to an accumulation of small HDL particles, which are also present in people who have apolipoprotein A-I Milano. These small HDL particles are thought to have strong heart-protective effects.

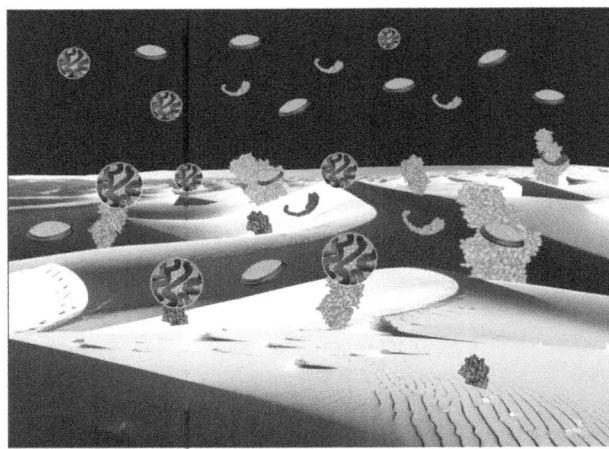

Improved function of cellular cholesterol transporters like ABCA1, ABCG1, and SR-BI can help remove cholesterol from cells. This can be done by increasing the presence of cholesterol acceptors like reconstituted HDL, apolipoprotein A-I or apolipoprotein A-I mimetics in the bloodstream. This not only raises HDL-cholesterol but can also provide protection against heart attacks.

Another way to boost HDL levels is by selectively removing the lipids from large HDL particles and then re-infusing them into the body. The blood is treated with sevoflurane, butanol, and other chemicals to convert large HDL particles rich in lipids into smaller, lipid-poor HDL. This method increases the pre-beta-1 HDL levels in the blood. This selective delipidation improves the ability of HDL to remove cholesterol from cells via the ABCA1 transporter. Delipidated plasma was used in animal studies to successfully reduce aortic atherosclerosis.

When HDL is reconstituted in a laboratory, it has all of the same biological properties as natural HDL. This includes removing cholesterol from the body, protecting against inflammation, oxidative damage, cell death, and infections. It also protects from blood clots and diabetes. Reconstituted HDL helps reduce plaque buildup in arteries. It does this by removing cholesterol quickly from arterial walls and sending it to the liver. A reconstituted HDL called CSL-111 (human apolipoprotein A-I combined with phosphatidylcholine) given intravenously over a period of four hours, can have immediate effects on the plaque characteristics of patients with peripheral arterial disease. This includes a reduction in lipid content and macrophages, as well as reducing local inflammation.

Reconstituted HDL has many potential benefits in reducing plaques. It can attenuate inflammation and increase blood vessel size. Reconstituted HDL has antioxidant properties, which can protect cells following a heart attack and improve heart function. It can also improve blood flow, neutralize toxins in those with high cholesterol, reduce platelet activation and improve sugar metabolism.

Reconstituted HDL therapy can be an effective way to reduce atherosclerotic plaque size. Apolipoprotein A-I Milano in reconstituted HDL can be particularly effective for treating cardiovascular disease. This is true in cases of acute conditions such as unstable plaques. Patients with acute coronary symptoms who received five weekly injections of apolipoprotein A-I Milano combined with phosphatidylcholine showed a reduction of 4.2% in plaque volume. This is a better result than a statin treatment, which is a gold standard approach available for atherosclerosis. In a study with coronary artery patients, CSL-111 also showed positive results on coronary atherosclerosis.

Later studies like the recent ApoA-I Event Reducing in Ischemic Syndromes II (AEGIS-II) trial of CSL-112 did not, however, find any benefit from reconstituted HDL in acute heart disease. These studies were performed on patients who had been treated with statins. This is the standard of care today. Statin treatment may have negated the benefits of reconstituted HDL that were observed in previous studies. High costs and the potential for liver damage are also limiting factors in developing these treatments.

Apolipoprotein Mimetic Peptides

Apolipoprotein A-I mimetic peptides are small molecules resembling apolipoprotein A-I that have similar effects on the body. These peptides can be taken orally or intravenously to increase HDL-cholesterol and reduce atherosclerosis risk. It is easier to produce and administer small peptides than the full-size apolipoprotein-A-I molecule. In the 1980s, these peptides first appeared to mimic certain parts of apolipoprotein A-I. They have evolved over time to better model apolipoprotein A-I and to better interact with lipids.

Apolipoprotein mimetic peptides can come from different proteins like apolipoprotein E and apolipoprotein J, but those derived from apolipoprotein A-I are most commonly used. These peptides were shown to reduce the atherosclerosis in animals such as mice, rabbits, and monkeys. They improve how HDL functions in the body by removing cholesterol from cells, increasing enzyme activities, reducing inflammation, and improving blood vessel function.

Apolipoprotein A-I mimetics like D-4F or 5A can activate various steps in the removal of cholesterol from the body. They can remove cholesterol from the arterial cells, aid in the processing of cholesterol by LCAT, and improve the delivery of cholesterol to the liver. They do this by forming pre-beta HDL particles with a low content of lipids. These particles play a key role in preventing plaque buildup. A strong anti-inflammatory effect is attributed to the 4F peptide's ability to bind oxidized lipids which provoke inflammation. Apolipoprotein A-I mimetics can also improve sugar processing and prevent blood clots.

Human trials of apolipoprotein A-I mimetics have been conducted after animal studies showed promising results. They were found to be safe and well tolerated. However, their bioavailability was low for clinical effects, with D-4F having a rate below 1%. Additionally, apolipoprotein A-I mimetic peptides are not stable in the digestive system. The molecules are rapidly removed from the body via the kidneys. This results in only temporary elevations in HDL-cholesterol. The mimetics can also cause high triglyceride due to the inhibition of lipoprotein lipase. . In order to administer reconstituted HDL or apolipoprotein A-I mimetic, the agents need to be injected directly into veins. It is expensive and difficult to perform this procedure, as it requires a large amount of biological material.

Statins

Statins, in addition to their effect on LDL-cholesterol, can also slightly increase HDL-cholesterol. The effects on HDL vary depending on the statin and the dose taken. Statins increase HDL-cholesterol by increasing apolipoprotein A-I and decreasing CETP activity. The reduction in LDL-cholesterol and increase in HDL-cholesterol are not directly related, indicating that statins work through different mechanisms to affect these markers of risk.

Statins can have a positive effect on HDL in the body. They increase the activity of LCAT which helps HDL to mature and decrease the PLTP activity which breaks down HDL.

Statins may also have other positive effects, such as reducing inflammation or oxidative stress. This is related to the increased activity of HDL-associated enzymes. Statins may also improve blood vessel function by activating the production of nitric oxide. It appears likely that beneficial effects of statins on HDL have a potential to negate the effects of other therapies directly targeting this lipoprotein.

Fibrates

Peroxisome proliferator-activated receptors (PPARs) alpha, delta, and gamma are a group of proteins that control many genes involved in how the body processes lipids, make adipose cells, respond to sugar, and deal with inflammation in blood vessels. PPARs act as switches to turn these genes on when activated by molecules called peroxisome proliferators.

Fibrates is a drug type that activates the PPAR alpha receptor. They have a moderate effect on the way the body processes lipids. Fibrates can increase HDL-cholesterol over a long period of time. This is unlike HDL mimetics or reconstituted HDL which only produce temporary increases. In studies that involved 53 large groups of patients, fibrates increased HDL-cholesterol by an average of 10%. Fibrates are thereby less effective than other drugs, such as CETP inhibitors and niacin, at increasing HDL-cholesterol. Fibrates can also alter the profile of HDL particles that are in the blood by increasing the small- and medium-sized HDL particles.

What It Is

Peroxisome proliferator-activated receptors, in short PPARs, are a group of proteins that regulate genes related to lipid processing, adipose cell function, insulin response, and inflammation in blood vessels.

PPAR alpha activation has a variety of effects on the body, including increasing the breakdown and processing of lipids. It also reduces inflammation. This increases the production of apolipoproteins A-I and A-II. Fibrates also reduce the levels of triglycerides. This may increase HDL-cholesterol by preventing the transfer of cholesteryl ester from HDL to triglyceride-rich lipoproteins.

Fibrates are a type of medication that can help reduce the risk of heart problems. However, the results of studies on their effectiveness have been mixed. Some earlier trials showed that drugs like gemfibrozil and clofibrate had positive effects in preventing heart issues. But other trials with drugs like bezafibrate, fenofibrate, and pemafibrate had negative results.

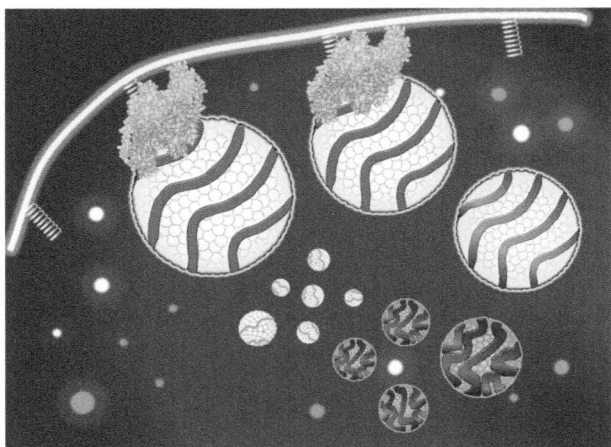

When triglyceride-rich lipoproteins like chylomicrons or VLDL are broken down more effectively by lipoprotein lipase, it leads to higher flux of cholesterol and other surface remnants to HDL. This helps to speed up the removal of cholesterol from the bloodstream, preventing it from building up in the arterial walls.

For example, in the Fenofibrate Intervention and Event Lowering in Diabetes (FIELD) trail, fenofibrate did not significantly reduce the risk of heart problems in patients with type 2 diabetes. In fact, it increased the chances of certain side effects like pancreatitis and pulmonary embolism. However, it lowered the overall number of cardiovascular events, mainly by reducing nonfatal heart attacks and procedures to open blocked arteries. As studies of reconstituted HDL, trials of fibrates were performed in patients treated with statins. This could reduce clinical benefits of fibrates observed in the pre-statin era.

Fibrates reduced heart problems in patients with metabolic syndrome. This was especially true for those who had high triglycerides, were overweight, and had low HDL-cholesterol. Fenofibrate, in the FIELD study specifically, was more effective with patients who had low HDL-cholesterol or hypertension. Fibrates may benefit people with metabolic symptoms.

Recent systematic reviews found that fibrates could reduce the risk for heart disease by 14% but did not appear to have any effect on the risk of death from heart problems. Fibrates do not appear to affect heart health when taken in conjunction with statins.

Niacin

Since almost 50 years, niacin (also known as nicotinic acid and vitamin B3) has been used to regulate lipid processing within the body. It was once the most powerful medication for increasing HDL-cholesterol. Niacin at a dose of 2 grams per day increases HDL-cholesterol by as much as 35%. According to studies, niacin can increase HDL-cholesterol by 16% on average and lower triglyceride by 20%. Niacin is also able to slightly lower LDL-cholesterol and Lp(a).

It works by temporarily decreasing the breakdown of lipids within adipocytes. This is done by blocking a receptor called HM74 that is found in human adipose tissues. It lowers the levels of fatty acids in the blood which are important for the production of triglycerides by the liver. Niacin is also shown to help break down lipoproteins rich in triglycerides. The effect of niacin is however only temporary.

Niacin increases HDL-cholesterol by reducing the levels of triglycerides in the blood. Niacin also stimulates macrophages to remove cholesterol through the ABCA1 membrane transporter, contributing to the increase in HDL-cholesterol.

Niacin can have multiple effects on HDL, leading to higher levels of large HDL2 particles and small HDL3 in the blood. Niacin, however, can temporarily decrease insulin sensitivity. This causes higher blood levels of glucose. In studies with a limited number of patients, niacin was shown to be beneficial for a variety of cardiovascular outcomes. The Coronary Drug Project was the first to test its use, and it showed promising results. Niacin could slow down the progression of atherosclerosis when used alone or in combination with statins. However, later large-scale studies of niacin in combination with statins provided essentially negative outcomes.

CETP Inhibitors

CETP inhibitors are small molecules that stop CETP from moving lipids between HDL and lipoproteins containing apolipoprotein B. These agents work really well in increasing HDL-cholesterol, and some can even raise it by more than 100%. The idea of using CETP inhibitors to raise HDL-cholesterol comes from studying people who have a genetic lack of CETP. These people often have high levels of HDL-cholesterol and low levels of LDL-cholesterol, which is considered to be good for preventing heart disease. Also, this lack is connected to having larger HDL particles, which typically lowers the risk of heart problems.

Several CETP inhibitors have been developed like torcetrapib, dalcetrapib, evacetrapib, anacetrapib, and obicetrapib. These inhibitors have the ability to increase HDL-cholesterol while decreasing LDL-cholesterol and apolipoprotein B. Each of these drugs works in a slightly different way to block the activity of CETP. Torcetrapib forms an inactive complex with HDL, dalcetrapib binds irreversibly to CETP through a chemical bond, and anacetrapib binds reversibly. Consequently, both torcetrapib and anacetrapib are good at stopping CETP from transferring cholesteryl ester and triglycerides. However, dalcetrapib is not as effective in this regard.

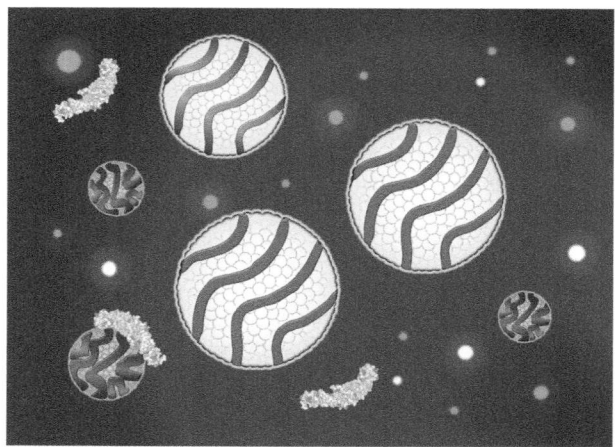

CETP has the ability to transfer cholesteryl esters from HDL to lipoproteins rich in triglycerides.Small molecules can decrease this transfer by preventing CETP from moving lipids between lipoproteins. As a result, more cholesteryl esters stay in HDL, leading to increased levels of HDL-cholesterol.

The first CETP inhibitor tested in clinical trials, torcetrapib, showed positive results for improving blood lipid levels. The molecule together with other CETP inhibitors was found to decrease CETP activity and increase HDL-cholesterol in healthy people. They also increased the size of HDL particles and levels of apolipoprotein A-I, prolonging its presence in the bloodstream.

CETP inhibitors can be particularly effective in reducing lipoproteins that are rich in triglycerides. The molecules help the body eliminate lipoproteins containing apolipoprotein B. Anacetrapib, however, is an exception. It does not affect triglyceride concentrations but reduces Lp(a) when taken with or without atorvastatin. CETP inhibitors can also change the composition of VLDL by increasing their content in triglycerides. This makes it easier to remove these particles from the blood by lipoprotein lipase. The LDL is then formed. CETP inhibitors also help break down chylomicrons, their remnants, and increase the levels of large HDL2 particles, which is considered beneficial for heart health.

There is ongoing discussion about whether CETP inhibitors can effectively prevent atherosclerosis by promoting the removal of cholesterol from tissues and arteries to the liver. Two ways are available to send HDL-cholesterol to the liver. It can go directly from HDL to the liver through SR-BI in a process called selective uptake. It can also be transferred indirectly to other lipoproteins, such as VLDL, IDL, and LDL, through CETP and then taken by the liver via the LDL receptor. The indirect transfer of HDL-cholesterol is the main way cholesterol returns from tissues to the liver in humans. This pathway is even more important if the LDL receptor activity increases, as it does when statins are taken. When CETP is blocked, however, this indirect path seems to slow down rather than speed up. Combining statin therapy and high LDL receptor activities with CETP inhibition may produce an even greater effect. Even if the indirect route is reduced, HDL may increase direct transport of cholesterol.

People with insufficient CETP activity show similar results as those who use CETP inhibitors. A person with a CETP deficiency has larger HDL2 particles and higher levels of LCAT and apolipoprotein E. The large HDL particles containing apolipoprotein A-I are very good at delivering cholesteryl esters to the liver through the SR-BI receptor. Inhibiting CETP may increase the ability of HDL to deliver cholesteryl ester to the liver by this route. In general, CETP inhibitors do not inhibit reverse cholesterol transport but they also do not accelerate it.

Clinical trials are important crash tests that help us understand how well a drug works and if it is safe. One trial called the Investigation Of Lipid Level Management To Understand Its Impact In Atherosclerotic Events (ILLUMINATE) was stopped early because more people were unexpectedly dying in the group taking the drug compared to those taking a placebo. Over 15,000 patients at high risk for heart disease participated in the trial. The patients were given either torcetrapib with atorvastatin or a fake pill with atorvastatin only over 550 days. Even though those taking torcetrapib had better cholesterol levels than the placebo group, the group that received it had more heart attacks and deaths. The death rate from cardiovascular causes increased by overwhelming +40%. The rate of cardiovascular events was up by +25%. And, the death due to other causes doubled. A second study, the Rating Atherosclerotic Change By Imaging with A New CETP Inhibitor (RADIANCE) trial, also showed that torcetrapib was linked to artery problems. These findings made people criticize using drugs like torcetrapib to treat heart disease.

Subsequent studies revealed that torcetrapib may have caused a rise in blood pressure, which could have negated the benefits of increasing HDL-cholesterol. These findings suggested torcetrapib could have unintended side effects by activating receptors in the body and causing hypertension. Other CETP inhibitors do not seem to have the same blood pressure effect, which indicates that this is not a problem common among drugs of this class.

These trials showed that we know little about how HDL protects against heart disease. In the ILLUMINATE study, the torcetrapib-treated group had more deaths from cancer (24 compared with 14 in the control group) and from infections (9 compared with 0). The drug seems to have caused these effects, as studies show that an increased HDL-cholesterol alone does not typically cause a higher death rate from non-cardiovascular diseases. Torcetrapib and other CETP inhibitors may impair endothelial functions or some other biochemical pathways.

Other clinical trials have further examined the clinical utility of CETP inhibitors. A Study of RO4607381 in Stable Coronary Heart Disease Patients With Recent Acute Coronary Syndrome (Dal-OUTCOMES) studied dalcetrapib on patients with recent heart problems. The studies did not show any significant effects of dalcetrapib on heart disease. In the Randomized EValuation of the Effects of Anacetrapib Through Lipid-modification (REVEAL) trial, patients were given either anacetrapib or a placebo, along with atorvastatin. The Determining the Efficacy and Tolerability of CETP Inhibition with Anacetrapib (DEFINE) study also looked at anacetrapib in patients with heart disease or at high risk for it. These studies showed that taking the drug slightly reduced cardiovascular events by 9%.

CETP inhibitors still have the potential to reduce cardiovascular risks in medical settings despite mixed results. They are considered promising because they can also reduce LDL-cholesterol, non-HDL-cholesterol, and triglycerides. Obicetrapib is a new drug of this class that can raise HDL-cholesterol and lower LDL-cholesterol, non-HDL-cholesterol, and triglycerides. Recently, a study showed that obicetrapib, when taken as a pill, can additionally reduce LDL-cholesterol by more than 50% when used with strong statin drugs. Ongoing trials are looking at the effectiveness of obicetrapib for different cardiovascular conditions.

History

Unexpected Therapeutic Failures

Who	When
Philip Barter and Pfizer	**2006**

Where	What
New York City NY, USA	Barter, P. J., M. Caulfield, M. Eriksson, S. M. Grundy, J. J. Kastelein, M. Komajda, J. Lopez-Sendon, L. Mosca, J. C. Tardif, D. D. Waters, C. L. Shear, J. H. Revkin, H. A. Buhr, M. R. Fisher, A. R. Tall and B. Brewer (2007) Effects of torcetrapib in patients at high risk for coronary events. New England Journal of Medicine 357: 2109-2022.

The link between low HDL-cholesterol and heart attacks was so well established by the end of the 20th century that researchers went to assess the HDL-raising hypothesis. They assumed that raising HDL-cholesterol in those with low levels would prevent cardiovascular disease. Some molecules in development led to high HDL-cholesterol levels. CETP inhibitors were the most powerful agents to increase HDL-cholesterol. There were few data on heart disease among people with very high HDL-cholesterol. Pharmacological companies, after the success of statins invested large sums of money in HDL, hoping for another breakthrough.

Researchers were stunned by the disappointing results of the ILLUMINATE study of torcetrapib, the molecule developed by Pfizer. Some predicted that CETP inhibitors could be useless, but the overall mood was positive and nobody anticipated such a failure. After the initial shock, scientists began asking the inevitable questions: What went wrong? Why and how did it happen?

Torcetrapib's ability to raise blood tension was the first thing that caught attention. Although initially considered to be clinically insignificant, the effects of torcetrapib were consistently observed across studies. Hypertension increases the risk of cardiovascular events. Torcetrapib's hypertensive effect could be responsible for the excess mortality.

Merck Research Laboratories observed that, in addition to elevated blood pressure, an increased concentration of aldosterone had a key physiological effect. Aldosterone, a hormone, is essential for the regulation of blood pressure. The human body's specialized adrenocortical cells secrete the molecule. Torcetrapib raised blood pressure in mice, dogs, and monkeys. The drug increased aldosterone production by isolated adrenocortical cell cultures, which suggested that hypertension was caused by the activation of this pathway. Anacetrapib was another CETP inhibitor studied by Merck. It did not however increase aldosterone production and could not raise blood pressure in animals. This leaves open the possibility of developing a CETP inhibitor that does not lead to hypertension.

The elevated blood pressure seen in studies with torcetrapib was not a sign of good things. The HDL field suddenly began to look ill. Studies on drug candidates that could raise HDL-cholesterol were still ongoing. HDL was viewed by many as an investment that was too big to fail and could not fail due to its strong epidemiology. In large-scale studies, several approaches to combating cardiovascular disease in people through HDL raising have been tested. Unfortunately, bad things never come alone. Within a few years, another HDL-raising drug failed spectacularly.

The failing agent was niacin. In the Atherothrombosis Intervention in Metabolic Syndrome With Low HDL/High Triglycerides and Impact on Global Health Outcomes (AIM-HIGH) trial, the molecule used to raise HDL-cholesterol in the past did not reduce the risk of heart issues in patients who had heart disease and were taking statins. The Heart Protection Study 2 – Treatment of HDL to Reduce the Incidence of Vascular Events (HPS2-THRIVE) trial, which included patients with atherosclerosis and statins, also found that niacin did not lower cardiovascular events. The drug also increased the risk for serious side effects. Endocrinologists have used niacin for years to correct abnormalities in lipoprotein metabolism. Niacin was highly effective in reducing cardiovascular disease during large-scale research in the 1970s. This included the Coronary Drug Project. However, it became almost useless by the 2000s. What has changed in this time period?

The studies from the 20th century and the 21st century were very different in terms of patients. While the pathology of heart disease was the same, the drug treatments had changed dramatically over the 30-year period between them. Statins were the key characteristic that distinguished modern patients from their predecessors. The statins developed in the 1980s were highly effective and soon became a gold standard in cardiovascular medicine. Researchers who wanted to develop novel drugs to reduce heart attacks had to do so only in patients taking statins. Some believed that statins were the reason why niacin did not provide any clinical benefit. At the time, there were no data available to support this hypothesis.

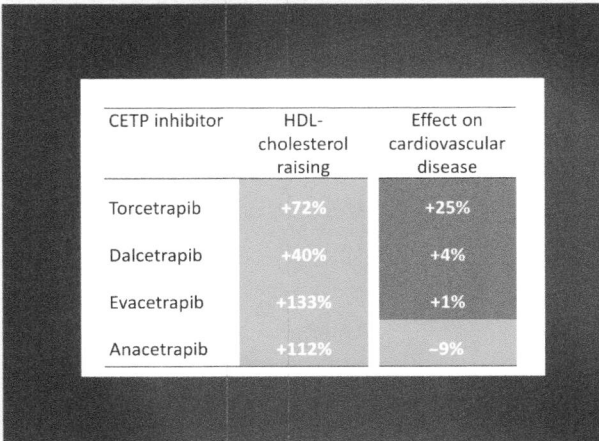

CETP inhibitor	HDL-cholesterol raising	Effect on cardiovascular disease
Torcetrapib	+72%	+25%
Dalcetrapib	+40%	+4%
Evacetrapib	+133%	+1%
Anacetrapib	+112%	−9%

The impact of CETP inhibitors on HDL-cholesterol levels and cardiovascular disease outcomes vary.

(Continued)

(Continued)

Meanwhile, the interest in the development of CETP inhibitors faded. Dalcetrapib was another drug in this class, developed by Hoffmann-La Roche. It did not reduce the risk of heart attacks for patients who recently suffered a heart attack and were on statins. Evacetrapib, another CETP inhibitor provided by Eli Lilly, did not lower the rate of heart problems among high-risk patients either.

In 2017, the last large-scale trial of CETP inhibitors evaluated anacetrapib, a drug developed by Merck, in a study involving more than 30,000 patients. The study finally showed positive results: a CETP inhibitor was able to reduce coronary heart diseases significantly. The long-term illness of this field seemed to be over. The clinical effect was however lower than expected, at only 9%, despite a 2.1-fold increase in HDL-cholesterol, accompanied by a reduction in LDL-cholesterol. This left the researchers puzzled.

Interesting Numbers

Pfizer projected annual sales for torcetrapib at 14 billion US dollars. However, the loss in total market value following the failure of this drug was 21 billion US dollars. The respective numbers for evacetrapib were 630 million and 6 billion US dollars.

The failure of torcetrapib came as a surprise. However, in the history of science, unexpected outcomes often prompt a reevaluation of our understanding. As anticipated, nature was about to introduce a new twist to the story.

This was epidemiologists who added a new piece to the HDL puzzle. In a few years, studies conducted in various countries revealed that the connection between HDL-cholesterol and heart attacks is not as straightforward as previously believed. Instead of a linear and inverse relationship, it was discovered that this association follows a U-shaped curve, which is quite unexpected. Interestingly, not only individuals with low HDL-cholesterol but also those with very high levels experienced elevated cardiovascular mortality. Despite this intriguing finding, no plausible biological explanation has been provided for this paradoxical curve.

Thermogenic Activators

Raising HDL-cholesterol can be good for the heart if it reflects using more fat to produce heat. This occurs in the brown adipose tissue in a process of thermogenesis.

A nice example is provided by mirabegron, a medication used to treat overactive bladder. Studies show that it might also help with obesity-related health issues by increasing the activity of brown adipose tissue and improving how the body processes sugars. When healthy women took mirabegron for four weeks, the medication increased the activity of the brown adipose tissue and also increased how much energy their bodies used while resting. It also improved their levels of HDL-cholesterol and control of blood sugar. This suggests that mirabegron and similar agents could be helpful in treating heart disease.

Open Question

Does an increase in HDL-cholesterol protect against heart disease when it is the result of activation of fat burning in adipose tissue?

Estrogens

Estrogens may be responsible for the elevated HDL-cholesterol in women. It is no wonder that estrogen supplements increase HDL-cholesterol between 5% and 15%. This increase is accompanied by an elevation of the number of HDL particles. Estrogens stimulate the liver to produce apolipoprotein A-I, which raises HDL-cholesterol. Also, they inhibit hepatic lipase which reduces HDL-cholesterol. Estrogens can also increase cholesterol efflux from cells.

Estrogens do not only alter HDL-cholesterol levels but also affect the type of HDL particles that circulate in the blood. They increase the size of HDL particles while decreasing remnant cholesterol.

How estrogens are administered can affect their effects on blood lipids. Oral estrogen replacement therapy increases HDL-cholesterol and apolipoprotein A-I, decreases total cholesterol and LDL-cholesterol, and increases triglyceride concentrations. Transdermal estrogen, however, does not affect lipoprotein concentrations but improves the ratio between triglycerides and HDL-cholesterol.

Oral estrogen therapy, despite raising HDL-cholesterol, does not reduce cardiovascular risk and can even increase it. Therefore, estrogens should not be used to prevent cardiovascular disease among postmenopausal females.

Others

Apabetalone, also known as RVX-208, is a medicine being developed by a company of Resverlogix. This molecule belongs to the bromodomain and extra-terminal domain (BET) reader inhibitors and helps increase the production of apolipoprotein A-I. The BET family proteins are important for controlling how genes are turned on and off in living organisms. The BET inhibitors stop these proteins from attaching to chromatin, which suppresses the transcription of certain genes. In studies, RVX-208 has been shown to raise apolipoprotein A-I levels in the blood of both healthy people and those with low HDL-cholesterol. In patients with heart disease who were already taking statins, RVX-208 increased apolipoprotein A-I and HDL-cholesterol even more. It also increased the size of HDL particles in the blood.

What It Is

BET family of proteins includes those that regulate the activity of genes and pathways related to the immune system of mammals.

A compound called R-138329 inhibited the uptake of cholesteryl ester by the liver through the SR-BI receptor. The agent temporarily increased HDL-cholesterol. However, it also caused a buildup of cholesterol within the liver and increased the formation of artery blockages in mice. These data clearly show that using HDL-cholesterol alone as a risk indicator for heart disease may not be accurate.

There are several ways to change the formation of HDL in the blood vessels. One way to do this is by giving a recombinant LCAT, a type of LCAT that is different from the normal LCAT. This type of LCAT is more powerful and stable than the normal type, and it may help increase HDL-cholesterol. Indeed, HDL-cholesterol levels rose when this recombinant LCAT, given as a single shot, was administered. In these studies, larger HDL particles containing more cholesteryl ester and apolipoprotein-A-I were formed. The removal of cholesterol and its conversion to bile acids which can be excreted through feces also increased. This suggests that the process for removing cholesterol from the blood was improved. When experimental animals were fed a diet high in cholesterol, they were also given recombinant LCAT. As a result of this, the artery-clogging lesions on their arteries were reduced. It is however important to use this approach with caution in humans, as recent research shows that there may be no direct link between LCAT, cholesterol removal from tissue, and the formation of lesions. Recent trial of this therapy was discontinued after some disappointung results.

Other ways exist to alter HDL processing in the body. One option is to activate lipoprotein lipase, which helps create more HDL by using remnant lipids from triglyceride-rich lipoproteins. This method has been shown to increase HDL-cholesterol in animals treated with a drug called ibrolipim. Ibrolipim not only raises HDL-cholesterol but also improves glucose metabolism, fatty liver, and the development of artery disease in animal models.

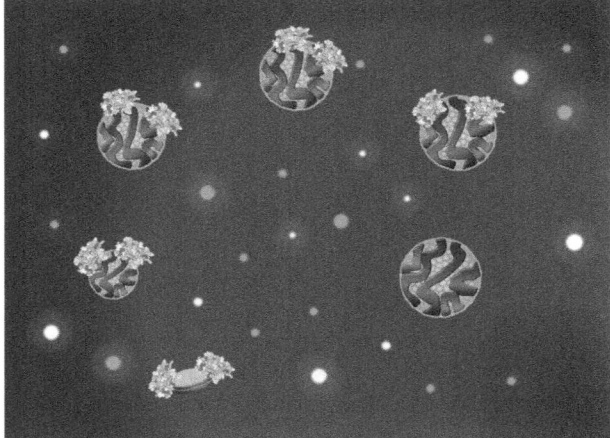

When the LCAT enzyme is more active, it makes more cholesteryl esters in HDL. This causes HDL to become larger and have more cholesteryl esters, which leads to higher levels of HDL-cholesterol.

History

U-shaped Epidemiology

Who	When
Børge Nordestgaard	**2017**

Where	What
Copenhagen, Denmark	Madsen CM, Varbo A and Nordestgaard BG. Extreme high high-density lipoprotein cholestrol is paradoxically associated with high mortality in men and women: two prospective cohort studies. Eur Heart J. 2017;38:2478–2486.

Since decades, it has been known that low levels of HDL-cholesterol in the blood increase the risk of cardiovascular disease. In 2010s, however, epidemiological studies showed that very high HDL-cholesterol levels can also be harmful. They increase the risk of heart diseases and death. This relationship is called U-shaped.

This observation was initially reported by researchers from Canada, Japan, and Denmark. The paradoxical finding was initially greeted with skepticism. However, similar observations were soon made in other populations around the world. Just recently, a very large study in China involving over 3 million adults found that there is a U-shaped association between HDL-cholesterol and risk of dying from heart disease and cancer. Some researchers still observed more death only at low HDL-cholesterol, calling the relationship L-shaped. Others found elevated mortality at extremely high HDL-cholesterol, resulting in the J-shaped curve. It is important to mention that the studies in which the U-shaped relationship was detected were large scale and included tens of thousands of participants followed over many years.

Børge Nordestgaard, Anne Tybjaerg-Hansen, and other researchers from the University of Copenhagen in Denmark were among the first to report this relationship. In large-scale studies that involved a large portion of the Danish population, the researchers observed an increase in heart disease mortality in those with extremely high HDL-cholesterol. They also found that extremely high HDL-cholesterol was related to mortality from other diseases. Unexpectedly, the extremely high HDL-cholesterol increased the risk of infectious diseases such as pneumonia and inflammatory diseases. Other researchers have even observed an increased risk for neurological disorders such as Alzheimer's.

This suggests that there is an optimal range of HDL-cholesterol levels, and having both low and extremely high levels may actually be risky. However, these extremely high levels are rare in the general population, affecting only a small percentage of people. Indeed, the frequency of extremely high HDL-cholesterol was estimated at 2–3% of the population in these studies.

(Continued)

(Continued)

It remains not clear why people with very high HDL-cholesterol have a higher risk of dying. These people are often women as well as those who drink a lot of alcohol, but they also tend to have better social and lifestyle factors.

There are data showing that certain changes in specific genes can cause both very high levels of HDL-cholesterol and an increased risk of heart disease. These changes may have negative effects on a person's health. For example, a rare variant in the *SCARB1* gene can raise HDL-cholesterol and increase the risk of heart disease. However, another mutation in the same gene can increase HDL-cholesterol without affecting atherosclerosis. So, having very high HDL-cholesterol due to a problem with the *SCARB1* gene does not always mean a higher risk of heart disease.

In a similar way, people with CETP deficiency may have very high HDL-cholesterol, but still be at risk of cardiovascular disease. Changes in the *CETP* gene that result in elevated HDL-cholesterol are however often associated with a lower risk of cardiovascular diseases, perhaps because they reduce triglycerides and LDL-cholesterol, as well as non-HDL levels. Genetic studies suggest that HDL-cholesterol and cardiovascular risk are not directly related in people who have a variant of the *LIPG* gene, which codes for endothelial lipase. However, these studies may only include certain genetic variations, and not all genes that influence HDL-cholesterol. They may not give a full understanding of the role that genetics plays in heart health.

Open Question

Can changes in lipoprotein metabolism help to protect against infectious disease?

Test Your Knowledge

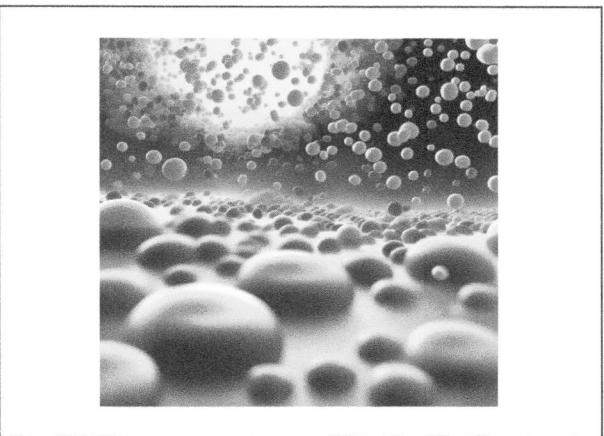

1   HDL concentration is approximately 30 micromoles per liter plasma. A typical adult contains about 3.5 liters plasma. Calculate the amount of HDL particles that an adult contains. One mole of any substance is composed of 6.02×10^{23} units of this substance.

   A 6.3×10^{19}

   B 6.3×10^{21}

   C 3.6×10^{19}

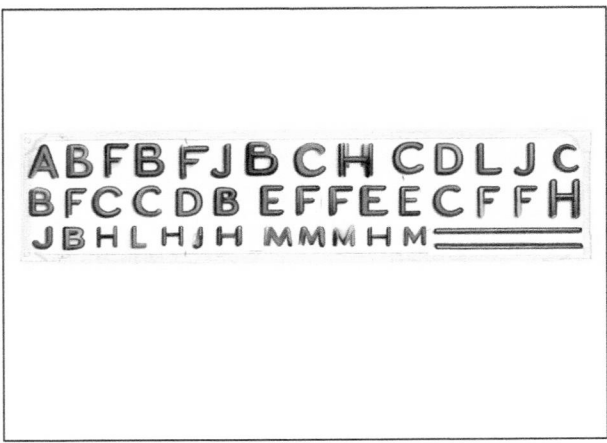

2 In the ABC nomenclature developed by Petar Alaupovic for apolipoproteins, which letters are unused?

   A the letters P to Z;

   B the letters G, I and K;

   C the letters G, I, K and P to Z.

3 Your HDL-cholesterol level was 37 mg/dl in January. Your doctor recommended that you engage in an exercise and diet program to increase your HDL-cholesterol. After six months, your HDL-cholesterol went up to a level of 40 mg/dl. Can you be satisfied?

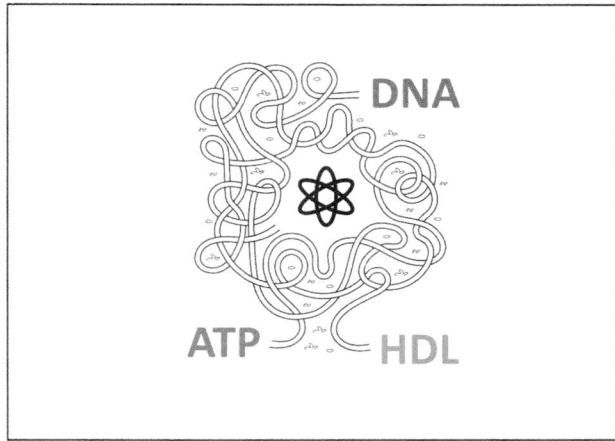

4   What molecule provides the body with its main source of energy?
    A ATP;
    B DNA;
    C HDL.

Answers

1, A; 2, C; 3, not necessarily. Assuming that HDL-cholesterol increases on the average from 37 to 39 mg/dl between winter and summer, your HDL-cholesterol could have naturally increased without any intervention, due to seasonal variations; 4, A.

Further Reading

Allard-Ratick, M. P., B. R. Kindya, J. Khambhati, M. C. Engels, P. B. Sandesara, R. S. Rosenson and L. S. Sperling (2021). "HDL: Fact, fiction, or function? HDL-cholesterol and cardiovascular risk." European Journal of Preventive Cardiology 28(2): 166–173.

Becher, T., S. Palanisamy, D. J. Kramer, M. Eljalby, S. J. Marx, A. G. Wibmer, S. D. Butler, C. E. S. Jiang, R. Vaughan, H. Schöder, A. Mark and P. Cohen (2021). "Brown adipose tissue is associated with cardiometabolic health." Nature Medicine 27(1): 58–65.

Bonacina, F., A. Pirillo, A. L. Catapano and G. D. Norata (2021). "HDL in immune-inflammatory responses: Implications beyond cardiovascular diseases." Cells 10(5): 1061.

Davidson, W. S., A. S. Shah, H. Sexmith and S. M. Gordon (2022). "The HDL proteome watch: Compilation of studies leads to new insights on HDL function." Biochimica et Biophysica Acta (BBA) – Molecular and Cell Biology 1867(2): 159072.

Franczyk, B., A. Gluba-Brzozka, A. Cialkowska-Rysz, J. Lawinski and J. Rysz (2023). "The impact of aerobic exercise on HDL quantity and quality: A narrative review." International Journal of Molecular Sciences 24(5): 4653.

Kjeldsen, E. W., L. T. Nordestgaard and R. Frikke-Schmidt (2021). "HDL cholesterol and non-cardiovascular disease: A narrative review." International Journal of Molecular Sciences 22(9): 4547.

Kjeldsen, E. W., J. Q. Thomassen and R. Frikke-Schmidt (2022). "HDL-cholesterol concentrations and risk of atherosclerotic cardiovascular disease – Insights from randomized clinical trials and human genetics." Biochimica et Biophysica Acta (BBA) – Molecular and Cell Biology 1867(1): 159063.

Kontush, A. (2020). "HDL and reverse remnant-cholesterol transport (RRT): Relevance to cardiovascular disease." Trends in Molecular Medicine 26(12): 1086–1100.

Kontush, A., M. Martin and F. Brites (2023). "Sweet swell of burning fat: Emerging role of high-density lipoprotein in energy homeostasis." Current Opinion in Lipidology 34(6): 235–242.

Marsche, G., J. T. Stadler, J. Kargl and M. Holzer (2022). "Understanding myeloperoxidase-induced damage to HDL structure and function in the vessel wall: Implications for HDL-based therapies." Antioxidants (Basel) 11(3): 556.

Ogura, M. (2022). "HDL, cholesterol efflux, and ABCA1: Free from good and evil dualism." Journal of Pharmacological Sciences 150(2): 81–89.

Ray, K. K., S. J. Nicholls, K. A. Buhr, H. N. Ginsberg, J. O. Johansson, K. Kalantar-Zadeh, E. Kulikowski, P. P. Toth, N. Wong, M. Sweeney and G. G. Schwartz (2020). "Effect of apabetalone added to standard therapy on major adverse cardiovascular events in patients with recent acute coronary syndrome and type 2 diabetes: A randomized clinical trial." JAMA 323(16): 1565–1573.

Rohatgi, A., M. Westerterp, A. von Eckardstein, A. Remaley and K. A. Rye (2021). "HDL in the 21st century: A multifunctional roadmap for future HDL research." Circulation 143(23): 2293–2309.

Tall, A. R. (2021). "HDL in morbidity and mortality: A 40+ year perspective." Clinical Chemistry 67(1): 19–23.

von Eckardstein, A., B. G. Nordestgaard, A. T. Remaley and A. L. Catapano (2023). "High-density lipoprotein revisited: Biological functions and clinical relevance." European Heart Journal 44(16): 1394–1407.

Zhang, G., J. Guo, H. Jin, X. Wei, X. Zhu, W. Jia and Y. Huang (2023). "Association between extremely high-density lipoprotein cholesterol and adverse cardiovascular outcomes: A systematic review and meta-analysis." Frontiers in Cardiovascular Medicine 10: 1201107.

Main Character Three: Triglyceride-Rich Lipoproteins, Carriers of Remnant Cholesterol

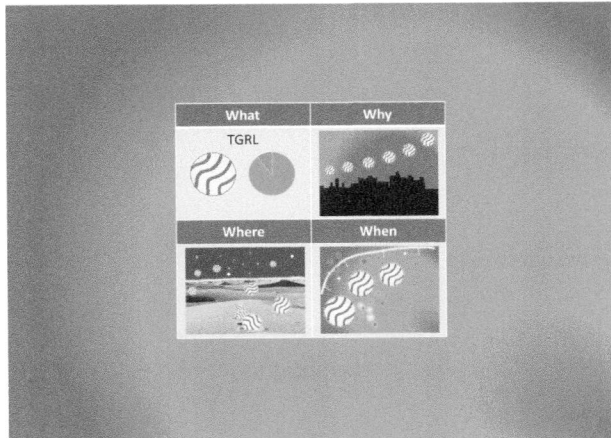

16

Remnant Cholesterol – Why It Is Important

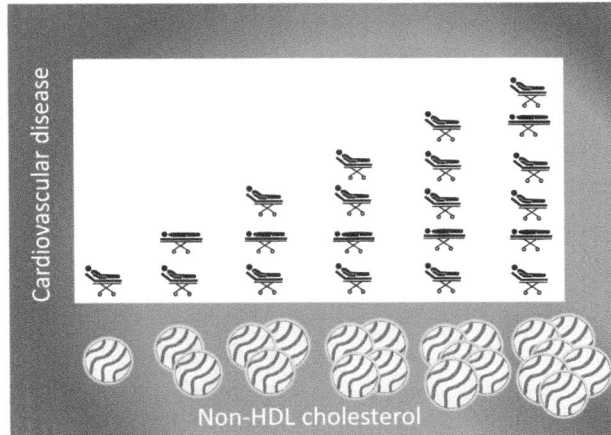

The relationship between non-HDL-cholesterol (which is a sum of remnant cholesterol and LDL-cholesterol) and mortality due to heart disease is depicted in the figure, with non-HDL-cholesterol levels plotted horizontally and the risk of developing heart disease plotted vertically. As non-HDL-cholesterol levels increase, the risk of developing heart disease also increases. Non-HDL-cholesterol is considered harmful.

Here is the final character in our story, the Remnant Cholesterol. This type of cholesterol is carried by large lipoproteins, whose primary purpose is to transport triglycerides through the body. They are therefore called triglyceride-rich lipoproteins.

Triglycerides, which are crucial for energy production, are arguably the most important lipids in circulation from a physiologic point of view. Triglyceride-rich lipoproteins have a larger size than HDL and LDL. In addition, because triglycerides weigh less than cholesterol, they are lighter.

The remnant cholesterol, along with the "bad" and "good" cholesterol, has been accused of a number of misdemeanors. It is true that it can be bad, but its primary function is useful. Lipoproteins, in particular, are essential to our lives – specifically for the production and use of energy within the body. Fats in the form triglycerides and sugars are the two main sources for energy in humans.

Despite being essential, triglycerides are not something we can consume in excess. In this situation, triglycerides begin to accumulate and their blood concentrations increase. Triglyceride-rich particles also increase the remnant cholesterol concentrations, which can lead to heart disease and

Cholesterol, Lipoproteins, and Cardiovascular Health: Separating the Good (HDL), the Bad (LDL), and the Remnant,
First Edition. Anatol Kontush.
© 2025 John Wiley & Sons, Inc. Published 2025 by John Wiley & Sons, Inc.

stroke. Research suggests that how our body processes fats after eating contributes to the heart disease, and it may also help predict the risk of the disease. Studies have shown that the amount of triglycerides in the blood when we have not eaten (also called fasting levels) can also predict the risk of heart disease. It does not matter if we have high or low blood sugar levels, there is a strong connection between fasting and non-fasting lipid levels.

Epidemiology

The genetics and epidemiology of triglyceride-rich lipoproteins show that high plasma triglyceride or remnant cholesterol levels can increase the risk of heart disease and stroke. This risk is determined by various factors, including LDL- and HDL-cholesterol. In the 1950s, John Gofman conducted groundbreaking studies on the role of lipoproteins in heart disease. Among other findings, his research proposed an atherogenic index that placed greater emphasis on the concentrations of triglyceride-rich lipoproteins rather than LDL. In the 1960s, the Livermore Study established a direct link between concentrations of triglyceride-rich lipoproteins like IDL and future heart disease. The exact cause of these diseases, whether it is abnormal triglyceride metabolism or its breakdown products, remains unclear. Remnant cholesterol may contribute to this relationship. Genetic studies have suggested that triglycerides could play a causal role in heart disease, although the specific contribution of triglyceride-rich lipoproteins is still unknown.

Interestingly, people with very high triglyceride levels in their blood (this condition is called chylomicronemia) do not develop heart disease as often as those with very high cholesterol levels. Acute pancreatitis is more likely to occur if they have triglycerides' levels in their blood that are higher than 900 mg/dl. These observations, at first glance, do not support a deleterious role for triglyceride-rich lipoproteins. This leads to doubts about the role of triglycerides in heart disease.

However, moderately high concentrations of plasma triglycerides are clearly detrimental to heart health. In the 1990s, studies showed that moderately high levels of triglycerides, whether fasting or not, are linked to an increased risk of heart disease. In fact, plasma triglyceride levels as low as 100 mg/dl can already cause heart disease. It is generally accepted that heart disease risk increases when fasting triglyceride concentrations exceed 150 mg/dl. This association was still present after adjusting for cholesterol levels.

These findings were later confirmed by other studies. Three studies based on the Copenhagen City Heart Study and the Women's Health Study showed in 2007 and 2008 that higher non-fasting levels of triglycerides are strongly associated with increased risk of heart attacks, coronary heart diseases, strokes, and deaths from any cause. Women with triglyceride concentrations above 440 mg/dl were 17 times more likely to suffer a heart attack than women with levels below 90 mg/dl over a 27–30-year period. The corresponding risk increase for men was five times. These higher risks for women were partly due to men consuming more alcohol than women. However, men with low alcohol intake had similar risks to women.

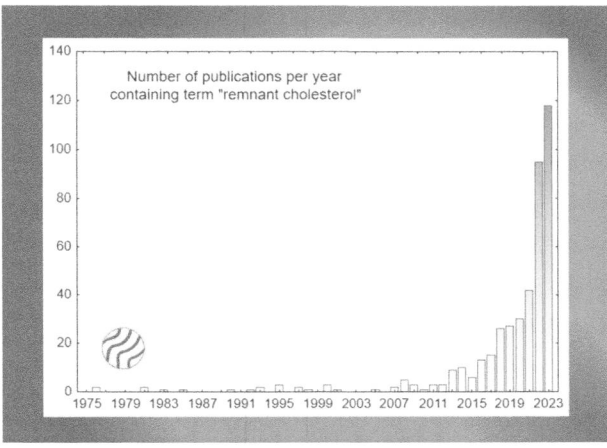

The study of remnant cholesterol has only recently gained attention, but it has experienced a rapid and significant growth.

In 2009, a group of researchers looked at more than 300,000 people in a study called the Emerging Risk Factors Collaboration. They found that having high levels of triglycerides in the blood increased the chances of getting heart disease. However, this link was not as strong when they considered other factors like HDL-cholesterol. The researchers also discovered that triglyceride levels can vary more than HDL-cholesterol on a daily basis, which might explain why HDL-cholesterol is more closely connected to heart disease. They also found that high triglyceride levels were linked to a higher risk of stroke. The risk of heart disease and stroke increased up to certain levels of fasting triglycerides.

Researchers combined two studies, the Copenhagen City Heart Study and the Copenhagen General Population Study, to find that higher non-fasting levels of triglycerides are associated with increased risk for various health problems. These risks were higher than those found by the Emerging Risk Factors Collaboration. Men and women who had an average non-fasting level of triglycerides at 580 mg/dl were at 5.1 times greater risk of having a heart attack than those with a 70 mg/dl level. Other health problems such as stroke and overall mortality were also associated with increased risks. Even if LDL-cholesterol levels were normal, high plasma triglyceride levels were associated with artery problems and inflammation. Some studies also showed that high triglycerides could increase cardiovascular risk after taking statins for lowering LDL-cholesterol. However, not all studies confirmed this association.

We do not know exactly which triglyceride-rich particles are causing heart disease. Measuring total plasma triglyceride levels is therefore a possibility to estimate the risk of the disease. We suppose that cholesterol makes these particles harmful to the heart. Measuring cholesterol in the particles is therefore a good way to evaluate the risk. The cholesterol present in remnant triglyceride-rich particles is called remnant cholesterol. Different methods are available to measure remnant cholesterol. However, there is no test that measures cholesterol in all types of triglyceride-rich particles at once. We can estimate remnant cholesterol by subtracting HDL-cholesterol and LDL-cholesterols from total cholesterol. This estimate is helpful because it does not require extra tests, and higher levels of the calculated remnant cholesterol have been linked to an increased risk of heart disease.

Indeed, numerous studies consistently show a strong link between high levels of remnant cholesterol in the bloodstream and the risk of heart disease. The Emerging Risk Factors Collaboration analyzed the data of more than 300,000 people without preexisting heart conditions. Researchers found that higher levels of remnant cholesterol, referred to as non-HDL cholesterol by them, were linked to a greater risk of heart attacks or fatalities due to heart disease. The study also concluded that measuring non-HDL-cholesterol (which includes LDL-cholesterol) yielded comparable outcomes to directly measuring LDL-cholesterol. This confirms that remnant cholesterol is a critical determinant when assessing heart disease risk.

A second notable example is The Prospective Studies Collaboration. This study examined an enormous cohort of more than 892,000 people who were initially free from cardiovascular disease. The study found that over a long period of time, there was a positive correlation between high levels of remnant cholesterol in the bloodstream and an increased susceptibility for heart disease. The accumulation of remnant particles increased the risk of developing heart disease.

Genetics

The link between remnant cholesterol and heart disease is confirmed by genetics. Mendelian randomization is like a lifelong experiment that helps us determine if high levels of triglycerides or remnant cholesterol are linked to inflammation, heart disease, and death. These studies are superior to regular observation because they do not have the same issues with confounding variables and reverse causality. In order to do these studies, we must choose genetic traits which only affect triglycerides and remnant cholesterol without affecting any other factors. This is not easy to do in practice.

In one study, a 39 mg/dl rise in remnant cholesterol was linked to a 2.8 times greater risk of heart disease. This could not be attributed solely to low HDL-cholesterol levels observed in parallel. Comparatively, observational studies revealed a risk that was only 1.4 times greater. In addition, high levels of triglycerides and remnant cholesterol determined by genes can cause chronic inflammation. This inflammation is associated with atherosclerosis. Comparatively, genes with high levels of LDL-cholesterol do not produce similarly strong association with inflammation.

What It Is

Remnant cholesterol stays for cholesterol present in remnant triglyceride-rich particles. It can be estimated as a difference between total cholesterol and HDL-cholesterol plus LDL-cholesterol. There is no test that measures cholesterol in all types of remnant triglyceride-rich particles at once.

Researchers found that using genetic variants of lipoprotein lipase, they could also observe that there is a twofold increase in risk for death from any cause, with every 90 mg/dl rise in triglycerides. Observational research showed a risk of only 1.2 times greater. Reducing triglyceride by 90 mg/dl was associated with a 50% reduction in the risk of death from any cause.

A study that used Mendelian randomization found evidence of the link between high levels of remnant cholesterol and coronary heart disease. The study examined a few genetic markers strongly associated with remnant cholesterol, LDL-cholesterol, and HDL-cholesterol. The genetic markers of remnant cholesterol and LDL-cholesterol were more strongly linked to the risk of disease than the actual levels of lipids. Particularly, remnant cholesterol was found to be more predictive of disease than LDL-cholesterol (hazard ratio of 2.82 for a 90 mg/dl increase in non-fasting remnant cholesterol compared to a 1.41 for LDL-cholesterol).

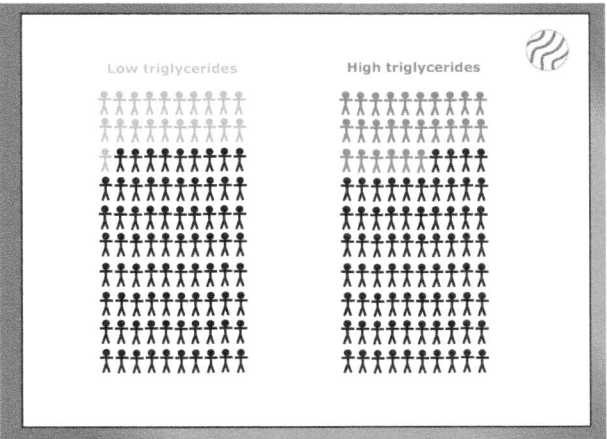

Approximately 20–25% of the population experiences imbalances in their triglyceride levels, either being too low or being too high.

Interestingly, lipoproteins rich in triglycerides are closely related to HDL. The links are inverse – as HDL-cholesterol increases, triglyceride levels in the blood decrease and vice versa. Elevated HDL-cholesterol (but not excessively) and low triglycerides are the ideal scenarios for cardiovascular health.

According to genetic studies, having high levels of triglyceride-rich lipoproteins (or remnant cholesterol) increases the risk of dying from cardiovascular disease. Low HDL-cholesterol levels may not cause the same problems directly, but they could be an indication of high triglycerides or remnant cholesterol. HDL-cholesterol may be a sign of cardiovascular health, but it might not directly cause problems.

In the last few decades, triglycerides have been less studied than HDL-cholesterol. When recent studies and trials show that low HDL-cholesterol levels are unlikely to be the cause of heart disease, this has led to a renewed interest in high triglyceride concentrations.

Practical Aspects: How to Calculate non-HDL-cholesterol

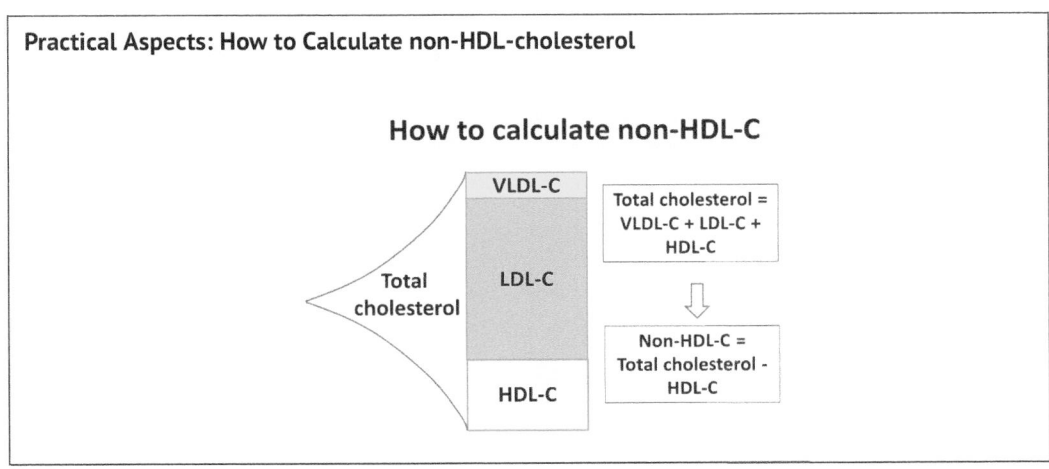

(Continued)

(Continued)

Remnant cholesterol (sometimes also called VLDL-cholesterol) is the cholesterol that can be found in triglyceride-rich lipoproteins such as chylomicron remnants, VLDL, and IDL. VLDL-cholesterol tends to be higher in most people than cholesterol in chylomicron remnants and IDL. Some laboratories may therefore use the term VLDL-cholesterol in place of remnant cholesterol. Non-HDL-cholesterol is the difference between total and HDL-cholesterol.

Non-HDL-cholesterol is the amount of cholesterol found in lipoproteins that are harmful, such as triglyceride-rich lipoproteins and LDL. Remnant cholesterol when combined with LDL-cholesterol can be measured. Both triglyceride-rich lipoproteins and LDL contain apolipoprotein B. Therefore, non-HDL cholesterol is also known as apolipoprotein B-containing lipoprotein-cholesterol. It is more useful than LDL-cholesterol because it also includes remnant cholesterol. Apolipoprotein B concentration in plasma can also provide information on harmful lipoproteins. Plasma triglycerides, on the other hand, only reflect remnant cholesterol and do not tell anything about LDL-cholesterol.

Thus, our understanding of the risk of heart disease associated with high levels of triglycerides has changed. We now know that triglycerides are just one component of a diverse group of particles called triglyceride-rich lipoproteins.

There is however currently no way to accurately measure the potential harm of triglyceride-rich lipoproteins or their remnants. Also, we do not know exactly how these particles' composition and number change after eating. It is possible to determine whether someone has high triglyceride levels by measuring the amount of triglycerides present in their blood, regardless of whether they are fasting or not. The risk of heart attack may be predicted better by non-fasting than fasting triglycerides. Since triglyceride-rich lipoproteins are known to contribute to inflammation and plaque buildup due to their cholesterol content, assessing triglyceride-rich lipoproteins' cholesterol may be an easier way to determine the risk of developing heart disease.

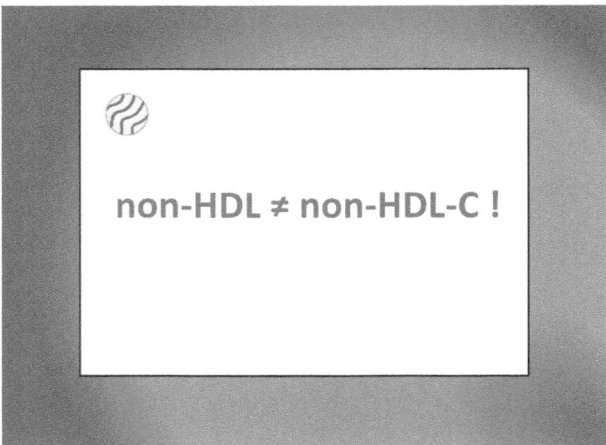

Non-HDL-cholesterol is a quick and easy way to determine the amount of blood cholesterol that can lead to heart problems. This includes cholesterol derived from LDL, remnants, and lipoprotein (a). It is recommended by experts and can help identify high levels of remnant lipoproteins. In clinical jargon, "non-HDL" is often used instead of "non-HDL-cholesterol," which is technically incorrect. Non-HDL is a term that comprises all lipoproteins except HDL, while non-HDL-cholesterol refers specifically to the cholesterol contained within these particles.

We can also estimate VLDL particle concentrations using NMR. This is related to remnant lipoproteins. VLDL-cholesterol is linked to heart disease in a similar way to LDL-cholesterol, according to studies. However, more research needs to be done to confirm this methodology.

The increase in non-HDL-cholesterol levels is linked to high levels of triglycerides. In this case, there is more VLDL-cholesterol and IDL-cholesterol. Parallel to this, the total apolipoprotein C-III and small LDL particles are also increased. There is also a decline in HDL-cholesterol because there are fewer large HDL particles. All of these changes can increase the risk of heart disease. However, it is still unclear which factors are directly responsible.

High triglycerides can be associated with diabetes and high blood sugar. Moderate hypertriglyceridemia can be a risk factor in developing problems with sugar processing. The way that adipose, muscle, and pancreatic cells work together is closely related to the triglyceride level in the blood. Heart disease risk can be predicted by elevated levels of fasting glucose as well as triglycerides. The triglyceride-glucose (TyG) index was created to better evaluate metabolic abnormalities among such patients. The index is calculated using the formula: ln [fasting triglycerides × fasting plasma glucose/2]. The index is an accurate biomarker for insulin resistance. The TyG index has been linked to coronary artery calcium and high heart disease risk.

There is an ongoing debate regarding the likelihood of apolipoprotein B-containing lipoproteins causing heart disease, and whether this likelihood increases gradually from VLDL to LDL. While most guidelines for treating high cholesterol prioritize LDL-cholesterol as the primary lipoprotein causing heart disease, several studies suggest that remnant cholesterol may be equally or even more likely to contribute to heart disease than LDL-cholesterol.

Studies consistently show that non-HDL-cholesterol is a similarly strong and even stronger indicator of heart disease risk than LDL-cholesterol in people with and without hypertriglyceridemia. For example, in a large meta-analysis of statin studies, patients with non-HDL-cholesterol levels over 130 mg/dl or LDL-cholesterol levels above 100 mg/dl were at the greatest relative risk in comparison to those who had both levels below target. Recent reseach shows that people who have high levels of non-HDL cholesterol from childhood to adulthood are more likely to have heart disease events.

The National Cholesterol Education Program Adult Treatment Panel III Report suggests that remnants from VLDL particles, IDL particles, and chylomicrons all induce atherosclerosis and that the cholesterol in VLDL is strongly correlated with the cholesterol in triglyceride-rich lipoproteins. Non-HDL-cholesterol can therefore be a better measure of the total burden of harmful lipoproteins.

The results of tests that measure remnant cholesterol specifically have been linked to heart disease. Some tests measure directly the number of particles carrying cholesterol. These methods measure the concentration of apolipoprotein B-100 and B-48.

Open Question

What is the best way to determine the risk of developing heart disease from lipoproteins? Is it LDL-cholesterol or non-HDL-cholesterol, apolipoprotein B, or some other metric?

The triglyceride level can differ greatly between people. As the levels rise, so normally does the risk of developing heart disease. It is therefore difficult to define what is normal and abnormal using specific percentages. Guidelines for "optimal" and "borderline" triglyceride levels are based on previous recommendations, as well as studies of triglyceride metabolic rates. The suggested cutoffs for high triglycerides were based on the current state of knowledge. However, more research is required to understand heart disease and pancreatitis risks in each category. As a temporary measure, plasma triglycerides can be used as a substitute marker because they are measured routinely in laboratories.

What It Is

High triglycerides in the blood are defined as 150 mg/dL (1.7 mmol/L) and higher. The levels of 150 to 199 mg/dL (1.8–2.2 mmol/L) are considered borderline high, those of 200–499 mg/dL (2.3–5.6 mmol/L) are high, and those of 500 mg/dL or above (5.7 mmol/L or above) are very high.

Practical Aspects: Should I Eat or Should I Not Eat?

In the past, doctors performed a lipid test when the patient was not fasting, that is, after a period of time without eating. The method was not considered reliable, as everyone's stomach empties differently. This affects the lipid levels in blood. Doctors now use fasting levels of lipids as an indication of how the body processes fats. In most countries, this is the norm. However, the recommendation to measure fasting levels of triglycerides is not based on studies that show it to be a better predictor of cardiovascular disease risk than non-fasting levels. Instead, after screening guidelines were implemented, researchers simply began using fasting measurements of triglycerides.

It is remarkable that lipoprotein profiles change only a small amount when healthy people consume normal food. This casts doubt on the concern about how much non-fasting triglyceride levels vary after eating. In fact, studies have shown that non-fasting triglyceride levels are just as good as or even better than fasting triglyceride levels at predicting future cardiovascular disease events. It makes sense, since people spend the majority of their day non-fasting (about 18 hours), that changes in triglyceride and lipoprotein concentrations after meals could lead to artery problems.

In some countries, such as Denmark, doctors will still perform the test if a patient has just eaten. The doctor may request a subsequent fasting test if the non-fasting levels of triglycerides are high. The lipid profile test can be done without fasting, which is easier for doctors, hospitals, and patients. The triglyceride level increases only slightly after eating a regular meal for a couple of hours. These small increases are not significant in terms of potential health risks.

Non-fasting lipid profile tests can still predict if someone is at risk of heart problems. They measure different types of lipids in the blood, including LDL-cholesterol. Since most people eat regularly throughout the day and only fast for a few hours at night, non-fasting tests might give a better idea of their average lipid levels than fasting tests. However, it is still unclear whether fasting or non-fasting plasma triglyceride levels are a better indicator of cardiovascular risk.

After eating, most people consume a lot of fat-rich meals throughout the day. After bedtime, triglycerides are at their highest level in the blood. They drop in the morning when we have not eaten overnight. In real life, triglyceride concentrations are typically 20–25% higher when eating than they were while fasting. However, this difference can be influenced by the initial fasting levels. Fasting triglyceride level can predict the response of triglyceride after eating.

Many studies have shown that high triglyceride concentrations after eight hours of fasting or longer are a risk factor for heart problems. This factor is often underestimated. It is caused by an excess of lipoproteins rich in triglycerides that are either produced too quickly or cleared too slowly. This accumulation can be caused by genetic variations, obesity, and insulin resistance.

After consuming a meal high in fat, about 80% of the increase in triglyceride levels comes from lipoproteins containing apolipoprotein B-48, while lipoproteins containing apolipoprotein B-100 account for most of the increase in particle number. Both during fasting and after eating, particles containing apolipoprotein B-48 are secreted as chylomicrons and smaller triglyceride-rich lipoprotein particles similar to VLDL.

17

Remnant Cholesterol – What It Is

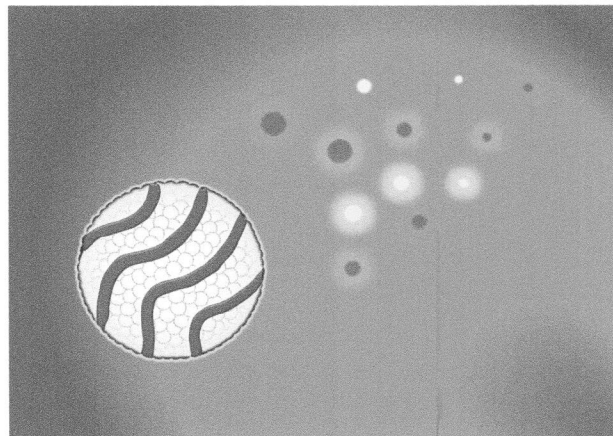

This image shows a VLDL particle in the blood. The molecule of the apolipoprotein B-100 is shown by reddish ribbons surrounding the particle. Phospholipids are represented by pink circles. Red blood cells (erythrocytes) and white blood cells (leucocytes) are depicted as red and white stars in the background. The size of VLDL is magnified 150 times larger than that of the cells.

Remnant cholesterol is carried by lipoproteins rich in triglycerides. The liver and intestines secrete a wide variety of them. The lipoproteins have a complex composition. Their concentration in the blood is associated with cardiovascular disease and pancreatitis.

More than other lipoproteins, triglyceride-rich particles are primarily composed of lipids. Triglyceride-rich lipoproteins are however different from LDL and HDL in many aspects. These include physical properties, chemical makeup, biological activity, and metabolic behavior. First, they are larger than other lipoproteins. Their size ranges from 30 nanometers up to more than 1 micrometer. Because they are so heterogeneous, it is hard to say what the average size is of triglyceride-rich particles. But, we can say with confidence that the particles are comparable in size to virus particles, like influenza viruses. They are also less dense than other lipoproteins.

Apolipoprotein B is the main protein found in triglyceride-rich lipoproteins. The apolipoprotein B comes in two different forms: the liver-produced apolipoprotein B-100 and the intestine-produced apolipoprotein B-48. The different forms allow us to classify lipoproteins more accurately than

Cholesterol, Lipoproteins, and Cardiovascular Health: Separating the Good (HDL), the Bad (LDL), and the Remnant, First Edition. Anatol Kontush.
© 2025 John Wiley & Sons, Inc. Published 2025 by John Wiley & Sons, Inc.

simply looking at their density or size. VLDL particles are primarily composed of apolipoprotein B-100. They are broken down into smaller particles called VLDL remnants, IDL, and LDL.

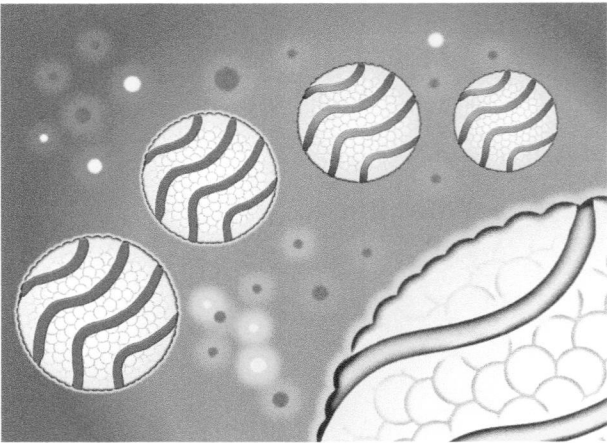

The main triglyceride-rich lipoproteins found in human plasma are chylomicrons, VLDL, and IDL.

Chylomicrons

The largest and lightest lipoproteins are chylomicrons. Apolipoprotein B-48 is found in them. The chylomicrons are broken down into smaller particles, called chylomicron remnants. These particles are not IDLs or LDLs. The intestine can also release smaller chylomicrons that are VLDL sized.

Chylomicrons are light because they contain a lot of triglycerides. Their density is less than 0.95 g/ml. They are lighter than water and under normal physiological conditions can float. The size of chylomicrons can range from 75 to 1,200 nm. They are spherical in shape and can be a variety of sizes. The apolipoproteins B-48 and other exchangeable apolipoproteins are also present in a small quantity.

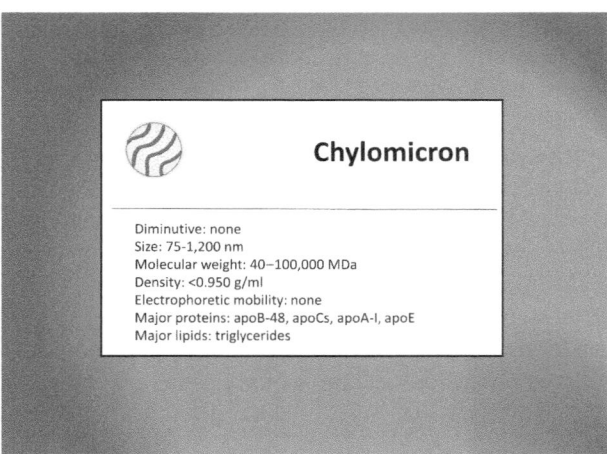

History

Discovery of Chylomicrons

Who	When
Simon Gage and Pierre Fish	**1924**

Where	What
Ithaca NY, USA	Gage, S. H. and P. A. Fish (1924) "Fat digestion, absorption, and assimilation in man and animals as determined by the dark-field microscope, and a fat-soluble dye." American Journal of Anatomy 34: 1-85.

Simon Gage and Pierre Fish at the Cornell University in Ithaca, New York, introduced the chylomicrons to the scientific community as early as 1924.

In the 17th century, the first evidence of the movement of lipids in the blood of animals was obtained. Gasparus Asellius, an Italian surgeon, observed that after eating food, the lacteals (lymphatic vessels) of animals became visible. The vessels turned milky. Robert Boyle, a British chemist, observed in 1665 that a milky, fatty fluid enters the bloodstream via the thoracic canal in the chest after a meal high in fat. William Hewson, a British surgeon, discovered in the 18th century that the fluid contained a lot of fat when it was dried. The particles were first observed by British and German scientists under a microscopy at the end of the 19th century. They called the particles "fat dust" and "blood dust."

Gage and Fish demonstrated in 1924 that the blood of humans taken after eating a fatty meal contained particles with a diameter of about 1 micrometer. The experiment involved a long fast of 18 to 24 hours followed by eating boiled rice sweetened in sugar. After the meal, tiny particles that were smaller than blood cell size appeared in the blood. Under a dark-field microscopy, the particles looked like falling snowflakes and appeared bright. After the fast, these particles were almost nonexistent. The particles were also found after eating the whites from four boiled eggs. The highest concentration of particles was observed after eating butter. This phenomenon lasted several hours.

Researchers concluded that particles are formed when fat is present in a meal. The amount of fat consumed determined the number of particles.

Gage and Fish called the particles chylomicrons, combining the Greek words "chylos", meaning chyle, and "micron", meaning any small object. The term was chosen to describe the origin of these particles, which were derived from chyle and had an average size of 1 micrometer, or micron at that time.

Researchers concluded that the particles derived from the digestion and absorption of fat. Gage and Fish then studied the appearance of these particles in the blood from different

(Continued)

(Continued)

individuals under various conditions. The researchers found that particles were higher at rest than when active. A higher body weight was also linked to more particles.

Using colored dyes, they also studied the deposition of fat in different organs. Finally, the appearance of chylomicrons in different animals such as dogs, horses, and hens was compared.

VLDL

VLDL is a type of lipoprotein that contains triglycerides. VLDL particles range in size from 35 to 100 nm. The particles also contain a lot of triglycerides, but less than the chylomicrons. VLDL has a higher protein content and a slightly higher density than chylomicrons. It ranges from 0.95 to 1.006 g/ml.

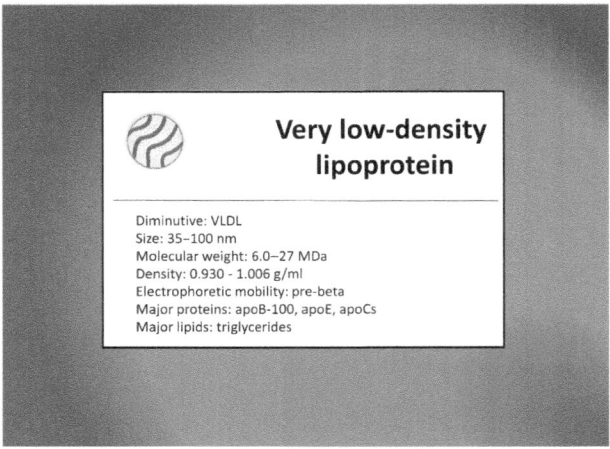

History

Discovery of Triglyceride-Rich Lipoproteins

Who	When
Frank Lindgren	**1951**

Where	What
Berkeley CA, USA	Lindgren, F. T., H. A. Elliott and J. W. Gofman (1951) "The ultracentrifugal characterization and isolation of human blood lipids and lipoproteins, with applications to the study of atherosclerosis." Journal of Physical and Colloid Chemistry 55: 80-93.

In 1951, Frank Lindgren and Hardin Jones discovered lipoproteins possessing a very low density at the Donner Lab in Berkeley. These particles floated at higher rates than Sf 17 and contained more triglycerides. Lipoproteins whose Sf was less than 17 contained higher amounts of cholesterol and lower amounts of triglycerides.

All of these particles were initially called LDLs. This definition allowed them to be distinguished from HDLs. Lindgren's studies separated this fraction further into LDL (1.006–1.050 g/ml, Sf 0–20) and VLDL (0.92–1.006 g/ml, Sf 20–100,000).

These studies used analytical ultracentrifugation. Preparative methods were developed later at the same lab as well as by Richard Havel and colleagues in 1955. The Lindgren method was not suited to precise chemical analysis, but the Havel method produced particles that could be studied in a variety of ways. The Donner Lab's preparative technique was also not simple. The results of both laboratories were however consistent despite the differences in methodology. VLDL was found to be high in triglycerides while LDL had a lot of cholesterol. The plasma concentrations of both lipoproteins also predicted future heart disease.

Richard Havel began working with lipoproteins in the New York Hospital in 1949, in a department headed by David Barr. They worked together with Howard Eder on the electrophoresis of alpha- and beta-lipoproteins. In 1953, he became a young researcher at the newly established National Heart Institute in Bethesda. He participated in studies on lipoprotein lipase in Christian Anfinsen's laboratory. Havel became interested in the metabolism of chylomicrons and triglycerides. At that time, there was no easy way to separate individual lipoproteins in plasma. John Gofman, along with his colleagues at the Donner Lab, developed a complex and expensive ultracentrifugation to separate lipoproteins. In the 1950s, an analytical ultracentrifuge was 30,000 US dollars. That was a lot of cash in those days. Havel and Eder, in collaboration with Joseph Bragdon, decided to develop a less sophisticated preparative ultracentrifugation.

(Continued)

(Continued)

They added concentrated salt solutions in a series of steps to plasma to increase its density. After each step, the plasma with added salt was ultracentrifuged overnight. This caused lipoproteins with a lower density than plasma with the salt to float at the top of test tube. This method separated lipoproteins with densities less than 1.019 g/ml, between 1.019 and 1.063 g/ml, and between 1.063 and 1.21 g/ml. The method was called sequential ultracentrifugation.

Preparative ultracentrifuges used in these studies were cheaper and became very popular. Beckman Instruments Spinco Model L, originally used by Havel, was sold in more than 3,000 exemplars.

This method was a breakthrough for the field of lipoproteins. According to Daniel Steinberg, a pioneering investigator of lipoproteins who also began his career at Anfinsen's lab, the Havel and Eder method revolutionized the field of lipoproteins, much like the Lowry protein method did for biochemistry. Even after almost 70 years, this method remains the most commonly used tool for isolating lipoproteins.

Joseph Bragdon, a pathologist who specialized in atherosclerosis, did not pursue a career in the field of lipoproteins. Richard Havel continued his lipoprotein studies and became one of the leading pioneers in the field.

Havel proposed the term VLDL in 1956 for lipoproteins with a density less than or equal to 1.019 g/ml. These were previously referred to as LDL along with those whose density was between 1.019 and 1.063 g/ml. The term LDL was reserved for this latter fraction.

The researchers further proposed that the term alpha- or beta-lipoproteins be restricted to particles separated via electrophoresis. The particles isolated by ultracentrifugation were the only ones that could be classified as VLDL, LDL, and HDL. The authors explicitly stated that fractions obtained by different methods were similar but not equivalent.

Later research defined another class of triglyceride-rich lipoproteins. Havel and coworkers separated IDL and VLDL using ultracentrifugation in 1960. IDL density ranged between 1.006 and 1.019 g/ml, with flotation rates Sf ranging between 12 and 20. The particles had a diameter between 28 and 35 nm, which was intermediate between VLDL and LDL. Their proteins were mainly composed of apolipoproteins B and E.

IDL and Other Remnants

The remnant lipoproteins are formed when triglycerides are partially removed from VLDL or chylomicrons. The number of remnants present in the blood depends on several factors, such as the extent of lipolysis, how well VLDL remnants turn into LDL, and how well VLDL and chylomicron remnants can be cleared by the liver.

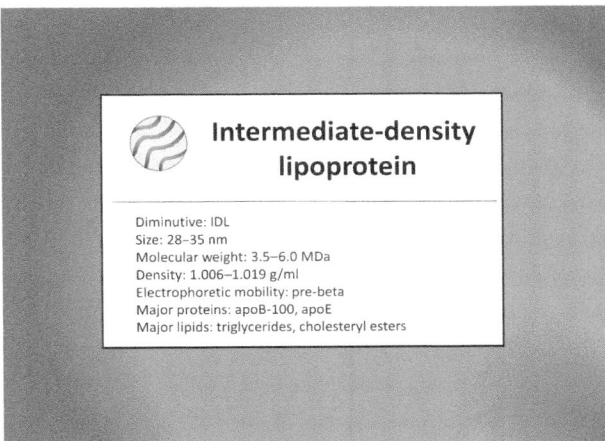

Remnant lipoproteins are a type of lipoproteins that are partially broken down by lipoprotein lipase. They have less triglyceride and more cholesteryl ester than non-modified chylomicrons or VLDL. Remnant particles have a much lower proportion of cholesterol than LDL. However, they are much larger and therefore can have up to four times more cholesterol per particle compared to LDL. People with slightly higher triglyceride concentrations have only small remnant particles.

Isolated by density, the IDL class covers remnant lipoproteins in general. IDL is mainly VLDL remnants, as chylomicrons or their remnants are usually quickly removed from the blood. IDL particles are the densest and smallest subtype of lipoproteins rich in triglycerides. They range in size between 28 and 35 nm and have a density of 1.006–1.019 g/ml. IDL particles have less lipid than VLDL, but more than LDL.

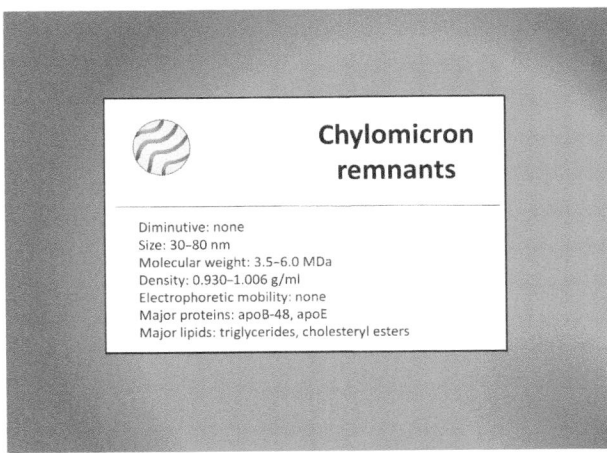

It is interesting that some data suggest that VLDLs and IDLs are not round-like spheres. The particles rather have flat surfaces and a polyhedral form. VLDL flat surfaces are similar to IDL, but IDL has sharper angles. This means that the flat surfaces of IDL are less hydrophobic than the curved surface of HDL. Because of this, IDLs do not bind as strongly to certain proteins as HDL does.

History

The Concept of Chylomicron Remnants

Who	When
Richard Havel	**1955**

Where	What
San Francisco CA, USA	Havel, R. J., H. A. Eder and J. H. Bragdon (1955) "The distribution and chemical composition of ultracentrifugally separated lipoproteins in human serum." Journal of Clinical Investigation 34: 1345-1353.

Dean Graham, John Gofman, and their colleagues were the first to observe a transient increase in VLDL following a fat meal. They also found a conversion of VLDL into IDL after heparin was administered to rabbits or humans. These studies revealed the first mutual conversion between lipoproteins.

Richard Havel proposed the conversion cascade of larger to smaller lipoproteins rich in triglycerides and then to LDL during the 1950s. In 1956, he moved from the NIH to the University of California San Francisco to continue his research on lipoproteins. These studies were based on the development of sequential ultracentrifugation for isolating and characterization of major classes of plasma lipoproteins. Researchers also used radioisotopes for labeling lipoproteins. This technique used differential labeling to track the movement of different lipid and protein components.

The key experiments were performed on rats called supradiaphragmatic, in which blood circulation occurs only above the diaphragm without involving the liver. In order to get enough remnant lipoproteins, it was necessary to exclude liver influence because the liver removes them quickly from the blood. The remnants that are produced by the breakdown of VLDL can be incorporated into the liver or converted to LDL. Researchers found that remnants of VLDL accumulate in the blood when Type III dyslipoproteinemia is present.

Richard Havel and Donald Fredrickson observed in 1956 that the removal from circulation of radiolabeled chylomicrons in dogs was parallel to the production of fatty acids. It was thought that the lipoprotein lipase may have caused this effect.

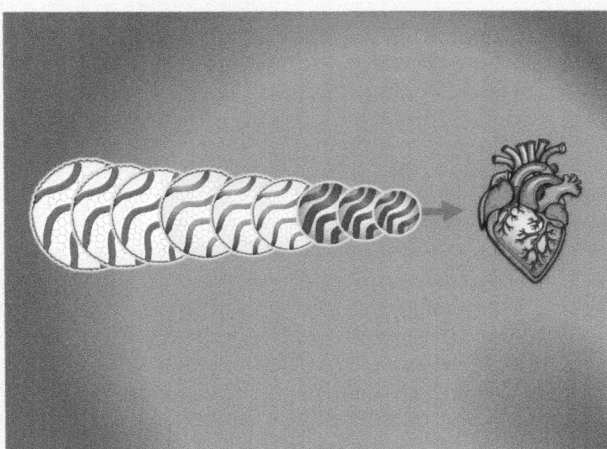

Havel was able to demonstrate that the accumulation in serum of triglycerides after fat intake could only be detected in VLDL and chylomicrons. The researchers also found that VLDL was involved in the transport of endogenous fatty acids produced by the liver. The studies showed that lipoprotein lipase hydrolyzed triglycerides in VLDL to fatty acids before the fatty acids entered cells. The extra-hepatic tissue absorbed most of the fatty acids released by lipolysis. The liver removed the rest of VLDL without lipolysis. Hydrolysis occured in the blood capillaries. VLDL was responsible for transporting fatty acids in the form of triglycerides from the liver into the tissues. First of all, these include adipose and muscle tissue.

Havel's laboratory showed subsequently that triglycerides from the intestine are transported into the adipose tissues by chylomicrons. Both pathways require lipoprotein lipase as well as local uptake by tissue of fatty acids. Albumin was responsible for transporting free fatty acids in the opposite direction, from adipose tissues to the liver. Apolipoprotein B was the major protein in both VLDLs and chylomicrons. Other protein components of VLDL and chylomicrons however differed, with apolipoprotein C and apolipoprotein E present in VLDL, and apolipoprotein A carried by chylomicrons.

The experiments that examined the assembly of VLDL within the liver cells nicely completed this work. Particles were found to be derived from triglycerides, which can come from either the liver or blood. The newly assembled particles are accumulated first in the secretory compartments of the cells and then secreted into the blood. The intestine produces chylomicrons in a similar manner.

Chylomicrons removed by the liver after the lipolysis were initially called the "chylomicron skeleton" by Havel in 1965. Trevor Redgrave proposed the term "chylomicron remnants" five years later. According to him, remnants are enriched in cholesteryl ester compared to non-modified chylomicrons. They also contain residual triglycerides that are less than 10% of their original level. It is likely that the neutral lipids, which form the core of the particle, were the reason for the abandonment of the term "skeleton." The remnants had a 90-nm diameter and were smaller than chylomicrons.

(Continued)

> **(Continued)**
>
> Havel finally developed the concept of chylomicron remnants after more than 10 years of hard work. He was a former boxer with Czech ancestry well known for his tough style of direction. His scientific rigor was a reflection of his clarity and attention to details. This concept was further solidified in the 1970s by demonstrating quantitative formation of LDL derived from VLDL. The lipoproteins, once discovered, were unified into a lipid-transport system. This was similar to the unification of elements in the immune or nervous systems.

Proteins

The proteins in triglyceride-rich lipoproteins and remnant particles are made up of either apolipoprotein B-48 or apolipoprotein B-100, along with other smaller proteins like apolipoproteins C and apolipoprotein E. The remnant particles have more apolipoprotein C-III and apolipoprotein E. The proteins affect how fast the lipids are broken down, how much cholesteryl ester is added, and how quickly the liver removes triglyceride-rich lipoproteins and remnants.

Apolipoprotein B-100 and Apolipoprotein B-48

VLDL is mainly composed of apolipoprotein B-100, apolipoproteins C, and apolipoprotein E. Their contributions to the total protein of VLDL are 40–50% for apolipoprotein B-100, 25–40% for apolipoproteins C, and 10–15% for apolipoprotein E. In chylomicrons, apolipoproteins C quantitatively prevail and large amounts of apolipoprotein A-I are also present, in addition to apolipoprotein B-48.

Apolipoproteins C

Apolipoproteins C-I, C-II, C-III, and C-IV are small apolipoproteins which play an important role in the triglyceride lipolysis. These proteins are linear in structure and have both lipid-binding domains and water-exposed ones. The smallest apolipoprotein is apolipoprotein C-I, which weighs 6.6 kDa. It can easily move between HDL and triglyceride-rich lipoproteins. Due to its strong positive charge, it can bind negatively charged free fatty acids and affect the activity of different proteins involved in lipoprotein processing.

Apolipoprotein C-II, an 8.8-kDa protein, activates several lipases that break down triglycerides. It is associated with HDL and VLDL and can move from one lipoprotein surface to another. Apolipoprotein C-II only functions when it is in lipoproteins rich in triglycerides.

Apolipoprotein C-III has the same mass of 8.8 kDa as apolipoprotein C-II, but the opposite effect of inhibiting the triglyceride lipolysis. It is present in small quantities in HDL, but it is mostly associated with triglyceride-rich lipoproteins where it plays an important role. Apolipoprotein C-III, in addition to inhibiting the triglyceride breakdown, also reduces hepatic uptake of triglyceride-rich remnants. The protein has a strong influence on the removal of remnant lipoproteins from the blood. The ratio of apolipoprotein C-III to apolipoprotein C-II is a key regulator for lipoprotein lipase and lipolysis.

Apolipoprotein C-IV is an 11 kDa protein. It is found in low amounts in HDL but is associated with particles rich in triglycerides. Apolipoprotein C-IV is less abundant in HDL than other apolipoproteins of this family.

Apolipoprotein E

Majority of the circulating apolipoprotein E is carried in triglyceride-rich lipoproteins. The apolipoprotein mediates their binding to receptors as well as removal by cells and tissues. Apolipoprotein E, which possesses the LDL-binding and heparin-binding domains, serves as a partner to the LDL-receptor, the LDL-receptor-related protein (in brief LRP), and the VLDL-receptor. It also ensures that lipoproteins bind to cell surface proteoglycans.

What It Is

LDL receptor-related protein (LPR) belongs to the LDL-receptor family and binds large triglyceride-rich particles.

Apolipoprotein E, like apolipoproteins C, is carried in the blood by both triglyceride-rich lipoproteins and HDL. The apolipoprotein molecule has multiple glycosylation sites, where sugar residues are able to bind covalently. This results in different isoforms. There are also three APOE alleles that are common: APOE2, APOE3, and APOE4.

Other Proteins

It is interesting that many proteins that were thought to only be found in HDL are actually also present in VLDL and chylomicrons. Some of these proteins are apolipoproteins A-I, A-II, A-IV, D, F, J, L-I, and M, PON1, and SAA. These proteins can move between HDL and lipoproteins rich in triglycerides.

History

Discovery of Apolipoprotein E

Who	When
Virgie Shore and Bernard Shore	**1973**

Where	What
San Francisco CA, USA	Shore, V. G. and B. Shore (1973) "Heterogeneity of human plasma very low density lipoproteins. Separation of species differing in protein components." Biochemistry 12: 502-507.

Apolipoprotein E was discovered in 1973 by Virgie Shores and Bernard Shores at the Lawrence Livermore Laboratory of the University of California. Two other laboratories independently made the discovery. It was found in VLDL isolated by ultracentrifugation.

(Continued)

(Continued)

The name of the protein was not given immediately. At that time, the ABC classification of apolipoproteins was not widely used and the new molecule's name was "arginine-rich protein" because of its high arginine content. There were then four apolipoproteins, ranging from apolipoproteins A to D. Two years later, the arginine-rich protein was given the letter E.

Gerd Utermann, a German scientist, discovered three isoforms of apolipoprotein E in the same year. The protein showed three bands when subjected to isoelectric focusing. These bands were named E-I, E-II, and E-III, according to the ABC nomenclature. Later, these isoforms became APOE2, APOE3, and APOE4. The isoforms are encoded by three APOE alleles and have different sugar residues that give them a different electrical charge.

Later studies revealed that isoforms differed by amino acids at residues 112 and 158. Apolipoprotein E2 contains a cysteine in both sites. Apolipoprotein E3 contains a cysteine in residue 112 with an arginine residue 158. And, apolipoprotein E4 contains arginines at both these sites. These amino acids are not located in the binding site of the LDL receptor, which is between residues 134 and 150. However, they do influence the structure of the apolipoprotein E to affect its affinity to the receptor.

Richard Havel, John Kane, and others found that apolipoprotein E levels were elevated in Type III dyslipoproteinemia shortly after its discovery. This defect caused abnormal plasma VLDL composition and metabolism. This disorder has been diagnosed since the characterization of apolipoprotein E. It was discovered that this condition was associated with apolipoprotein E2 isoform. The altered protein decreased VLDL binding, which delayed VLDL removal and increased concentrations of VLDL in the blood.

The work of Thomas Innerarity and Robert Mahley at the NIH in the late 1970s showed that apolipoprotein E is required for lipoproteins to interact with cellular receptors. Apolipoprotein E interaction with receptors such as the LDL-receptor or the LRP removes remnant lipoproteins from the blood.

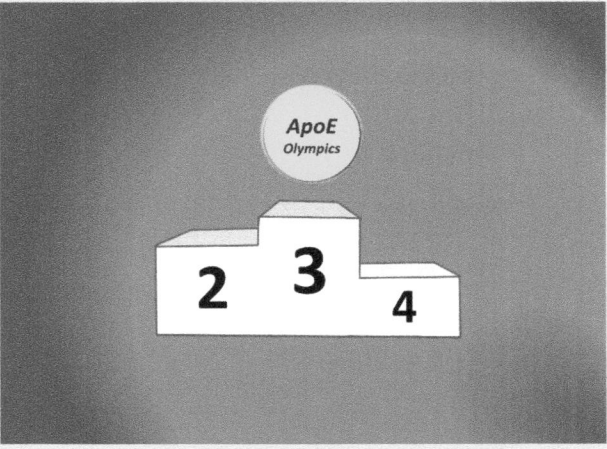

Jan Breslow, at Rockefeller University New York, sequenced the APOE gene in 1985. Apolipoprotein E is a 34-kDa protein with 299 amino acid residues that is encoded by chromosome 19. In 1992, Nobuyo Mieda, at the University of North Carolina, and Jan Breslow, independently, removed the gene in mice. They created apolipoprotein E knockout mice. This was the most popular mouse model of atherosclerosis for many years. It was a difficult task to complete, as mice are naturally resistant against this disease. This breakthrough was therefore hailed as a new golden age in experimental atherosclerosis. Homozygous mice were hypercholesterolemic when fed standard chow and developed spontaneous atherosclerosis by the age of three to four months. The remnant lipoproteins were the main carriers of cholesterol in their plasma. Western diets accelerated the development of atherosclerotic lesions.

Drs Shores' discovery of apolipoprotein E was significant, as it has led to better understanding of the role of this protein in various health conditions. This includes its relationship with cardiovascular disease and Alzheimer's disease. In the 1990s, it was found that the APOE4 gene is strongly associated with Alzheimer's. This allele became a major risk factor in this debilitating disorder. In the end, APOE3 remained the allele associated with the lowest risk of several diseases.

The discovery of apolipoprotein E also paved the path for future research on its functions and implications to human health. Since then, it has become an important target to study genetic predispositions to diseases and develop potential therapeutic interventions.

Lipids

Enzymes influence the lipids of triglyceride-rich lipoproteins, as well as their remnants. The remnant particles contain more cholesteryl ester and less triglyceride than intact chylomicrons or VLDL.

In triglyceride-rich lipoproteins, lipids play an important role. Lipidomic analysis identifies numerous individual species of lipid in triglyceride-rich particles, just as it does in other lipoproteins. Similar to other lipoproteins, their lipidome is primarily composed of triglycerides, phospholipids, cholesterol, and cholesteryl esters. Triglycerides typically account for 55–95% total mass of triglyceride-rich lipoproteins, phospholipids make up 3–18%, cholesteryl esters account for 1–14%, and free cholesterol is 0.5–8%.

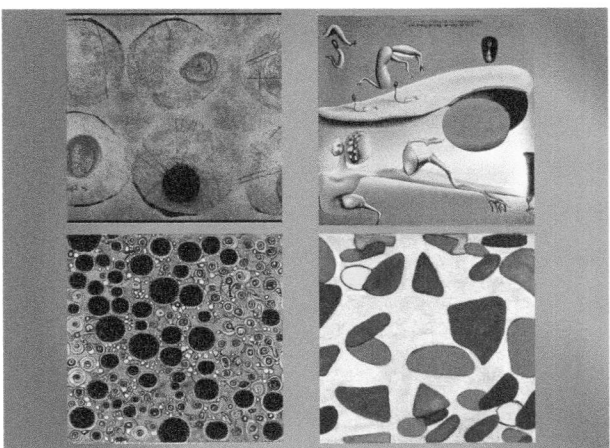

Here are artistic representations of triglyceride-rich lipoproteins in the styles of Leonardo da Vinci, Salvador Dali, Juan Miro, and Gustav Klimt (clockwise starting at the top left) generated by Stable Diffusion AI. The images were generated based on prompts that included the name of the artist and the term "triglyceride-rich lipoprotein." Source: Black Technology LTD/https://stablediffusionweb.com/ (accessed 9 February 2024).

Triglycerides are the main lipid class in triglyceride-rich lipoproteins. The triglycerides are added to the lipoproteins within the cells prior to secretion. By contrast, cholesteryl ester largely originates from CETP-mediated exchange with HDL. Both triglycerides and cholesteryl ester are hydrophobic. They are both located in the lipid core of triglyceride-rich lipoproteins.

Similarly to other lipoproteins, phospholipids make up the majority of the lipids on the surface of lipoproteins rich in triglycerides. They are essential for maintaining the particle structure and ensuring proper packaging of lipids. As in other lipoproteins, the major types of phospholipids in triglyceride-rich lipoproteins are phosphatidylcholine and sphingomyelin. There are also small amounts of sterols, mostly cholesterol, in the surface layer. These lipids regulate fluidity.

The triglyceride-rich lipoproteins carry minor bioactive lipids, such as ceramides and other classes of phospholipids. They also contain lysosphingolipids, glycosphingolipids, diacylglycerides, monoacylglycerides, free fatty acids, and other molecules.

Heterogeneity

Like other lipoproteins, triglyceride-rich particles are also divided into subtypes with different characteristics.

Ultracentrifugation has been the standard method for separating triglyceride-rich lipoproteins. This technique can be used to isolate chylomicrons as well as VLDL, IDL, and chylomicron remnants. This technique can be further extended to fractionate VLDL into subtypes VLDL1 and VLDL2. Like other lipoproteins, large particles rich in triglycerides are less dense, while smaller particles are denser. As the density of the particles increases, their size decreases. This is accompanied by an increase in proteins and a decrease in lipids. Parallel to this, the particle surface-to-core ratio also increases. VLDL is the most prevalent type of triglyceride-rich lipoproteins in individuals with normal levels of lipids. The particles rich in triglycerides have a density below 1.013 g/ml.

NMR, size exclusion chromatography, and ion mobility can be used to separate triglyceride-rich lipoproteins based on their size. There are three types of triglyceride-rich lipoproteins in people with normal levels of lipids: large-, medium-, and small-sized particles.

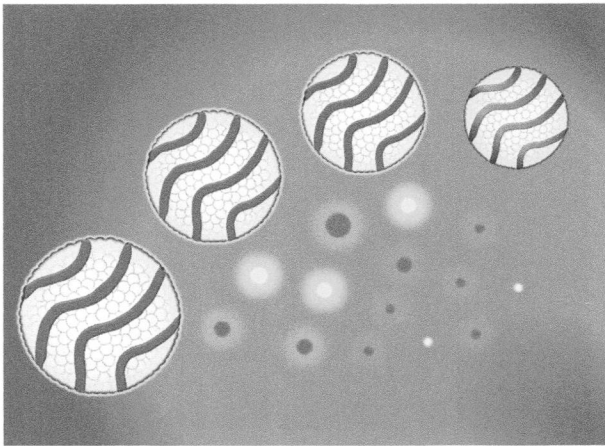

There are several subtypes of triglyceride-rich lipoproteins, which include chylomicrons, VLDL1, VLDL2, and IDL. Red blood cells (erythrocytes) and white blood cells (leucocytes) are depicted as red and white stars in the background. The size of the lipoproteins is magnified 150 times larger than that of the cells.

The primary reason for the heterogeneity of triglyceride-rich lipoproteins lies in the structure and length of apolipoprotein B. This protein has the ability to alter its shape depending on the amount of lipid it has bound, resulting in the formation of particles of varying sizes. It is important to remember that each lipoprotein particle contains only one apolipoprotein B molecule.

18

Remnant Cholesterol – Where It Comes From

Lipoproteins rich in triglycerides are responsible for moving triglycerides from the liver and intestines to the muscles to be used for energy and to the adipose tissues for storage. Cholesterol is a part of these particles which become remnants after their lipolysis by lipoprotein lipase.

The processing of triglyceride-rich lipoproteins involves several steps, illustrated in a clockwise direction starting from the top left. After a meal, chylomicrons are released by the intestine and then broken down into chylomicron remnants by lipoprotein lipase. In between meals, VLDL is released by the liver and broken down into IDL and LDL by lipoprotein lipase. Both chylomicron remnants and LDL bind to the LDL receptors and are removed from circulation. However, they can also enter the arterial wall and lead to cholesterol buildup in that area. CETP facilitates the exchange of cholesteryl esters and triglycerides between triglyceride-rich lipoproteins and HDL.

Production of Triglyceride-Rich Lipoproteins

When we consume fat, chylomicrons form in the small intestine and are secreted into the lymph. They then quickly move to the bloodstream and travel throughout the body.

Chylomicrons form primarily in the jejunum, which is the second part of the small intestine, and are unique to the intestine. Enterocytes produce apolipoprotein B-48, which helps make these particles using the triglycerides and cholesterol we consume. VLDL particles are made in the liver using triglycerides and cholesterol from different sources, such as those the body produces or obtains. These lipoproteins contain apolipoprotein B-100.

Cholesterol, Lipoproteins, and Cardiovascular Health: Separating the Good (HDL), the Bad (LDL), and the Remnant, First Edition. Anatol Kontush.
© 2025 John Wiley & Sons, Inc. Published 2025 by John Wiley & Sons, Inc.

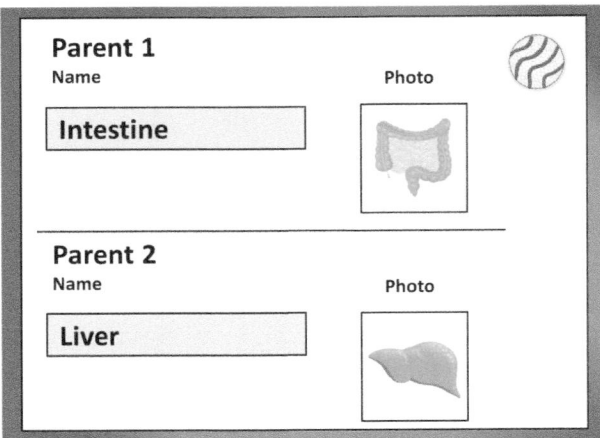

The liver and the intestine are vital organs that determine the processing of triglyceride-rich lipoproteins.

Chylomicrons are assembled from lipids that are absorbed at mealtime. Intestinal lipases break down triglycerides in the stomach and small intestinal tract when we consume about 100 grams of them per day. They then form fatty acids and other substances. The majority of fatty acids and cholesterol in the intestines are derived from food. The cells of the small intestinal tract absorb these lipids by a process known as diffusion, or with the help of specialized proteins that transport lipids. The bacteria that live in the gut, called microbiota, can reduce the amount of cholesterol the body takes up. This is done, for instance, by converting cholesterol into coprostanol which the body easily eliminates through the intestine.

After a meal, the intestine releases chylomicrons which are then broken down into smaller chylomicron remnants by lipoprotein lipase. In between meals, the liver releases VLDL (shown in the figure) which is also broken down into smaller particles of IDL and LDL by lipoprotein lipase.

In the enterocytes, triglycerides undergo a resynthesis from fatty acids to be stored for a brief time in structures known as lipid droplets before they combine with the apolipoprotein B-48, phospholipids, and cholesterol. This combination occurs in a part of cells called the endoplasmic reticulum and is facilitated through a protein known as microsomal triglyceride transfer protein, which aids in the formation of chylomicron particles. The chylomicrons leave the endoplasmic reticulum and enter the Golgi apparatus, a secretory part of the cell. The chylomicrons are then sent to the lymphatic system via a system called lacteals in the intestine before entering the bloodstream through thoracic duct, a main vessel linking the lymph to the blood.

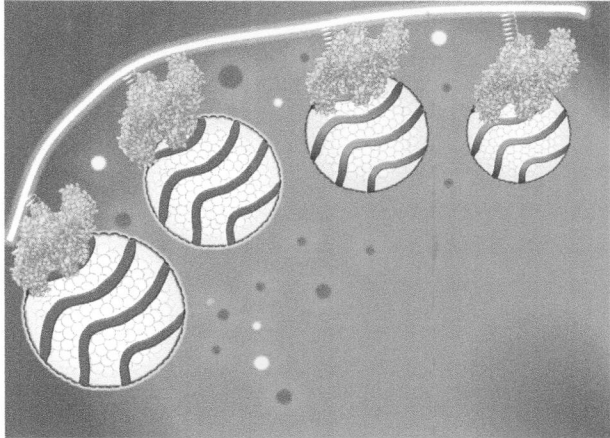

Dimeric lipoprotein lipase (in ligh blue and green) is responsible for breaking down chylomicrons and VLDL (shown in the figure) into smaller remnant particles, such as chylomicron remnants and IDL.

Chylomicrons greatly increase the amount of fat in the blood following a meal. The rate at which chylomicrons are formed is determined by the availability of apolipoprotein B-48, which is controlled by insulin and gut peptides. It is also influenced by neural networks and signals from nutrients like fatty acids and glucose. When there is more or less protein and lipid available, chylomicrons can have different sizes and compositions.

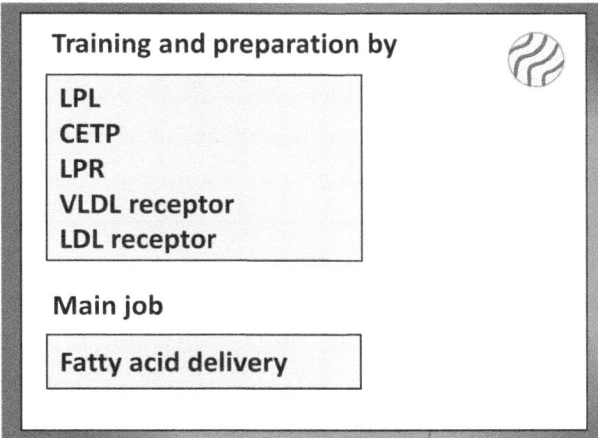

Various proteins and enzymes, including lipoprotein lipase (LPL), CETP, LRP, VLDL receptor, and LDL receptor, are responsible for processing triglyceride-rich lipoproteins in the blood. These lipoproteins play a main role in delivering fatty acids to tissues for energy production.

VLDL is produced in liver cells by a two-step procedure. With the aid of microsomal triglyceride transfer protein, lipids and apolipoprotein B-100 are first combined to form a basic VLDL particle.

In the endoplasmic reticulum, triglycerides are added to this particle. The liver makes triglycerides in different ways, such as creating them from scratch, using stored fats, or getting them from fatty acids in the blood. When the liver makes more new lipids, it also makes more VLDL. The liver also releases more VLDL when it receives more triglycerides from remnants in the bloodstream because there is less breakdown of chylomicrons and VLDL in the blood.

The liver produces different types of particles containing apolipoprotein B-100. The particles are sized from large VLDL1 to small VLDL2. The amount of triglycerides produced in the liver determines the type of particles. When there is an abundance of triglycerides, the liver produces VLDL1 that are large and rich in triglyceride. When triglycerides become scarce, smaller VLDL2 is secreted.

In healthy individuals during fasting, the liver releases approximately 20–70 grams of triglycerides in VLDL each day, associated with one gram of apolipoprotein B-100. The intestine absorbs, and packages, in the range of 50–200 grams of triglyceride every day, depending on the amount of fat in the diet.

Scientists have discovered new proteins in the past 20 years that can affect how our bodies handle triglyceride-rich lipoproteins. Examples include apolipoprotein A–V, angiopoietin-like protein 3 (ANGPTL3), and GPIHBP1.

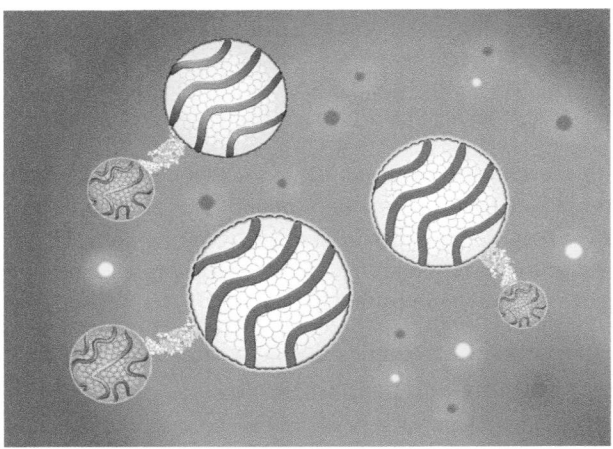

CETP facilitates the exchange of triglycerides from triglyceride-rich lipoproteins with cholesteryl esters from HDL. This exchange occurs through a tunnel connecting these particles.

History

Discovery of Lipoprotein Lipase
In 1943, lipoprotein lipase enzyme was discovered.

Paul Hahn, a biochemist at the University of Rochester (USA), was researching how to measure the red blood cell volume in dogs. Heparin was injected into his catheters to prevent them from clotting. He noticed by chance that lipemia quickly cleared and blood plasma became transparent instead of milky when heparin was injected in the dogs which ate a meal containing fats and then had lipemic blood. Even after placing the plasma on a laboratory bench and removing it from the dog, he found that the lipemia cleared. Adding heparin to plasma did not clear the diet-induced lipemia. This indicated that the heparin was releasing a factor to remove fat from plasma. Hahn published a 1943 paper in *Science* describing a lipemia-clearing factor caused by heparin. He noted that "this was a striking phenomenon, even when the level of lipemia was so high, the plasma looked like light cream."

Who	When
Paul Hahn	**1943**
Where	**What**
Rochester NY, USA	Hahn, P. F. (1943) "Abolishment of alimentary lipemia following Injection of Heparin." Science 98: 19-20.

What It Is

Lipemia occurs when there is a high level of lipids present in the blood.

In the 1950s, experiments with heparin were carried out at the Donner Lab under the directorship of John Gofman. The researchers found that the decrease in turbidity correlated with the conversion of VLDL into LDL. In 1952, Christian Anfinsen's laboratory at NIH showed that an enzyme was catalyzing a conversion from larger to smaller VLDL with the formation of HDL.

(Continued)

(Continued)

Edward Korn, at the NIH in 1955, discovered that an enzyme (the clearing factor) was responsible for the removal of circulating fatty acids after a meal. He named it lipoprotein lipase. Hahn had observed the same effects. Korn found that lipoprotein lipase is present in the heart and is released from blood vessels when heparin is used. He proposed that lipoprotein lipase breaks down triglycerides to glycerol and free fatty acids which are not cloudy, explaining why heparin produces a "clearing effect." He suggested that lipoprotein lipase breaks down triglycerides along the blood vessel surface in the heart even without heparin.

In 1956, Richard Havel, Robert Gordon, and their colleagues at the University of California San Francisco discovered a family suffering from severe hyperchylomicronemia. The researchers found that plasma taken from family members after heparin was injected did not degrade chylomicrons. Havel and Gordon then linked this condition (also known as familial chylomicronemia or Type I hyperlipoproteinemia) with a defective lipoprotein lipase enzyme. This was the very first genetic lipoprotein-lipase deficiency to be observed.

A few years later, another interesting part of this story occurred. Researchers found that plasma taken from children with lipoprotein lipase deficiency could still be used to break down triglyceride in emulsion particle form. It was surprising to see that this went against the original hypothesis that lipoprotein lipase deficiency causes hyperchylomicronemia. The only difference was that Havel and Gordon used chylomicrons rather than lipid emulsions. This conflicting evidence led to the discovery of a new enzyme called hepatic lipase. This enzyme breaks down remnants, IDL, and HDL phospholipids, but not chylomicrons. The enzyme was active against triglycerides found in smaller emulsions rich in triglycerides.

A third enzyme, capable of destroying lipids within the plasma, was also found later after heparin administration. This enzyme was called endothelial lipase.

19

Remnant Cholesterol – What It Does

Lipoproteins rich in triglycerides transport the triglycerides, whose lipolysis plays a vital role in energy production. The breakdown of triglycerides within the blood results in fatty acids. The body tissues easily absorb the fatty acids, and they are used to produce energy. In a process called beta-oxidation, fatty acids are oxidized to synthesize ATP. This is a form of energy that cells use ubiquitously. Energy supply from lipoproteins rich in triglycerides can protect cells from damage and help them stay alive.

Triglycerides are derived from chylomicrons produced by the intestine, or VLDL produced by the liver. Lipoprotein lipase, attached to the endothelium in capillaries that nourish muscle and adipose tissue cells, converts triglycerides into fatty acids and glycerol upon their entry into the circulation. Locally, the cells absorb the free fatty acids. Triglycerides mainly reach the cells after a meal. Free fatty acids can be used as energy in muscle cells. These molecules can also be converted back into triglycerides in adipocytes and stored for future use. A hormone-sensitive lipase in adipocytes breaks down cellular triglycerides into fatty acids during fasting. The fatty acids are released from the adipose tissues and enter the plasma, providing fuel to all tissues.

The body creates remnant particles when it breaks down chylomicrons. The liver then clears these particles using receptors such as LDL receptor, LRP and VLDL receptor. This process also occurs with large VLDL1, which becomes small VLDL2, as well as with remnants and IDL. They are either converted to LDL or removed by liver receptors. Lipid transport occurs via delivery of lipoproteins through their receptors into targeted tissues. The liver can also produce small VLDL2 that can be converted into IDL and LDL. The breakdown of VLDL and chylomicrons in muscle and adipose tissue is not random. Insulin plays a central role in controlling this process. It stimulates the breakdown of fats in adipose tissues while inhibiting it in skeletal muscles. Additionally, insulin enhances glucose uptake by muscles and promotes the use of fat for energy by adipose tissues.

Remnant cholesterol is transported to cells and tissues along with the rest of the remnant particles. It plays a role as a source of cholesterol necessary for proper cell function. Any remnant cholesterol that is not absorbed by the tissues is taken up by the liver, where it can be utilized to generate new VLDL particles. Additionally, it may also be excreted in the bile along with the remaining liver cholesterol.

Lipolysis as a Source of Energy

Lipoprotein lipase, an enzyme of great importance, breaks down triglycerides found in chylomicrons or VLDL into free fatty acids. The enzyme is produced primarily in tissues such as the heart and muscle, which use triglycerides as energy. It is also made in adipose cells where triglycerides

Cholesterol, Lipoproteins, and Cardiovascular Health: Separating the Good (HDL), the Bad (LDL), and the Remnant, First Edition. Anatol Kontush.
© 2025 John Wiley & Sons, Inc. Published 2025 by John Wiley & Sons, Inc.

can be stored. Lipoprotein lipase is attached to molecules called proteoglycans on the inner surface of arterial walls. A special protein, GPIHBP1, transports it to the blood vessels. This allows lipoprotein lipase to break down triglycerides found in chylomicrons or VLDL. This causes fatty acids and other substances to be released, which lead to the formation of smaller particles of chylomicron and VLDL remnants.

Lipoprotein lipase is active as dimers composed of two identical protein molecules. The enzyme makes VLDL and chylomicron particles less stable. The core, which is made of triglycerides in the particles, is destroyed. It is necessary to adjust the particle surface because it becomes too large for the reduced core. The extra surface of chylomicrons or VLDL is removed, releasing phospholipids as well as free cholesterol and apolipoproteins. These molecules can be either taken up by HDL particles already present or used to form new HDL. Lipoprotein lipase also helps cells bind lipoproteins rich in triglycerides and remove them from the blood.

The activity of lipoprotein lipase is closely controlled on different levels because it plays such a vital role in maintaining the balance of energy in the body. Several proteins, such as apolipoproteins C-I, C-II, C-III, and A-V as well as ANGPTL3, ANGPTL4, and ANGPTL8, help regulate lipoprotein lipase. ANGPTL3, 4, and 8 play a part in stopping the action of this enzyme, while apolipoprotein A-V enhances it.

The lipolysis increases the concentrations of HDL particles within the blood. As well, many proteins are transferred between HDL, on the one side, and VLDL or chylomicrons, on the other. This exchange may affect the activity of lipoprotein lipase. Apolipoprotein C-II is a lipoprotein-lipase activator, while apolipoprotein C-III is an inhibitor. Apolipoprotein A-II is also important for controlling lipoprotein lipase activity. It does this by dislodging apolipoproteins C from HDL to triglyceride-rich lipoproteins. This is because apolipoprotein A-II has a higher affinity for the HDL surface than other apolipoproteins.

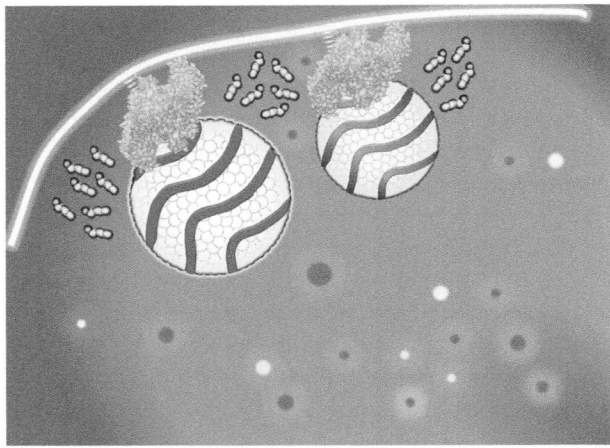

Triglyceride-rich lipoproteins in the blood are responsible for delivering fatty acids to tissues. These fatty acids are then utilized to produce energy in the form of ATP or stored inside cells in the form of triglycerides.

Hepatic lipase degrades lipids, especially phospholipids on the surface of lipoprotein remnants. Hepatic lipase action leads to a lower level of IDL particles in the blood because they are cleared more quickly.

In the past, scientists believed that the breakdown of triglycerides is a simple process controlled only by lipoprotein lipase, and apolipoprotein C-II and apolipoprotein C-III on triglyceride-rich lipoproteins. Now, we know that other proteins are involved. Mutations can make them not function properly, resulting in very high or low triglyceride levels. These proteins are partially inactive in some people who have high triglycerides.

Apolipoprotein C-III influences how effectively triglycerides are removed from the blood. Lower levels of apolipoprotein C-III can lead to a more effective removal of triglycerides, which in turn leads to their lower level. Higher levels of apolipoprotein C-III in conditions such as diabetes and obesity can lead to less effective removal and higher levels of triglycerides.

ANGPTL3, 4, and 8 are angiopoietin-like proteins that inhibit lipoprotein lipase. They also play a part in the direction of chylomicrons or VLDL toward adipose tissues or skeletal muscles. It depends on the nutritional and hormonal state at the time.

VLDL particles produced in the liver are broken down into smaller particles, including VLDL remnants, IDL, and LDL. Similarly, chylomicrons produced in the intestine are also broken down into smaller particles, but not into IDL or LDL. During this breakdown process, the size of the lipoproteins decreases. The amount of triglycerides is reduced while the amount of cholesteryl ester increases. CETP plays a main role in this process by facilitating the exchange of triglycerides from triglyceride-rich lipoproteins with cholesteryl esters from LDL and HDL. The apolipoprotein B remains associated with the lipoproteins it was initially released with. Since there is one apolipoprotein B molecule per particle, its concentration can be measured to determine the number of particles present. Apolipoprotein Cs and other proteins are primarily transferred to HDL when triglyceride-rich lipoproteins are broken down.

Lipoprotein lipase breaks down newly secreted VLDLs into a mixture of particles. These particles, called apolipoprotein B-100 remnants or IDL, are further broken down into LDL by lipoprotein lipase and hepatic lipase. These remnants can be seen as temporary or transient. Other apolipoprotein B-100 remnants and all apolipoprotein B-48 remnants undergo a process that makes them resistant to further breakdown. They remain in the VLDL and IDL density range as end-product remnants until the liver removes them from circulation. The reason for the existence of different pathways is unclear.

The longer a remnant stays in circulation, the higher its cholesterol content becomes while its triglyceride content decreases. End-product remnants can contain over 7,500 molecules of cholesterol per particle, compared to the 2,000–2,700 found in LDL. The presence of apolipoproteins C-II and C-III on the surface of triglyceride-rich lipoproteins, as well as the availability and activity of lipoprotein lipase, determine how long remnants remain in circulation and whether they reach the sub-endothelial area of the arterial wall. The longer these particles stay in the bloodstream, the greater their chances of being retained in the arterial intima. After most of the lipids from chylomicrons are removed, apolipoprotein B-48 returns to the liver as part of chylomicron remnants. In the liver, it is taken up by cells and broken down.

History

Links Between Triglyceride-Rich Lipoproteins and HDL
Many of the proteins found in chylomicrons in the 1950s were also present in HDL. The studies used terminal residues to analyze the proteins but could not yet reveal their structure. Based on these data, researchers were able to develop a hypothesis about links between HDL and chylomicrons.

(Continued)

(Continued)

Who	When
Esko Nikkilä	**1978**

Where	What
Helsinki, Finland	Nikkilä EA, Taskinen MR and Kekki M. Relation of plasma high-density lipoprotein cholestrol to lipoprotein-lipase activity in adipose tissue and skeletal muscle of man. Atherosclerosis. 1978;29:497-501

Richard Havel's studies in the 1960s revealed that LDL and HDL could be the result of the action of lipoprotein lipase on VLDL. This involved the breakdown of triglycerides into fatty acids. These studies observed the transfer of proteins from lipoproteins rich in triglycerides to HDL after lipolysis. Parallel to this, phospholipids were also transferred. Apolipoproteins C were found to be transferred from HDL into VLDL while phospholipids moved in the opposite direction. The researchers observed that HDL is the source of the apolipoprotein C found in chylomicrons. The intestine cannot produce this protein.

In the 1970s, studies showed that HDLs and triglyceride-rich lipoproteins were also linked by the transfer of cholesterol when triglycerides are broken down by lipoprotein lipase. This process produces energy for the human body. The amount of HDL-cholesterol in the blood is increased by the transfer of cholesterol from lipoproteins rich in triglycerides to HDL. Clinical tests confirmed that low triglycerides are linked to high HDL-cholesterol.

Esko Nikkilä, at the University of Helsinki, Finland, Josef Patsch, and Antonio Gotto, Jr., at the Baylor College of Medicine, Houston, and Alan Tall, at Columbia University, New York, have all studied this process. Lipoprotein lipase breaks down triglycerides into free fatty acids when triglyceride-rich lipoproteins are injected into the bloodstream. This causes them to shrink and become denser like a deflated soccer ball. Some of the surface components, such as proteins, phospholipids, and cholesterol, are shed during this process. These surface remnants can form larger vesicles. The proteins and lipids that are on the surface triglyceride-rich lipoproteins can also be used to form small disc-shaped pre-beta HDL particles. In the presence of HDL, however, these structures prefer to fuse with existing HDL particles.

During the 1970s, these studies experienced rapid development. However, in the early 1980s, another type of interaction between HDL and triglyceride-rich lipoproteins was discovered. This pathway involves the exchange of specific lipids through CETP, resulting in the formation of HDL with more triglycerides and triglyceride-rich lipoproteins with more cholesteryl esters. Additionally, during this time, Nikkilä tragically passed away in a car accident, causing the focus of research in the HDL field to shift toward CETP.

Removal

When lipoproteins rich in triglycerides bind to cellular receptors such as the LDL receptor, LRP, and VLDL receptor, they are removed from the bloodstream. These receptors are mostly located on the surface of liver cells. LRP is a member of the LDL family and binds large particles rich in triglycerides. VLDL receptors are found in the heart muscle and adipose tissues. They help transport fatty acids to tissues from VLDLs. The heart and other muscles use the fatty acids for energy while adipose tissues store them.

The liver directly removes chylomicron remnants as well as varying amounts of VLDL remnants from the blood. The LDL receptor is the main player in this process, but other proteins such as LRP or proteoglycans can also be involved. For chylomicron remnants to be removed from the blood, they must have functional apolipoprotein E. It is interesting that their major structural protein, apolipoprotein B-48, does not allow them to interact with the LDL receptor.

Different pathways are used by the liver to eliminate VLDL remnants. Around 25–75% of these particles do not turn into LDL but are instead removed directly. Apolipoprotein C-III prevents remnant particles in the VLDL and IDL range from being taken by the liver while apolipoprotein E aids with their removal. According to studies in mice, endothelial and hepatic lipases may help remove remnants as well without the need for lipoprotein receptors and lipoprotein lipase.

Apolipoprotein E helps clear several types of particles in the blood, like chylomicron remnants. It interacts with LRP receptors and VLDLs. Apolipoprotein C-III, on the other hand, interferes with VLDL binding to these receptors. This leads to the formation of smaller LDL particles and counteracts the effects of apolipoprotein E.

It is essential to remove all remnants in order to prevent inflammation, which can be caused by triglyceride-rich lipoproteins. When chylomicrons accumulate, they can cause inflammation in the pancreas. This increases the risk of acute pancreatitis. VLDL, IDL, and VLDL remnants as well as chylomicrons can also cause inflammation and allow immune cells to move into the arterial wall.

When triglyceride-rich lipoproteins in the blood are broken down by lipoprotein lipase on the surface of blood vessel cells, this releases substances that can lead to atherosclerosis. These substances activate genes that cause inflammation, blood clotting, and cell death. The breakdown of triglyceride-rich lipoproteins also releases some fatty acids that can cause inflammation in blood vessel cells and produce harmful substances like reactive oxygen species. This is especially true for saturated fats compared to polyunsaturated fats like omega-3.

20

Remnant Cholesterol – How It Stops Working

Causes

Hypertriglyceridemia

Triglyceride levels can vary greatly between people. Plasma triglyceride concentrations can range between 30 and 10,000 mg/dl. When fasting triglyceride level is above 100 mg/dl, a buildup of triglyceride-rich lipoproteins and their remnants occurs in the blood. In the Copenhagen General Population Study, 27% of participants had slightly elevated triglyceride levels in their blood (between 180 and 350 mg/dl), and only 0.1% of them had extremely high levels (above 900 mg/dl). When it came to remnant cholesterol measurements, almost half of the individuals (45%) had slightly elevated levels between 45 and 90 mg/dl, and about one-fifth (21%) had levels higher than 90 mg/dl.

Variability in plasma triglyceride concentrations is due to the wide variation of clearance and secretion rates between individuals. This variation is caused by differences in the way that the body produces triglycerides, apolipoproteins B, and chylomicrons. How much triglyceride the liver can remove, how effectively it does so, and the amount of lipoproteins that are triglyceride-rich is what determines the concentrations in the blood. Lipolysis is controlled by different enzymes, proteins, and remnants of triglyceride-rich lipoproteins.

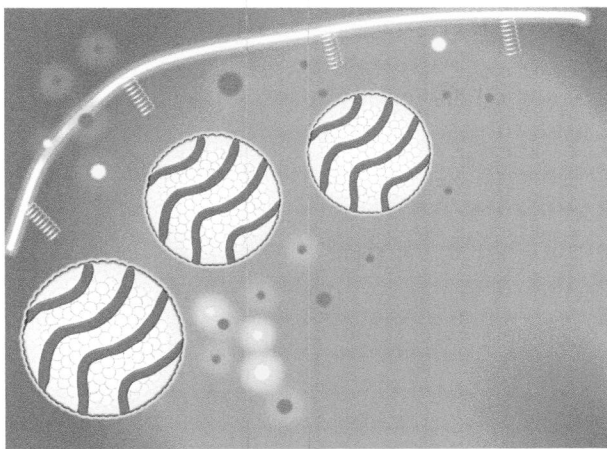

The absence of lipoprotein lipase leads to the accumulation of large triglyceride-rich particles in the bloodstream, resulting in highly elevated levels of triglycerides and remnant cholesterol.

Cholesterol, Lipoproteins, and Cardiovascular Health: Separating the Good (HDL), the Bad (LDL), and the Remnant, First Edition. Anatol Kontush.
© 2025 John Wiley & Sons, Inc. Published 2025 by John Wiley & Sons, Inc.

Overproduction is the most common cause of borderline or moderately high triglyceride concentrations. When triglyceride concentrations are very high, it is usually due to a reduced triglyceride lipolysis by lipoprotein lipase. In people with moderately high triglycerides, the liver produces more VLDL. This causes a higher level of lipoproteins that are rich in triglycerides and their remnants. When triglycerides are even higher, it is because the body does not break down them well. This leads to both VLDLs and chylomicrons building up. Those with severely high triglycerides – especially those that have genetic defects – may have normal production but a very low breakdown. In severe hypertriglyceridemia, the main problem is therefore that triglyceride-rich lipoproteins are not cleared from the blood well enough.

The process of triglyceride turnover is affected in dyslipidemias. High triglycerides in atherogenic dyslipidemia are associated with low HDL-cholesterol and high levels of small, dense LDL. This increases the risk of cardiovascular disease. Atherogenic dyslipidemia, which is associated with high triglycerides and low HDL-cholesterol, can also include high levels of LDL-cholesterol and total cholesterol.

A combination of genetic factors as well as lifestyle choices can cause high triglycerides. Genetic studies have revealed that certain components in triglyceride-rich lipoproteins and processes involved in the production and breakdown of these lipoproteins are related to heart disease. Genetic factors can increase the levels of triglyceride-rich lipoproteins and cause heart disease. Rare genetic disorders can increase triglyceride concentrations in the blood. This includes defects in genes coding for lipoproteins lipase, apolipoprotein C-II, apolipoprotein A-V, and others. Excessive alcohol consumption, diabetes, and obesity are the most common lifestyle causes of high triglycerides. In fact, both dietary sugar and alcohol can increase plasma triglycerides.

In some conditions, the removal of VLDL and chylomicrons from circulation may be delayed. Remnants can form when there are too many triglyceride-rich lipoproteins or certain genetic or physiological factors that limit the breakdown. The liver and small intestinal tract produce a greater amount of large triglyceride-rich lipoproteins, and chylomicrons which are high in triglycerides when there is a moderately elevated level of triglycerides. The accumulation of remnants is caused by this, in combination with the poor lipoprotein-lipase activity that occurs when people have diabetes or obesity. Even when the production of triglyceride-rich lipoproteins is normal, certain variations in the *LPL* or other genes that influence lipoprotein lipase activity can cause an increase in triglycerides and remnant accumulation. In the absence of lipoprotein lipase, triglyceride breakdown is not possible. This results in a significant increase in large triglyceride-rich lipoproteins without any increase in remnants. Pancreatitis, rather than cardiovascular disease, is the major risk in this situation. This possibly results from elevated blood viscosity in chylomicronemia.

The type of particles that contain the apolipoprotein B-100 and B-48 varies depending on the level of triglycerides. These particles are mostly small VLDL or IDL when triglycerides levels are low. When triglycerides increase, chylomicron remnants and VLDLs are accumulated. These particles contain a high amount of cholesterol, less apolipoprotein C, and more apolipoprotein E. They are either removed by the liver or converted into IDL and LDL. These particles are very different from each other in most people.

People with high triglyceride levels have VLDL and IDL particles that contain a variety of different lipids. Among these, triglyceride-rich lipoproteins and their remnants are the main carriers of neutral lipids like triglycerides, diglycerides, and cholesteryl esters, specific phospholipids like phosphatidylethanolamine and phosphatidylinositol, and ceramides. Remnant particles may therefore have a particular pattern of lipids. This pattern seems to be more specific for chylomicrons than VLDLs.

Type III dyslipidemia is characterized by a genetic variant of the apolipoprotein E, called apolipoprotein E2, which prevents VLDLs from binding to receptors on cells. These receptors are responsible for removing remnant particles which accumulate in the blood. Plasma concentrations of VLDL that contain apolipoprotein E are elevated in such individuals. Apolipoprotein E is also enriched in these VLDL when compared to normal individuals. In this condition, plasma concentrations of VLDL and triglycerides are elevated. VLDL shows abnormal beta migration instead of pre-beta migration on paper electrophoresis and is known as beta-VLDL. Only some of Type III patients will, however, develop remnant dyslipidemia. Other factors like insulin resistance are needed to increase the production of VLDL or chylomicrons, or to impair their breakdown. High levels of lipids are the result. One characteristic of this condition is a very high ratio of cholesterol to triglycerides in the remnants. This happens because there is a significant buildup of VLDL and chylomicron remnants that are rich in cholesterol. This dyslipidemia is associated with an increased risk of heart disease, even though LDL-cholesterol is usually low. Plasma concentrations of apolipoprotein B, and in particular apolipoprotein C-III, are elevated in Type IV dyslipidemia. In Type V dyslipidemia, apolipoproteins C-III and E are also higher. As a consequence, increases in plasma apolipoprotein C-III typically reflect the presence of hypertriglyceridemia.

Excessive secretion of VLDL by the liver can result in elevated levels of triglycerides and remnant cholesterol, thereby increasing the likelihood of heart attack and stroke. This occurrence is commonly observed in individuals with diabetes and metabolic syndrome.

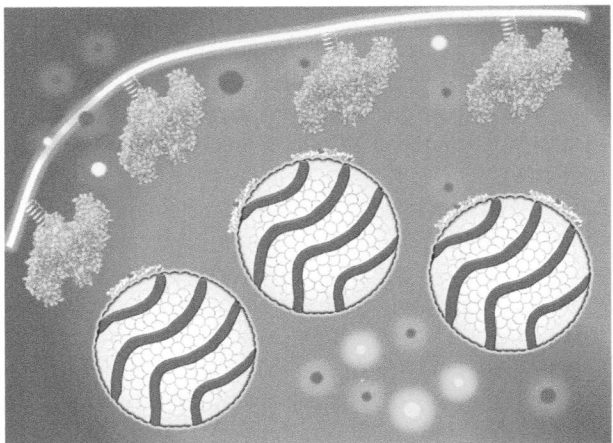

High levels of apolipoprotein C-III (shown at the surface of VLDL) in the bloodstream hinder the breakdown of lipoproteins rich in triglycerides by lipoprotein lipase. They also reduce uptake of remnant particles by the liver. Consequently, this causes the buildup of triglyceride-rich particles and increased levels of both triglycerides and remnant cholesterol.

History

Atherogenicity of Triglyceride-Rich Lipoproteins

Who	When
Donald Zilversmit	**1973**

Where	What
Ithaca, NY, USA	Zilversmit, D. B. (1973) "A proposal linking atherogenesis to the interaction of endothelial lipoprotein lipase with triglyceride-rich lipoproteins. Circulation Research 33: 633-638.

John Gofman published a study in 1951 that showed elevated concentrations of IDL, a remnant lipoprotein rich in triglycerides, in patients with atherosclerosis. Rabbits fed cholesterol revealed similar results. These concentrations could also be used to predict heart disease in population studies.

In 1962, studies by Richard Havel from the United States and Lars Carlson from Sweden suggested that elevated levels of VLDL, LDL, or both can cause atherosclerosis.

Donald Zilversmit first proposed the hypothesis that triglyceride-rich particles are atherogenic in 1973. He suggested that atherogenesis was linked to the interaction between lipoprotein lipase and triglyceride-rich lipoproteins such as VLDL or chylomicrons. The remnant particles that are formed during this reaction may enter the arterial walls and cause cholesterol buildup.

It was confirmed that the high cholesterol levels of rabbits that were fed a diet rich in cholesterol are due in part to an accumulation of chylomicron remnants. Lipoprotein lipase is found in the aorta, which suggests that remnants are formed on the inner surface of the arteries. Even without releasing them into the bloodstream, they can be deposited in deeper layers of the arterial wall.

Zilversmit's hypothesis on the atherogenicity of triglyceride-rich lipoproteins suggests that elevated blood triglycerides can contribute to atherosclerosis. Important to note that triglycerides do not cause atherosclerosis on their own – they are broken down by lipoprotein lipase and do not accumulate in the arterial walls. Their concentrations in plasma are a sign of the dangerous remnant particles that are rich in cholesterol.

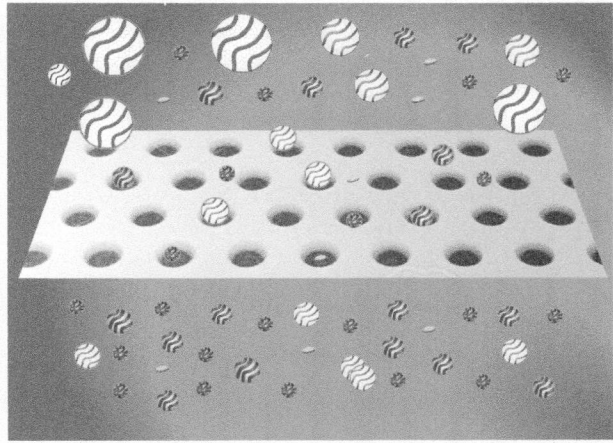

Large lipoproteins are unable to penetrate the endothelium of the arterial wall, whereas smaller lipoproteins have the ability to do so.

According to this theory, high triglyceride values are a sign of either increased VLDL production in the liver or a delayed removal of VLDL from the blood. VLDL particles are transformed into IDL when triglycerides are removed from VLDL by lipoprotein lipase. IDL particles are highly atherogenic. They have a capacity to cause plaque to form within the arterial walls.

IDL particles are also capable of being further modified to LDL which is particularly atherogenic. LDL can easily penetrate arterial walls, causing the formation of fatty streaks. IDLs are removed quickly from the bloodstream while LDLs circulate for days. LDL is therefore a major contributor to the accumulation of cholesterol within lesions.

Zilversmit's hypothesis suggests that high levels of triglycerides can contribute to atherosclerosis through the production and conversion of VLDL and IDL particles. This pathway ultimately leads to an increase in cholesterol within the arterial wall.

Later studies like those performed at the laboratory of Zilversmit by Børge Nordestgaard confirmed this hypothesis.

Genetics

About half of the differences in triglycerides between people can be explained by differences in their DNA. We can determine if high levels of triglycerides are a cause or just a reflection of heart disease by studying genetics.

Abnormalities in how the body processes triglycerides are linked to certain genes. Some of the genes associated with these abnormalities are those of apolipoproteins A-I, A-IV, A-V, C-III, and E, as well as lipoprotein lipase, microsomal triglyceride transfer protein, and SR-BI receptor. The *APOA5* gene has the strongest link to an increased risk of heart disease caused by high levels of triglycerides. Apolipoprotein A-V plays a role in various steps of lipid processing, such as producing VLDL, clearing remnant lipoprotein particles from the blood, and controlling the activity of lipoprotein lipase. A change in the *APOA5* gene can cause elevated levels of plasma triglycerides and increase the risk of heart disease. Other genes that may affect how our body responds to dietary fats include *ANGPTL4* and *PPARA*.

Familial hyperchylomicronemia can lead to high levels of chylomicrons, which may cause pancreatitis. The triglycerides are also very high, and the HDL- and LDL-cholesterol levels are low. Mutations in the *LPL* gene are responsible for this condition. Symptoms are milder in those who have only one copy of this gene mutation. This condition does not however increase the risk of heart disease, because lipoproteins lipase can have both positive and adverse effects on cardiovascular health.

Heart disease can also be caused by other genes that are related to triglyceride processing. Mutations of the *APOC3* can result in lower levels of triglycerides and LDL-cholesterol. They also reduce the risk of atherosclerosis. In studies that examine the entire genome, the gene cluster involving *APOA1*, *APOC3*, *APOA4*, and *APOA5* is strongly related to triglyceride levels.

Two studies have shown that apolipoprotein C-III plays a major role in heart disease and triglyceride levels. In one study, rare mutations of the *APOC3* gene were detected in 1 in 150 individuals. These mutations were linked to a 39% reduction in triglycerides and a 40% decrease in heart disease risk.

In Denmark, a second study was conducted using data collected from the general population. Low levels of non-fasting triglycerides were associated with a reduced risk of cardiovascular disease. Those who had triglycerides less than 90 mg/dl were 60% less likely to develop heart disease compared with those who had levels higher than 350 mg/dl. The *APOC3* mutations were associated with a reduced triglyceride level (by 56%) and a decreased risk of ischemic vascular diseases (by 59%).

In the 1970s, early studies found that mutations in apolipoprotein C-II were responsible for elevated triglycerides. In a patient with severe hypertriglyceridemia, the lipoprotein lipase activity was normal but there was no apolipoprotein C-II in VLDL. As apolipoprotein C-II is an activator for lipoprotein lipase, its absence resulted in hypertriglyceridemia.

Most people have LDL particles that are 3–10 times more abundant than lipoproteins rich in triglycerides. LDL is also found to stay in the bloodstream for a longer period of time than VLDL and chylomicrons. While chylomicrons and VLDL are removed from the blood within 4–13 hours, LDL remains in the blood for 2.5-3.5 days. This is similar to the long residence time of HDL, which is from four to six days. The average turnover time for chylomicron triglycerides in the blood is 7 minutes, whereas that for VLDL triglycerides is about 20 minutes. The capacity of these lipoproteins to induce heart disease thereby increases by prolonging their stay in the blood. This still makes LDL the main target for preventing heart disease. Although it is difficult to measure the cholesterol content of remnant particles with accuracy, studies have shown that they can carry up to 30% of plasma cholesterol.

Consequences

Cholesterol Buildup and Inflammation in Arterial Walls

Long time ago, it was believed that atherosclerosis occurs primarily after eating when blood vessels are exposed to high levels of lipoproteins rich in triglycerides. These lipoproteins have been shown to cause heart disease. In fact, remnants of these lipoproteins are found in plaques that block blood vessels.

CETP accelerates the process of cholesterol transfer from HDL into triglyceride-rich lipoproteins after eating. It can lead to more cholesterol-rich remnant particles. Parallel to this, HDL particles become richer in triglyceride and poorer in cholesterol. This is the same process that occurs in people who have high levels of blood triglycerides. In the hours following a meal, HDL-cholesterol may slightly drop. This decrease is small when eating normal foods. After eating, both people with diabetes and healthy individuals experience a slight increase in inflammation.

The release of fatty acids, and other molecules, occurs when lipoproteins lipase is activated. This can be on the blood vessel lining or inside the arterial wall. These substances may cause inflammation and localized damage. This could explain why high remnant cholesterol levels from birth can cause mild inflammation. The remnants are directly taken up by macrophages and foam cells, unlike LDL which needs to be modified before uptake. High triglycerides can therefore indicate a greater amount of remnants containing cholesterol. These remnants trigger mild inflammation, foam cells and plaque formation, and ultimately heart disease.

Triglycerides are broken down in most cells. Cholesterol cannot be broken by any cell. This means that cholesterol in remnant triglyceride-rich lipoproteins is more likely to cause cardiovascular disease and atherosclerosis than a simple increase in triglycerides. In the arteries, cholesterol, and not triglycerides, builds up in foam cell plaques. Chylomicrons cannot enter the arterial walls, unlike remnant lipoproteins such as IDL and LDL. Because remnant lipoproteins are larger than LDL, they may be trapped in the arterial walls more easily.

At high triglyceride concentrations of above 900 mg/dl, triglyceride particle sizes are too large to cause atherosclerosis. When triglyceride concentrations are moderately elevated, however, the particles become small enough to enter the arteries. This can lead to atherosclerosis.

As triglyceride-rich lipoproteins break down, they create different subsets of smaller particles with varying abilities to cause heart disease. Chylomicron and VLDL remnants quickly go into the walls of the arteries and add cholesterol to the buildup of plaque. When there are too many of these particles in the blood, they build up in the space under the inner layer of the artery. Particles smaller than 70 nm like chylomicron remnants, VLDL remnants, and IDL can pass through the endothelium and get stuck in the arterial wall, leading to the formation and progression of atherosclerotic lesions. This happens because these particles contain a lot of cholesterol and can cause inflammation and blood clotting.

Remnant particles in the small VLDL and IDL range can have 5–20 more cholesterol per particle compared to LDL. These particles can cross the walls of the blood vessels and contribute to the formation of atherosclerotic plaques. They can stick to proteoglycans in the arterial wall especially when there is apolipoprotein C-III and apolipoprotein E on their surface. All these factors contribute to the buildup of cholesterol in the plaque, with remnant particles being more likely to enter than leave the arterial wall compared to LDL particles. Unlike LDL, these particles can be taken up by certain cells in the blood vessels without regulation, leading to the formation of foam cells.

Elevated levels of remnant particles for a long time after eating can cause damage to the arteries. Remnants containing apolipoprotein B-48 or apolipoprotein B-100 can be found in atherosclerotic lesions in humans. This suggests that both types of remnants play a role in the heart disease. Since it was first proposed, there has been growing recognition that post-meal lipoproteins from the intestines are important in this disease.

When remnant particles are broken down, lipid molecules are released that can cause blood vessel problems and inflammation. Monocyte-derived macrophages are cells that come from the blood to remove these particles from the arterial intima. However, when they take up too much cholesterol, they become foam cells, which start to form early signs of damage to the arterial walls. The process continues until a streak of fatty substances is formed. As smooth muscle cells move from the middle layer into the damaged area, more changes occur. A dangerous plaque can form that may break or be damaged on its surface. It can result in a blood clot blocking the artery and causing serious health issues.

The remnant lipoproteins that are rich in triglycerides can cause more inflammation than LDL. They contain more cholesterol per particle compared to LDL. Unlike LDL, these remnants are able to induce foam cells without any further modification. A recent study found a strong correlation between high blood triglycerides and inflammation of the vascular system, something that was not seen with LDL-cholesterol.

VLDL composition is a key factor in determining the risk of cardiovascular disease. In large-scale studies, only VLDL particles containing apolipoprotein C-III were associated with an increased risk of coronary artery disease. This was also true for their products which are small LDL particle. Apolipoprotein C-III prevents VLDL from attaching to cell receptors, which results in the formation of small, dense LDL particles that are especially harmful. This reverses the effects caused by apolipoprotein E.

History

Discovery of CETP

Who	When
Christopher Fielding	**1978**

Where	What
San Francisco CA, USA	Chajek, T. and C. J. Fielding (1978) "Isolation and characterization of a human serum cholesterylester transfer protein." Proceedings of the National Academy of Sciences USA 75: 3445-3449.

Alex Nichols and Lester Smith at the Donner Laboratory were the first to observe plasma CETP activity. They reported in 1965 that when human serum was kept for 16 hours, the cholesteryl ester content increased across all lipoprotein classes, including VLDL, LDL, and HDL. In the same time period, triglycerides from VLDL were transferred to LDL and HDL. In lipoproteins, the levels of phospholipids and nonesterified cholesterol decreased. This reflected the activity of LCAT. These results showed that cholesteryl ester exchange for triglycerides occurred in serum lipoproteins.

Zilversmit and colleagues studied this phenomenon in the 1970s. They discovered a protein in human plasma that helps move cholesteryl esters between lipoproteins. By separating plasma through different processes, they were able to purify the protein 3,500 times compared to the starting plasma. They found that the protein is a type of glycoprotein with specific characteristics, making it different from other known proteins. The exchange protein does not seem to have any effect on the process of cholesterol esterification in lipoproteins.

CETP was isolated from HDL by the laboratory of Christopher Fielding at the University of California, San Francisco in 1978. They showed that CETP helps move cholesteryl ester from HDL to VLDL or LDL. The process of transporting cholesteryl ester is linked to the movement of triglyceride from VLDL and LDL back to the HDL fraction.

In the 1980s, the human gene of CETP was studied and cloned, which led to research into how it is controlled. CETP was found to be a glycoprotein consisting of 476 amino acids, with a molecular mass of 66–74 kDa. The purified protein contained a high percentage (45%) of hydrophobic residues, as determined by amino acid analysis.

(Continued)

(Continued)

In 1982, Ying Chea Ha and Philip Barter in Australia discovered that CETP was active in fish, birds, reptiles, and mammals but not rats or mice. CETP was also present in primates. It was produced primarily in the liver.

Soon after, it was discovered that Japanese families with CETP deficiency had extremely high levels of HDL-cholesterol. However, there was ongoing debate about whether these levels actually provided protection against heart disease.

The researchers then studied its structure and function using recombinant forms of CETP. After cloning and studying the human *CETP* gene, researchers began to study how it was regulated. Transgenic mice have been created to investigate how CETP influences atherosclerosis. Researchers found that animals lacking CETP also had higher HDL-cholesterol levels. Inhibitors of CETP were soon discovered which improved markers for cardiovascular risk in both humans and animals. The inhibitors first increased HDL-cholesterol. They also reduced the levels of LDL-cholesterol and triglycerides. The clinical effects were however questioned in the subsequent trials.

In 2007, the molecular structure of CETP was finally resolved, which helped researchers understand how it works and develop better drugs for treatment. The protein showed a 6-nm long tunnel filled with two molecules of cholesteryl esters and blocked at each end by a molecule of phospholipid. The openings of the tunnel are big enough for the lipids to get through, helped by a flexible helix and possibly also by a moving flap. The shape of CETP matches the shape of HDL particles, and it might change to fit bigger lipoprotein particles. Mutations in the middle of the tunnel stop lipids from being transferred, which suggests that neutral lipids like cholesteryl ester and triglyceride move through this tunnel.

The discovery and understanding of CETP's function paved the way for further research into its potential as a therapeutic target for managing cardiovascular diseases. Scientists continue to investigate the complex mechanisms underlying CETP activity and explore potential interventions to modulate its function for therapeutic purposes

21

Remnant Cholesterol – What to Do to Correct It

Data suggest that a reduction in blood triglycerides is associated with a reduction in cardiovascular events. It is also true in patients who are treated with statins, after adjusting for the LDL-cholesterol reduction caused by these drugs. In statin trials, the effects of lowering LDL-cholesterol and triglycerides are similar. There is however more evidence to support the reduction of LDL-cholesterol than triglycerides.

Lifestyle changes and therapy can reduce triglycerides. Clinical trials have shown that therapies aimed at lowering triglyceride levels affect not only triglyceride-rich lipoproteins but also LDL, HDL, and inflammation. High levels of triglycerides, whether fasting or not, in the blood are also associated with heart disease and other conditions. The role of triglycerides in clinical effects is therefore difficult to single out.

A large study has not been done to determine the effect of reducing triglycerides on heart disease in those with high levels of triglycerides. Statin trials and other studies did not specifically include individuals with high triglycerides. These studies do not allow us to say with certainty if lowering triglycerides and remnant cholesterol helps prevent heart disease. Some reviews and analyses looked at the effects of lowering triglycerides but may have led people to believe that this was not helpful. More research is needed to determine if lowering high triglycerides reduces the risk of heart attack.

The ideal study would involve a group of people with moderately or mildly high triglycerides but normal LDL-cholesterol, taking a powerful cholesterol-lowering drug along with another to specifically reduce triglycerides. The study would compare the participants to a control group. Clinically, triglyceride-rich lipoproteins are as mysterious as HDL because there is no effective therapy that targets them.

According to the 2019 guidelines of the European Society of Cardiology (also known as ESC) and the European Atherosclerosis Society, the risk of cardiovascular disease increases when triglyceride concentrations are higher than 150 mg/dl. They only recommend medication for high-risk patients when the triglyceride level is above 200 mg/dl. The goal of lowering triglycerides and other remnant lipoproteins is not defined because studies have shown that there are limited benefits to lowering them.

According to recent guidelines issued by professional societies, it is best to focus on non-HDL-cholesterol when treating patients with dyslipidemia. Some recent studies have shown non-HDL-cholesterol levels to be more closely linked to cardiovascular events than LDL-cholesterol or apolipoprotein B. In a meta-analysis, it was found that reducing non-HDL-cholesterol is associated with lower risks of heart attacks and cardiovascular death. Statin therapy seems to have a greater impact on non-HDL-cholesterol. This metric may therefore be the better treatment target, particularly for people with high levels of triglycerides. Researchers have suggested that

Cholesterol, Lipoproteins, and Cardiovascular Health: Separating the Good (HDL), the Bad (LDL), and the Remnant, First Edition. Anatol Kontush.
© 2025 John Wiley & Sons, Inc. Published 2025 by John Wiley & Sons, Inc.

apolipoprotein B could be used as an alternative for non-HDL-cholesterol, because it is found in both LDL and VLDL.

Hypertriglyceridemia affects a number of processes in the body. Due to this complexity, many drugs that lower triglycerides have multiple effects. These drugs can alter the way certain genes function, making them more effective but also causing side effects. It is difficult to control the blood levels of lipoproteins rich in triglycerides and their remnants. It is hard to find treatments that will specifically lower these particles while reducing the risk of heart diseases.

It is not clear whether it is more effective to focus on how triglyceride-rich particles are broken down and then removed from the body rather than trying to reduce their production. Clinical trials using reliable tests should be conducted in order to determine if the current treatments are effective. In addition, it is important to determine which lipoproteins are more likely than LDL to cause heart disease. It will be necessary to perform specific tests on each particle type.

Lifestyle

People with moderately to slightly high levels of triglycerides can benefit from lifestyle changes. Lifestyle can help reduce triglyceride levels. Physical activity, reduced alcohol intake, reduced sugar consumption, and weight loss are all part of the lifestyle changes that can reduce blood triglycerides. Losing weight through eating less and exercising is the most important change in lifestyle.

Exercise

Aerobic exercise is strongly recommended to people with moderately high triglycerides. A sedentary lifestyle and lack of exercise are associated with high triglycerides. Exercise may however have a different effect on each individual. For example, those with high triglycerides such as diabetes or heart disease might respond better to it. Exercise can reduce plasma triglycerides, which in turn protects the heart.

Exercise reduces triglyceride levels, likely because it decreases the activity of CETP. Regular and short-term exercise both increase the activity of the lipoprotein lipase, which breaks down the triglycerides.

Nutrition

To lower triglycerides, it is recommended to make dietary changes and lose weight. Avoiding foods high in refined carbohydrates and eating more fish (especially fatty ones) is recommended. Also, it is recommended to increase the intake of fruits, vegetables, and whole grains, while reducing alcohol consumption.To achieve and maintain healthy weight, it is important to adjust calorie intake.

The Mediterranean diet is effective in reducing plasma triglycerides. Diets high in polyunsaturated fats can help lower triglycerides. This is done by reducing the amount of triglycerides in the liver and downregulating genes that are involved in triglyceride synthesis. Polyunsaturated fats are abundant in fish oils. It is strange, but true, that eating foods high in fish oil can lower triglycerides. However, fish oils were not shown to reduce heart disease risk when combined with statins.

It is also important to cut back on alcohol if triglycerides are high, as this can lead to liver and other alcohol-related problems.

Therapy

Many treatments are available to help reduce the level of triglycerides. However, each treatment must be evaluated individually to determine how it will affect the lipids.

Statins

Theoretically, a more effective removal of lipoproteins that are harmful should reduce the risk of heart attack. Statins can increase the number of LDL receptors and, by removing remnant particles faster, they can reduce their concentrations. Statins can lower LDL-cholesterol as well as triglycerides. The statin's effect on triglycerides is dependent on the dosage and the starting triglyceride level.

When triglycerides are close to a healthy level, statins do not work as well. They are more effective at higher levels. Statins can also be used to clear up remnants from a meal high in fat. Statins reduce remnant particles. This is beneficial for those at risk of heart disease. However, relying solely on the LDL receptor to remove remnants may not be enough to reduce the risk of heart disease. Other receptors, such as the VLDL receptor or LRP, are also important in removing remnants. Statins do not activate these receptors, unlike the LDL receptor.

Statins do not lower triglycerides specifically but can help patients with moderately high levels of triglycerides. Statins' ability to lower LDL-cholesterol has a direct impact on plasma triglyceride concentrations. Statins taken for several years can reduce the risk of pancreatitis among people with moderately high triglyceride levels.

PCSK9 inhibitors work in a similar manner to statins. They increase the activity of the LDL receptors as well as help eliminate IDL. While statins are more effective at lowering triglycerides and clearing VLDL in those with high levels of triglycerides, PCSK9 inhibitors are able to reduce the smaller remnants of triglyceride-rich lipoproteins in people who have lower levels. However, they cannot do so in those with high levels. PCSK9 inhibitors do not seem to have a great deal of effect on apolipoproteins B-48 and chylomicronemia.

Fibrates

If triglyceride levels are still high on a statin treatment, doctors may use a fibrate. Fibrates lower triglyceride levels to a different extent. However, they have only a small impact on apolipoprotein B. Fibrates also affect LDL-cholesterol levels differently depending on the initial concentration of triglycerides. Combining statins with fibrates can cause negative health effects, some of them quite dangerous. Rhabdomyolysis, a serious condition, can occur in rare cases. It happens when damaged muscle releases proteins and electrolytes in the bloodstream. These substances may cause permanent kidney and heart damage.

Open Question

Can the acceleration of lipoprotein breakdown protect against heart disease? Can activating lipoprotein lipase or hepatic lipase protect against heart disease?

Lipolysis is a way to lower the triglyceride levels in the blood. Fibrates accelerate the lipolysis to improve the breakdown of the triglycerides and remove them from the blood. Fibrates are able to reduce triglyceride levels in the blood almost by half. It is not fully clear however if the improved breakdown of triglycerides, and conversion of VLDL to LDL are beneficial. Several studies have found that genetic variations linked to this process may lower the risk of heart disease.

In a study that examined the effects of lowering triglycerides in fibrate trials, a reduction of 10 mg/dl led to a 5% decrease in coronary events. This was especially true for those with triglyceride values of at least 180 mg/dl. In a study of 555 heart attack victims, giving them both fibrate and niacin reduced their risk of death by 26% and heart disease-related deaths by 36%. In other studies, fibrate was given to participants with high levels of triglycerides. This resulted in a reduction of 54% in the overall number of heart events and 43% in those who had high triglycerides. In the studies where a fibrate is added to a statin, this reduction in triglycerides has been statistically significant for patients with high levels of triglycerides.

In the 1990s, when their mechanism of action was understood, a great deal of interest in fibrates grew. The promising results of clinical trials, such as the Veterans Affairs High-Density Lipoprotein Cholesterol Intervention Trial (VA-HIT) study, that demonstrated the efficacy and safety of gemfibrozil in cardiovascular events, were evident. Fenofibrate's beneficial effects on microvascular problems in the Fenofibrate Intervention and Event Lowering in Diabetes (FIELD) study, which confirmed earlier findings for clofibrate in a previous study, helped to consolidate interest in these drugs. There was also some benefit to patients with high triglycerides and low HDL-cholesterol, particularly those with elevated remnants. The study of pemafibrate, which was part of the recent Pemafibrate to Reduce cardiovascular OutcoMes by reducing triglycerides IN diabetic patiENTs (PROMINENT) trial, found that the drug, despite its ability to lower triglycerides, did not, however, reduce cardiovascular disease risk among people with type 2 diabetes, high triglycerides, and low HDL-cholesterol treated with statins. It is interesting that pemafibrate increased LDL-cholesterol and apolipoprotein B in this trial. The fibrates are therefore no longer widely used as lipid-lowering medications.

History

Discovery of Fibrates

Who	When
Jeffrey Thorp and Wilson Waring	**1962**

Where	What
Macclesfield, UK	Thorp, J. M. and W. S. Waring (1962) "Modification of metabolism and distribution of lipids by ethyl chlorophenoxy isobutyrate." Nature 194: 948-949.

The first fibrates were made in the 1950s.

Scientists in France discovered in 1953 that certain acids derived from dehydrocholic and other acids similar to it could lower plasma cholesterol in humans and rats. Researchers from Imperial Chemical Industries in England discovered a few years later that certain fatty acids can also lower cholesterol and lipid levels in rats. The best compound, discovered by Jeffrey Thorp and Wilson Waring at the Imperial Chemical Industries in Macclesfield in 1962, was called ethyl-alpha-4-chlorophenoxyisobutyrate, or clofibrate for short. This compound is derived from fibric acid (also known as phenoxy-isobutyric) which was described by Italian researchers in 1947 and by French researchers in 1956.

Clofibrate reduced the levels of lipids found in animal blood. Initially, this effect was attributed to seasonal variations in hormones such as androsterone. Later, however, studies of patients with high levels of cholesterol revealed that clofibrate could reduce lipids without the involvement of androsterone. Clofibrate also reduced triglycerides and VLDL more than cholesterol and LDL. It was found that the drug is safe to use for a long time and does not require androsterone in order to work.

Researchers took 30 years to fully understand how clofibrate works. They were worried because long-term clofibrate use caused liver enlargements in rats, due to an increase of certain intracellular structures known as peroxisomes. In 1967, the United States approved clofibrate for treating high levels of lipids. The exact mechanism of how clofibrate lowered lipid levels and its ability to grow peroxisomes was not known until later studies. In rats and mice, other compounds, such as Wy-14,653 or tibric acids, that lowered lipids, were more effective in causing peroxisome development, even though they were different from clofibrate.

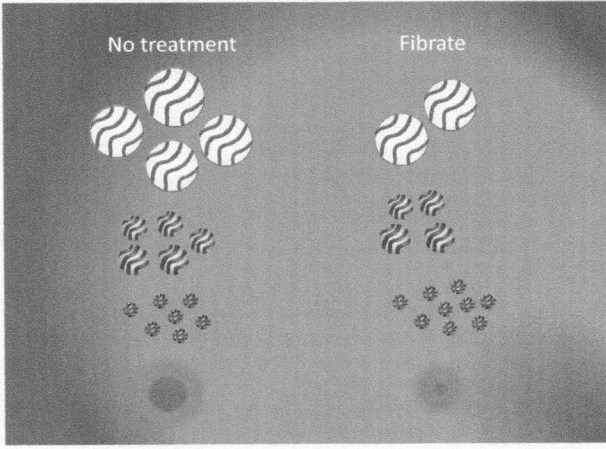

Fibrates lower triglyceride and LDL-cholesterol levels, while also raising HDL-cholesterol.

Thorp and his coworkers were concerned about the liver enlargement caused by clofibrate. They were looking for a safer and more effective drug to lower cholesterol. The researchers tested two drugs that were similar, methyl-clofenapate and clobuzarit. They found that these drugs had a greater ability to lower cholesterol than clofibrate. Nevertheless, methyl-clofenapate was found to cause liver enlargement and cancer in rats. It was therefore

(Continued)

(Continued)
not used in any further studies. Clobuzarit did not lower cholesterol as well as clofibrate but it had anti-inflammatory effects and decreased levels of an inflammatory protein fibrinogen. Several other pharmaceutical companies have also made efforts to enhance the structure of clofibrate in order to develop more effective cholesterol-lowering drugs. However, despite various modifications attempted, none have proven successful in achieving the desired results. Procetofen, which was manufactured by Groupe Fournier SA and first sold in France in 1974, was the only exception. Later, it was renamed to fenofibrate in order to comply with the naming guidelines of the World Health Organization. It was found that fenofibrate had better effects on humans than clofibrate. This is due to the way it is processed in the body, and also its effect on cholesterol levels. In the early 1980s and the late 1970s, other fibrates like bezafibrate or ciprofibrate were introduced to Europe and the United States. Their use was, however, restricted because of studies that showed some of these drugs cause liver cancer in mice and rats. The drugs were also responsible for the increased growth of peroxisomes. In 1983, it was found that both humans and primates did not suffer from this. Further epidemiological studies on humans confirmed that there was no increased cancer risk. Fibrates are now safe for use. In the 1990s, large studies demonstrated that fibrates are effective in reducing plasma triglycerides and thereby reducing cardiovascular complications in patients with diabetes. However, in 2022, the PROMINENT trial that was conducted in conjunction with statins and looked at people with high triglyceride levels and type 2 diabetes found that fibrates did not reduce heart disease risk in patients with both conditions. The use of fibrates to lower lipids decreased as a result.

CETP Inhibitors

CETP increases remnant cholesterol when it moves cholesteryl esters from HDL into triglyceride-rich lipoproteins. Remnant cholesterol can be reduced if CETP activity is blocked. Indeed, researchers found that CETP inhibitors such as evacetrapib and anacetrapib led to a significant decrease in cholesterol in VLDL (which includes remnants). CETP inhibitors also reduce blood triglyceride levels.

A recent large-scale study revealed the potential of CETP inhibitors to fight heart disease. Anacetrapib treatment reduced cardiovascular events by only 9% in patients with cardiovascular disease treated with a statin. This effect may have been caused by the reduction of non-HDL-cholesterol. However, even though evacetrapib, another CETP inhibitor, decreased non-HDL-cholesterol more, it did not provide any cardiovascular benefit. Clinical trials for obicetrapib, a novel CETP inhibitor, are currently underway.

Open Question
Can CETP inhibitors protect against heart disease by lowering LDL-cholesterol or non-HDL-cholesterol?

Polyunsaturated Fatty Acids

Researchers noticed in the 1970s that Greenland Eskimos, who consumed a high amount of omega-3 fatty acids, had a lower risk of heart disease. The treatment with omega-3 fatty acids helps lower the level of triglycerides. This is done by decreasing the amount of triglycerides released by the liver and increasing the rate at which triglycerides leave the blood.

It was found that fish oil contains a variety of active components. Scientists have since studied the benefits of two main omega-3 fatty acids, which are docosahexaenoic and eicosapentaenoic acids. Each omega-3 fatty acid has similar, but slightly different effects. High doses of omega-3 fatty acid (usually a mixture of docosahexaenoic and eicosapentaenoic acids) can reduce the secretion of VLDL triglycerides and apolipoprotein B by 25–30%. A clinical trial called the EpanoVa fOr Lowering Very high triglyceridEs (EVOLVE) showed that omega-3 fatty acid in a certain form is safe and effective for those with very high blood lipid levels.

However, the effect of the acids on VLDL clearance varies. In some studies, the amount of VLDL converted to IDL or LDL increased but plasma IDL or LDL levels did not change. This means that omega-3 fatty acid may not have an impact on the number of harmful lipoproteins, or on remnant populations.

Open Question

Does the ability of individual polyunsaturated fatty acids to protect against heart disease differ?

The Reduction of Cardiovascular Events with EPA – Intervention Trial (REDUCE-IT) found that high-dose of eicosapentaenoic acid reduced the risk of cardiovascular event. Only a small portion of the benefit was attributed to changes in triglyceride-rich lipoprotein levels. Ethyl eicosapentaenoic acid, also known by the names icosapent ethyl, E-EPA, and Vascepa, was recommended to adults with triglyceride concentrations of at least 150 mg/dl. On the other hand, the Outcomes Study to Assess STatin Residual Risk Reduction with EpaNova in HiGh CV Risk PatienTs with Hypertriglyceridaemia (STRENGTH) did not show any benefit with a combination of docohexaenoic and eicosapentaenoic acids, indicating that they may have different effects on cardiovascular disease.

The medium-chain triglyceride oils contain other fatty acids that are also helpful to reduce triglycerides. They are used to treat severe cases of chylomicronemia, when triglyceride concentrations are very high (above 1,000 mg/dl).

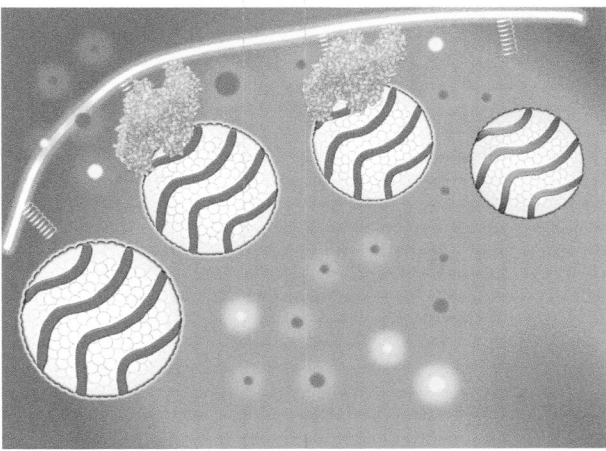

When the breakdown of triglyceride-rich lipoproteins in the blood is sped up, it lowers the levels of both triglycerides and remnant cholesterol, which helps reduce the chances of heart diseases.

Some data reveal that consuming alpha-linolenic acid, another polyunsaturated fat, can reduce the risk of developing heart disease, and the likelihood of dying of a heart attack. Eating foods containing the acid can improve cholesterol levels. It lowers triglycerides and total cholesterol while increasing HDL-cholesterol.

Niacin

Niacin is a powerful triglyceride-lowering agent. It also helps reverse the buildup of liver fat. Niacin in high doses can however cause side effects for humans. Niacin activates a specific pathway through a receptor called GPR109A, leading to the release of prostaglandins. This can induce flushing, which is a reddening and warming of the skin. Its use is often limited because it causes this reaction. Extended-release niacin is generally better tolerated but can be more harmful to the liver compared to immediate-release niacin. In some studies, extended-release niacin has been linked to higher infection rates. Early trials showed that niacin, when used alone to treat cardiovascular disease, was effective. The drug, however, was not effective when combined with statins.

These results are contrary to the notion that high triglyceride concentrations alone are a risk factor for cardiovascular disease. However, trials using fibrates showed a reduction in heart disease risk in people with fasting levels of triglycerides of 200 mg/dl and higher, particularly in those with low HDL-cholesterol. Therefore, it is possible that there is a threshold below which lowering the triglyceride level could be beneficial. More trials are needed to resolve this issue.

Other Agents

Genetic studies have found potential targets for reducing triglycerides, which are proteins that have a strong effect on triglyceride levels in the blood. These proteins include apolipoprotein C-III, apolipoprotein A-V, and lipoprotein lipase. Based on the genetic findings, it is suggested that treatments targeting lipoprotein lipase function, reducing apolipoprotein C-III function, decreasing ANGPTL4 function, or enhancing apo A-V function could be beneficial for heart health.

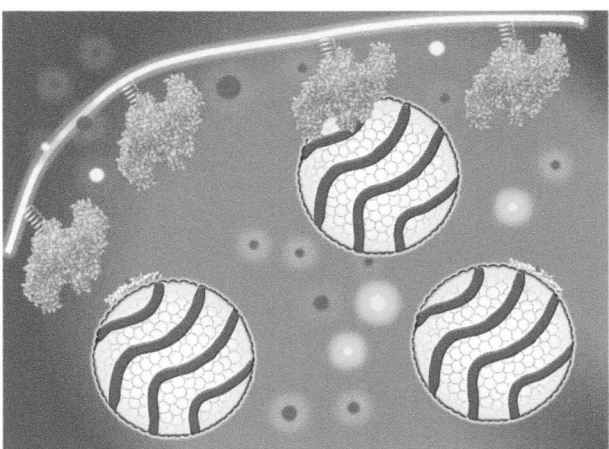

Lowering levels of apolipoprotein C-III (shown at the surface of VLDL), which stops lipoprotein lipase from working properly, helps break down triglyceride-rich lipoproteins in the blood. This strategy reduces levels of triglycerides and remnant cholesterol in the blood and may help prevent heart disease.

Apolipoprotein C-III, ANGPTL3, and other treatment targets have been identified to stimulate lipolysis. Apolipoprotein C-III contributes to high triglyceride levels by mainly slowing their breakdown. Apolipoprotein C-III also interferes with the way triglyceride-rich lipoproteins bind to liver receptors, making it more difficult for these particles to leave the blood. Apolipoprotein C-III levels are often correlated with the severity of high triglycerides.

Fish oil containing omega-3 fatty acids can lower apolipoprotein C-III levels in those with high levels of triglycerides. An experimental treatment (an antisense oligonucleotide) targeting the *APOC3* gene significantly reduced levels of apolipoprotein C-III, fasting plasma triglycerides, and non-HDL-cholesterol in people with severe hypertriglyceridemia, even when lipoprotein lipase activity was absent. Volanesorsen is a treatment that stops the production in the liver of apolipoprotein C-III by binding to the specific part of apolipoprotein C-III mRNA. Two more drugs targeting apolipoprotein C-III are currently being developed.

It is important to keep in mind that blocking apolipoprotein C-III could lead to an increase in fat accumulation in the liver. Apolipoprotein C-III deficiency in humans can lead to a greater breakdown of triglycerides, but not a higher liver uptake of remnants. These results suggest that improving triglyceride degradation without increasing the removal of remnants might not be enough to reduce heart disease risk in people with high triglyceride levels.

ANGPTL3 inhibits the lipoprotein lipase enzyme. People with ANGPTL3 variants that result in low levels of ANGPTL3 also have low levels of triglycerides, LDL-cholesterol, and HDL-cholesterol. Early trials of evinacumab (a monoclonal anti-ANGPTL3 antibody) showed that it had a positive effect on LDL-cholesterol, HDL-cholesterol, and apolipoprotein B in people with high triglycerides. One dose of the monoclonal antibody reduced triglyceride by 80% for patients with the high levels. This effect lasted 90 days and suggests that it reduced triglycerides independent of lipoproteins lipase. In addition, inhibition of ANGPTL3 has the potential to reduce LDL-cholesterol by limiting LDL particle formation.

The alipogene tiparvovec AAV1-LPLS447X gene replacement therapy, also known as Glybera, is a gene replacement therapy that was developed to treat familial chylomicronemia due to certain *LPL* gene mutations. This therapy injects additional copies of the gene that creates a powerful form of this enzyme into the muscles of patients affected. This therapy is used to treat patients with two copies of a mutated lipoprotein lipase or other mutations that lead to lipoprotein deficiency. The drug, which was injected into muscles in important clinical trials, was well tolerated. The patients showed signs of improvement in symptoms and reduced the incidence of pancreatitis up to five years after treatment.

History

The Response-to-Retention Hypothesis of Atherosclerosis
Robert Virchow observed in 1850 that inflamed areas of the arterial wall often contained cells with lipid deposits. This showed him that the changes must have occurred during inflammation. Virchow believed that lesions caused by atherosclerotic disease were the result of damage to the arterial walls and the body's inflammatory and growth responses.

Virchow's idea has been expanded and revised over the years by many researchers. Nikolai Anitschkow first proposed the lipid filtration theory in 1913. According to him, the atherosclerotic lesions were caused by the lipids that invaded the arterial wall. Irving Page developed this hypothesis in 1954. He believed that lipids and their derivatives were foreign bodies irritating arterial tissue, causing a proliferative reaction of removal around the tissue in a type of

(Continued)

(Continued)

chronic inflammation. According to him, the passage of lipid-laden lipoproteins through the arterial wall caused the lipoproteins to be disrupted, resulting in an unsoluble residue that triggered tissue reaction.

Who	When
Kevin Williams and Ira Tabas	**1995**

Where	What
Philadelphia and New York, USA	Williams, K. J. and I. Tabas (1995) The response-to-retention hypothesis of early atherogenesis. Arteriosclerosis, Thrombosis, and Vascular Biology 15: 551-561.

In other studies, the damage caused to the arterial walls by LDL was examined. This was because studies revealed LDL in the atherosclerotic intima. Elspeth Smit from Aberdeen University revealed in the 1970s that a large amount of LDL could be found in the inner layer of aortas in humans. Lipoprotein concentration in the intima had a strong correlation with LDL-cholesterol levels in plasma. Certain enzymes can release a portion of the LDL that was stuck in the tissue. This amount of stuck LDL is closely related to how much cholesterol has built up within the tissue. This suggests that the retention of lipoproteins in the tissue is an important step in the buildup of cholesterol in atherosclerotic lesions.

In the 1970s, Russell Ross and John Glomset also focused on the growth of muscle cells in the inner layer of blood vessels. The researchers believed that damage to the inner lining of the vessels caused platelets to gather at the affected area. These cells then released substances that stimulated muscle cell growth. This response-to-injury hypothesis of atherosclerosis became widely accepted by the medical community as the primary cause of the disease.

Further research revealed that even minor damage in the blood vessel lining can lead to atherosclerosis. This suggests that dysfunction of endothelial cells is more important than complete removal of these cells. Russell Ross refined the response-to-injury hypothesis in 1993. He stated that atherosclerosis is a protective reaction to different types of damage to arterial walls. The protective response can however become excessive and lead to a disease over time, depending on the type of harm and the duration. Ross suggested in 1999 that atherosclerosis is an inflammatory condition caused by both lipid accumulation and inflammation.

Studies revealed that there was no evidence to support the idea that damage to the blood vessels' lining is necessary or sufficient to form lesions. The vessel cells were intact throughout most of the stages of lesion formation. In more severe cases, they were damaged. The severe damage to these cells could be an important factor in the restenosis that occurs after balloon injury, or in advanced complex plaques. However, it does not seem to play a major role in early atherosclerosis.

To resolve this contradiction, the response-to-retention hypothesis was developed by Kevin Williams and Ira Tabas in 1995. They suggested that the development of atherosclerosis is influenced by small injuries to endothelium which do not completely remove cells but still affect their function. Increased permeability to lipoproteins, in particular harmful ones, is a principal change that occurs in the endothelial cells. This idea is connected to the lipid filtration hypothesis. Atherosclerosis is not caused by changes in permeability or the loss of endothelial cells. This is because a healthy endothelium transports many molecules and particles, including lipoproteins. The early accumulation of dangerous lipoproteins in vulnerable areas of the arterial wall is not due to increased permeability, but rather retention. All studies have agreed that the more susceptible sites retain harmful lipoproteins.

LDL undergoes changes when it is retained in the arterial walls. These changes have important effects on the body. LDL is retained by interaction with proteoglycans – branched molecules on the cell surface. In laboratory tests, LDL bound to proteoglycans forms structures and clumps that resemble what is seen in the arterial walls in the body. Under certain conditions, these complexes have a higher likelihood of being oxidized. Certain enzymes producing reactive oxygen species within the arterial wall can also affect retained LDL. If modified in laboratory tests by arterial cells, this lipoprotein has been readily taken up by macrophages and smooth muscle cells, leading to foam cell formation. Retained lipoproteins are responsible for all early signs of atherosclerotic lesion, whether they cause them directly or indirectly. Also, they can accelerate further retention of lipoproteins and their clumping by stimulating the local production of substances that are involved in this process.

Atherosclerotic lesions can be affected by the flow of blood in arteries. Lesions are more common in artery segments with turbulent blood flow, which are under low shear stress. This can happen at branch points, or when blood pressure is high. The development of lesions is influenced by changes in arterial wall cells due to turbulent blood flow. Shear stress to the blood vessels is not however the direct cause of atherosclerosis. High levels of lipoproteins are also required for the development of lesions within the blood vessels. Atherosclerosis does not occur when the vessels are under low shear stresses if LDL-cholesterol does not exceed a certain threshold. Even at sites with high shear stresses, lesions may still form if the blood contains high levels of harmful lipoproteins. Shear stress in the blood vessels may contribute to atherosclerosis. However, increased lipoprotein retention is the primary factor leading to the development of lesions. Low shear stress stimulates the production of molecules which promote lipoprotein accumulation.

The oxidation of lipoproteins is thought to play a role in atherosclerosis. Current evidence indicates that oxidation occurs after lipoproteins have been trapped in the walls of arteries. The oxidation that occurs is a natural result of trapping the lipoproteins and keeping them away from antioxidants, which are abundant in the blood. Healthy cells nearby may also contribute to the oxidation. LDL becomes more susceptible to oxidation when it adheres to arterial proteoglycans.

(Continued)

(Continued)

Injecting rabbits with LDL causes the arteries to undergo a change in a matter of hours. The LDL forms clumps and gets stuck in the arterial wall This occurs in areas that are more prone to developing lesions of the arteries. It is not that more cholesterol enters the artery but rather, it is not leaving properly. Evidence suggests that certain molecules found in the arterial wall like proteoglycans or enzymes such as lipoprotein lipase and sphingomyelinase, are involved in clumping LDL. These proteins, which are present in the normal arteries, can contribute to atherosclerosis. LDL also forms a close bond with the proteoglycans of the arteries. Even without its main function as an enzyme, lipoprotein lipase helps LDL adhere to arterial walls.

Atherosclerosis is a complex process, with multiple factors at play. However, one event is responsible for the onset of the disease. Lipoproteins are retained in arterial walls. Lipoprotein retention is more important than other factors like high cholesterol levels and changes to the blood flow and to the cells and tissues in the artery. The disease cannot develop without lipoprotein retention. It is also enough to cause other areas of the artery to react and start developing atherosclerosis. These early changes are responsible for all of the subsequent events that occur in the disease.

Not only LDL but also other lipoproteins are capable of causing atherosclerosis. This process is largely influenced by the size of lipoproteins. Lipoproteins that are small enough, such as the small-sized remnant lipoprotein rich in triglycerides called IDL, can easily pass through arterial walls and be caught by proteoglycans. VLDL, chylomicrons, and other large lipoproteins cannot however pass through the blood cell lining. It is also important to note that these lipoproteins must have apolipoprotein B in order to adhere to proteoglycans. Consequently, remnants of triglyceride-rich particles can cause cardiovascular disease.

The response-to-retention hypothesis opens exciting possibilities to combat atherosclerosis and cardiovascular disease at their roots. Recent years have seen improvements in the understanding and treatment of atherosclerotic heart disease. However, the pace of progress has slowed, and the risk for heart attacks or strokes remains high. The majority of studies are focused on the advanced plaques. Nevertheless, new efforts have been made to prevent arterial disease or treat it before it becomes severe. Kevin Williams, one of the authors of the hypothesis, has recently stated that new evidence is needed to support intensive short-term treatment for early plaques which may be able to heal. Patients and doctors may be motivated to act sooner if they are screened for these plaques at midlife. In order to reduce deaths due to this disease, the focus should be shifted from the end stage to the earlier stages of atherosclerosis that are likely to be successfully treated.

Open Question

Can the atherosclerosis process be stopped by blocking the retention of lipoproteins within arterial walls?

Test Your Knowledge

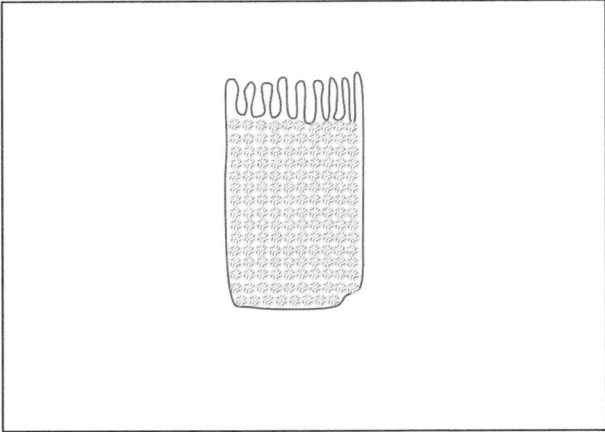

1 Enterocytes secrete and produce chylomicrons. How many chylomicrons with an average size 500 nm will fit in a cell of an enterocyte that has an average volume 1400 cubic micrometers?

 A About two thousand.

 B About two million.

 C About two billion.

2 What proportion of the surface area is lost when VLDL particle undergoes lipolysis and becomes LDL? These surface components can be used to make how many HDL particles? Assume that the average size for VLDL is 66 nm. The average size for LDL is 22 nm. And, the average size for HDL is 11 nm.

   A 8/9 its initial surface equivalent to 32 HDL particles;

   B 1/9 its initial surface equivalent to 3 HDL particles;

   C 1/2 its initial surface equivalent to 15 HDL particles.

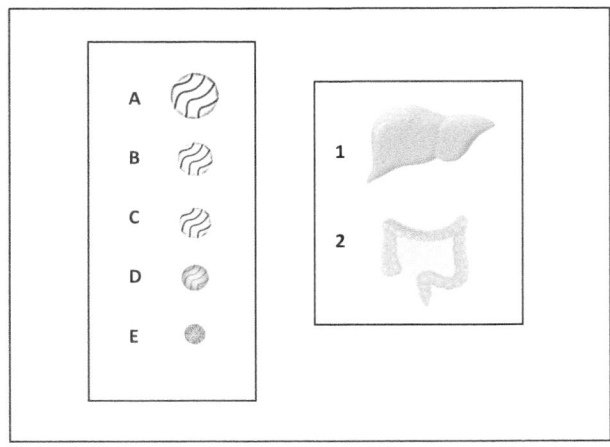

3 What organs produce what lipoproteins? Make connections.

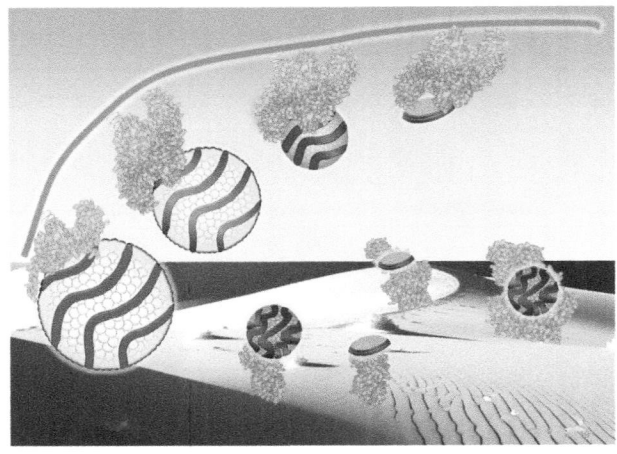

4 Find 7 errors.

Answers

1, C; 2, A; 3, A2, B1, C1, D1, E1, E2; (1) VLDL is not broken down by lipoprotein lipase monomer; (2) LDL is not broken down by lipoprotein lipase; (3) discoid HDL is not broken down by lipoprotein lipase; (4) LPL dimer does not directly bind to the endothelium; (5) spherical HDL does not bind to ABCA1; (6) discoid HDL does not bind to ABCA1 monomer; (7) discoid HDL does not bind to SR-BI.

Further Reading

Balling, M., S. Afzal, G. D. Smith, A. Varbo, A. Langsted, P. R. Kamstrup and B. G. Nordestgaard (2023). "Elevated LDL triglycerides and atherosclerotic risk." Journal of the American College of Cardiology 81(2): 136–152.

Bjornson, E., C. J. Packard, M. Adiels, L. Andersson, N. Matikainen, S. Soderlund, J. Kahri, A. Hakkarainen, N. Lundbom, J. Lundbom, C. Sihlbom, A. Thorsell, H. Zhou, M. R. Taskinen and J. Boren (2020). "Apolipoprotein B-48 metabolism in chylomicrons and very low-density lipoproteins and its role in triglyceride transport in normo- and hypertriglyceridemic human subjects." Journal of Internal Medicine 288(4): 422–438.

Boren, J., C. J. Packard and M. R. Taskinen (2020). "The roles of ApoC-III on the metabolism of triglyceride-rich lipoproteins in humans." Frontiers in Endocrinology (Lausanne) 11: 474.

Boren, J., M. R. Taskinen, E. Bjornson and C. J. Packard (2022). "Metabolism of triglyceride-rich lipoproteins in health and dyslipidaemia." Nature Reviews Cardiology 19(9): 577–592.

Boren, J. and K. J. Williams (2016). "The central role of arterial retention of cholesterol-rich apolipoprotein-B-containing lipoproteins in the pathogenesis of atherosclerosis: A triumph of simplicity." Current Opinion in Lipidology 27(5): 473–483.

Budoff, M. J., D. L. Bhatt, A. Kinninger, S. Lakshmanan, J. B. Muhlestein, V. T. Le, H. T. May, K. Shaikh, C. Shekar, S. K. Roy, J. Tayek and J. R. Nelson (2020). "Effect of icosapent ethyl on progression of coronary atherosclerosis in patients with elevated triglycerides on statin therapy: Final results of the EVAPORATE trial." European Heart Journal 41(40): 3925–3932.

Castaner, O., X. Pinto, I. Subirana, A. J. Amor, E. Ros, A. Hernaez, M. A. Martinez-Gonzalez, D. Corella, J. Salas-Salvado, R. Estruch, J. Lapetra, E. Gomez-Gracia, A. M. Alonso-Gomez, M. Fiol, L. Serra-Majem, E. Corbella, D. Benaiges, J. V. Sorli, M. Ruiz-Canela, N. Babio, L. T. Sierra, E. Ortega and M. Fito (2020). "Remnant cholesterol, not LDL cholesterol, is associated with incident cardiovascular disease." Journal of the American College of Cardiology 76(23): 2712–2724.

Chait, A., H. N. Ginsberg, T. Vaisar, J. W. Heinecke, I. J. Goldberg and K. E. Bornfeldt (2020). "Remnants of the triglyceride-rich lipoproteins, diabetes, and cardiovascular disease." Diabetes 69(4): 508–516.

Das Pradhan, A., R. J. Glynn, J. C. Fruchart, J. G. MacFadyen, E. S. Zaharris, B. M. Everett, S. E. Campbell, R. Oshima, P. Amarenco, D. J. Blom, E. A. Brinton, R. H. Eckel, M. B. Elam, J. S. Felicio, H. N. Ginsberg, A. Goudev, S. Ishibashi, J. Joseph, T. Kodama, W. Koenig, L. A. Leiter, A. J. Lorenzatti, B. Mankovsky, N. Marx, B. G. Nordestgaard, D. Pall, K. K. Ray, R. D. Santos, H. Soran, A. Susekov, M. Tendera, K. Yokote, N. P. Paynter, J. E. Buring, P. Libby, P. M. Ridker and I. Prominent (2022). "Triglyceride lowering with pemafibrate to reduce cardiovascular risk." New England Journal of Medicine 387(21): 1923–1934.

Cholesterol, Lipoproteins, and Cardiovascular Health: Separating the Good (HDL), the Bad (LDL), and the Remnant, First Edition. Anatol Kontush.
© 2025 John Wiley & Sons, Inc. Published 2025 by John Wiley & Sons, Inc.

Dron, J. S. and R. A. Hegele (2020). "Genetics of hypertriglyceridemia." Frontiers in Endocrinology (Lausanne) 11: 455.

Duran, E. K., A. W. Aday, N. R. Cook, J. E. Buring, P. M. Ridker and A. D. Pradhan (2020). "Triglyceride-rich lipoprotein cholesterol, small dense LDL cholesterol, and incident cardiovascular disease." Journal of the American College of Cardiology 75(17): 2122–2135.

Gaudet, D., P. Clifton, D. Sullivan, J. Baker, C. Schwabe, S. Thackwray, R. Scott, J. Hamilton, B. Given, S. Melquist, R. Zhou, T. Chang, J. S. Martin, G. F. Watts, I. J. Goldberg, J. W. Knowles, R. A. Hegele and C. M. Ballantyne (2023). "RNA interference therapy targeting apolipoprotein C-III in hypertriglyceridemia." NEJM Evidence 2(12): EVIDoa2200325.

Gaudet, D., E. Karwatowska-Prokopczuk, S. J. Baum, E. Hurh, J. Kingsbury, V. J. Bartlett, A. L. Figueroa, P. Piscitelli, W. Singleton, J. L. Witztum, R. S. Geary, S. Tsimikas, L. S. O'Dea and I. Vupanorsen Study (2020). "Vupanorsen, an N-acetyl galactosamine-conjugated antisense drug to ANGPTL3 mRNA, lowers triglycerides and atherogenic lipoproteins in patients with diabetes, hepatic steatosis, and hypertriglyceridaemia." European Heart Journal 41(40): 3936–3945.

Ginsberg, H. N. and I. J. Goldberg (2023). "Broadening the scope of dyslipidemia therapy by targeting APOC3 (Apolipoprotein C3) and ANGPTL3 (Angiopoietin-Like Protein 3)." Arteriosclerosis Thrombosis and Vascular Biology 43(3): 388–398.

Ginsberg, H. N., C. J. Packard, M. J. Chapman, J. Borén, C. A. Aguilar-Salinas, M. Averna, B. A. Ference, D. Gaudet, R. A. Hegele, S. Kersten, G. F. Lewis, A. H. Lichtenstein, P. Moulin, B. G. Nordestgaard, A. T. Remaley, B. Staels, E. S. G. Stroes, M.-R. Taskinen, L. S. Tokgözoğlu, A. Tybjaerg-Hansen, J. K. Stock and A. L. Catapano (2021). "Triglyceride-rich lipoproteins and their remnants: Metabolic insights, role in atherosclerotic cardiovascular disease, and emerging therapeutic strategies—A consensus statement from the European Atherosclerosis Society." European Heart Journal 42(47): 4791–4806.

Gouni-Berthold, I., V. J. Alexander, Q. Q. Yang, E. Hurh, E. Steinhagen-Thiessen, P. M. Moriarty, S. G. Hughes, D. Gaudet, R. A. Hegele, L. S. O'Dea, E. S. G. Stroes, S. Tsimikas and J. L. Witztum (2021). "Efficacy and safety of volanesorsen in patients with multifactorial chylomicronaemia (COMPASS): A multicentre, double-blind, randomised, placebo-controlled, phase 3 trial." Lancet Diabetes & Endocrinology 9(5): 264–275.

Ilich, J. Z., P. Y. Liu, H. Shin, Y. Kim and Y. Chi (2022). "Cardiometabolic indices after weight loss with calcium or dairy foods: Secondary analyses from a randomized trial with overweight/obese postmenopausal women." Nutrients 14(5).

Johannesen, C. D. L., M. B. Mortensen, A. Langsted and B. G. Nordestgaard (2021). "Apolipoprotein B and Non-HDL cholesterol better reflect residual risk than LDL cholesterol in statin-treated patients." Journal of the American College of Cardiology 77(11): 1439–1450.

Langsted, A., C. M. Madsen and B. G. Nordestgaard (2020). "Contribution of remnant cholesterol to cardiovascular risk." Journal of Internal Medicine 288(1): 116–127.

Laufs, U., K. G. Parhofer, H. N. Ginsberg and R. A. Hegele (2020). "Clinical review on triglycerides." European Heart Journal 41(1): 99–109c.

Packard, C. J. (2022). "Remnants, LDL, and the quantification of lipoprotein-associated risk in atherosclerotic cardiovascular disease." Current Atherosclerosis Reports 24(3): 133–142.

Quispe, R., S. S. Martin, E. D. Michos, I. Lamba, R. S. Blumenthal, A. Saeed, J. Lima, R. Puri, S. Nomura, M. Tsai, J. Wilkins, C. M. Ballantyne, S. Nicholls, S. R. Jones and M. B. Elshazly (2021). "Remnant cholesterol predicts cardiovascular disease beyond LDL and ApoB: A primary prevention study." European Heart Journal 42(42): 4324–4332.

Rosenson, R. S., D. Gaudet, C. M. Ballantyne, S. J. Baum, J. Bergeron, E. E. Kershaw, P. M. Moriarty, P. Rubba, D. C. Whitcomb, P. Banerjee, A. Gewitz, C. Gonzaga-Jauregui, J. McGinniss, M. P. Ponda, R.

Pordy, J. Zhao and D. J. Rader (2023). "Evinacumab in severe hypertriglyceridemia with or without lipoprotein lipase pathway mutations: A phase 2 randomized trial." Nature Medicine 29(3): 729–737.

Rosenson, R. S., A. Shaik and W. Song (2021). "New therapies for lowering triglyceride-rich lipoproteins: JACC focus seminar 3/4." Journal of the American College of Cardiology 78(18): 1817–1830.

Salinas, C. A. A. and M. J. Chapman (2020). "Remnant lipoproteins: Are they equal to or more atherogenic than LDL?" Current Opinion in Lipidology 31(3): 132–139.

Tardif, J. C., E. Karwatowska-Prokopczuk, E. St Amour, C. M. Ballantyne, M. D. Shapiro, P. M. Moriarty, S. J. Baum, E. Hurh, V. J. Bartlett, J. Kingsbury, A. L. Figueroa, V. J. Alexander, J. Tami, J. L. Witztum, R. S. Geary, L. S. O'Dea, S. Tsimikas, D. Gaudet and I. Olezarsen Study (2022). "Apolipoprotein C-III reduction in subjects with moderate hypertriglyceridaemia and at high cardiovascular risk." European Heart Journal 43(14): 1401–1412.

Wadström, B. N., A. B. Wulff, K. M. Pedersen, G. B. Jensen and B. G. Nordestgaard (2021). "Elevated remnant cholesterol increases the risk of peripheral artery disease, myocardial infarction, and ischaemic stroke: A cohort-based study " European Heart Journal 43(34): 3258–3269.

Williams, K. J. (2024). "Eradicating atherosclerotic events by targeting early subclinical disease: It is time to retire the therapeutic paradigm of too much, too late." Arteriosclerosis, Thrombosis, and Vascular Biology 44: 48–64.

Epilogue

The Story Never Ends

Our exploration of the hidden realms of cholesterol, lipids, cells, and diseases is coming to an end. Throughout our journey, we have followed three key lipoproteins as they flow through arteries, passing by organs in the human body. Our aim has been to understand how these lipoproteins contribute to cholesterol buildup in the arteries, atherosclerosis, and heart disease in both the extracellular and intracellular spaces. We have seen how they can both aggravate and protect against these conditions through a complex network of pathways within cells and tissues.

Now, it is time to answer the ultimate question – which outcome will prevail? How will the story end? Will there be a heart attack or will it be prevented?

It is clear that there is no simple answer. Each lipoprotein has a range of effects on health that can vary. Additionally, the lipoproteins interact with each other in a continuous manner. The outcome depends on many factors. In some cases, lipoproteins consistently increase cholesterol delivery to arterial walls leading to disease progression. In other situations, cholesterol delivery decreases, temporarily halting the disease.

Like any tale of life and death, this one never truly ends. Lipoproteins continue their journey, providing us with vital energy while leaving behind a small amount of cholesterol in arterial walls. Just as death is an inevitable consequence of life, atherosclerosis arises as a side effect of energy production – an essential component of life itself. The body's response to infection, also crucial for survival, further worsens the disease.

As long as we live, it is unlikely that we can completely stop this process. However, genetic disorders like familial hypercholesterolemia remind us that our biological makeup can shape the future. Our future is not predetermined and can be altered. Although we may not fully prevent the disease, we can delay its onset and reduce its threat.

When considering the risk factors for heart disease, there are numerous possibilities for reducing its likelihood. By addressing each individual factor, we have the potential to greatly improve our overall health and well-being. By combining these efforts, we can minimize suffering and promote healing.

Dyslipidemia, referring to issues with plasma lipoproteins, remains the primary risk factor. This emphasizes the potential for targeting of these particles. Focusing on remnant cholesterol may provide additional benefits beyond lowering LDL-cholesterol. While the exact role of HDL in atherosclerosis remains unclear, improving the processing of this lipoprotein could be another piece to solving the puzzle of cholesterol.

Cholesterol, Lipoproteins, and Cardiovascular Health: Separating the Good (HDL), the Bad (LDL), and the Remnant, First Edition. Anatol Kontush.
© 2025 John Wiley & Sons, Inc. Published 2025 by John Wiley & Sons, Inc.

However, all of this is in the future. For now, countless lipoprotein particles continuously glide in darkness along smooth walls, resembling yellowish riders against a backdrop of a reddish sunset. It is our responsibility to keep this road safe.

At the end of our journey with bad cholesterol, good cholesterol, and remnant cholesterol, we understand that atherosclerosis can be prevented.

The death of cardiovascular diseases is not inevitable.

Glossary

Abetalipoproteinemia is a hypolipidemia involving reduced concentrations of LDL-cholesterol.

Adenosine triphosphate (ATP) is a molecule that gives energy to living cells for different processes like muscle movement, nerve signals, and making other biomolecules like proteins, DNA and RNA. It is found in all living organisms and is called the "molecular unit of currency" for moving energy inside cells. When it is used up, it turns into adenosine diphosphate (ADP) or adenosine monophosphate (AMP).

Alpha-helix is a spiral-shaped structure formed by a single polypeptide chain. The backbone of the chain forms the inner part of the helix, while the side chains extend outward. Internal hydrogen bonds stabilize the helix. Alpha-helixes are often found in membrane proteins and in regions of proteins that need to be rigid.

Amphiphilic (amphiphatic) means having both hydrophobic (attracted to lipids, or nonpolar) and hydrophilic (attracted to water, or polar) properties.

Angiopoietin-like proteins (ANGPTLs) is a family of proteins that is structurally similar to the angiopoietins – a type of blood vessel growth factors that help create new blood vessels in embryos and after birth.

Antisense oligonucleotide is a short, synthetic piece of DNA that can change how RNA works and decrease, restore, or change how proteins are made.

Apolipoproteins are specialized proteins present in lipoproteins which have specific shapes and functions.

ATP-binding cassette transporter A1 (ABCA1) is a large membrane protein with a molecular mass of 254 kDa that helps with the formation of HDL and keeps cell cholesterol balanced. ABCA1 exports cholesterol to small discoid HDL and to apolipoprotein A-I.

ATP-binding cassette transporter G1 (ABCG1) is another membrane protein that helps with the formation and remodeling of HDL. ABCA1 exports cholesterol to large spherical HDL.

Beta-sheet is formed by multiple polypeptide chains or segments of a single chain that align side-by-side. The backbone forms a zigzag pattern, with each strand connected by hydrogen bonds

Cholesterol, Lipoproteins, and Cardiovascular Health: Separating the Good (HDL), the Bad (LDL), and the Remnant, First Edition. Anatol Kontush.
© 2025 John Wiley & Sons, Inc. Published 2025 by John Wiley & Sons, Inc.

between strands. Beta-sheets can be parallel (with strands running in the same direction) or antiparallel (with strands running in opposite directions). Beta-sheets are often found in globular proteins and provide structural stability.

Bile, also known as gall, is a yellow–green liquid constantly made by the liver that helps break down fats in the small intestine. In people, bile is mostly made of water in which are dissolved a number of solid constituents including bile salts, lipids and other substances. It is stored and concentrated in the gallbladder and released into the small intestine after eating.

Body-mass index (BMI) is a measure of body fat based on height and weight that applies to adult men and women. BMI is defined as the body mass divided by the square of the body height, and is expressed in units of kg/m^2.

Bromodomain and extra-terminal (BET) family of proteins includes proteins that control the activity of many genes and pathways related to the immune system in mammals.

Cholesteryl ester is an ester that has a fatty acid part taking the place of the hydroxyl group in cholesterol.

Cholesteryl ester transfer protein (CETP) is a plasma lipid transfer protein that exchanges cholesteryl ester and triglyceride between different types of lipoproteins in the blood. It normally retrieves cholesteryl ester from HDL and transfers it to other lipoproteins, such as VLDL, in return for triglyceride.

Chylomicron is the largest and lightest lipoprotein of human plasma. Its size can reach more than 1 micrometer. Chylomicron is largely composed of triglyceride.

Dalton, or Da, is a unit used to measure the mass of atoms and molecules. It is defined as 1/12 of the mass of a carbon atom. The mass of a carbon atom is 12 daltons. Big molecules can have masses of thousands or even millions of daltons, which are often written as kDa (kilodalton) or MDa (megadalton).

Disability-adjusted life years (DALYs) measure the total impact of a disease or health condition on a population by adding up the years lost due to early death and the years lived with a disability.

Dyslipidemias (also called dyslipoproteinemias) are medical conditions involving abnormal concentrations of lipids (triglycerides and/or cholesterol) in the blood plasma.

Electrophoresis is a method to separate biomolecules or their complexes according to their mobility in the electric field. The mobility is a measure of charge and size of the molecules.

Ester is a substance made from a precursor where the hydrogen atom (H) of at least one hydroxyl group (OH) is replaced by a different group (−R).

Gallbladder is a small organ that holds bile before it goes into the small intestine. In people, it is shaped like a pear and sits under the liver. The gallbladder gets bile from the liver and sends it to the small intestine to help digest fats. Sometimes, gallstones can form in the gallbladder and cause pain.

High-density lipoprotein (HDL) is a small and dense lipoprotein of human plasma. An average size of HDL is around 10 nm. HDL carries a lot of protein and phospholipid.

High triglycerides in the blood are defined as 150 mg/dL (1.7 mmol/L) and higher. The levels of 150 to 199 mg/dL (1.8–2.2 mmol/L) are considered borderline high, those of 200–499 mg/dL (2.3–5.6 mmol/L) are high, and those of 500 mg/dL or above (5.7 mmol/L or above) are very high.

Homozygous and **heterozygous** indicate whether a living organism has two of the same copies of a gene for a particular trait (homozygous), or whether it has two different ones (heterozygous).

Hydrophilic, derived from the Greek for water (*hydros*) and love (*philia*), means having an affinity for water and being capable of interacting with it.

Hydrophobic, derived from the Greek words for water (*hydros*) and fear (*phobos*), means having little or no affinity for water.

Hydroxymethylglutaryl coenzyme A (HMG-CoA) reductase is a key enzyme that regulates synthesis of cholesterol in cells.

Hyperalphalipoproteinemia is a condition where the circulating levels of HDL-cholesterol are higher than what is typically seen in a healthy population of the same age and sex. This is usually defined as being higher than the 95th or sometimes 90th percentile.

Hypercholesterolemia is a dyslipidemia involving elevated concentrations of cholesterol.

Hyperlipidemia (also called hyperlipoproteinemia) is a dyslipidemia involving abnormally elevated concentrations of lipids.

Hypoalphalipoproteinemia is a condition where the circulating levels of HDL-cholesterol are lower than what is typically seen in a healthy population of the same age and sex. This is usually defined as being lower than the 5th or sometimes 10th percentile.

Hypocholesterolemia is a dyslipidemia involving reduced concentrations of cholesterol.

Hypolipidemia (also called hypolipoproteinemia) is a dyslipidemia involving abnormally reduced concentrations of lipids.

Intermediate-density lipoprotein (IDL) is a remnant lipoprotein of human plasma. An average size of IDL is around 30 nm. IDL is rich in triglyceride and cholesterol.

Intestinal villi are tiny hair-like parts that stick out into the inside of the small intestine.

Ischemia is a medical condition that occurs when there is a reduced blood flow to a particular part of the body, usually due to a blockage or narrowing of an artery.

Isoelectric focusing is a method to separate and analyze proteins and peptides in protein samples. It uses gel electrophoresis to separate the proteins based on their isoelectric point, which is the pH at which a molecule has no net electrical charge.

Lacteals are lymphatic capillaries in the villi of the small intestine.

LDL-cholesterol is cholesterol carried by LDL in the blood. Its concentration is expressed as milligram per deciliter (mg/dl) or millimole per liter plasma (mmol/l). High LDL-cholesterol is associated with elevated risk of heart disease and is typically defined as higher than 130 mg/dl in people who are not treated by lipid-lowering drugs. High levels of LDL-cholesterol require treatment to reduce cardiovascular risk.

LDL receptor is a protein present at the surface of the liver and other organs and tissues. The protein specifically interacts with apolipoprotein B-100 on LDL and helps clear a large part of LDL from the blood.

LDL receptor-related protein (LPR) is a protein present at cell surfaces that belongs to the family of the LDL receptor and binds large triglyceride-rich particles in the circulation.

Lecithin-cholesterol acyltransferase (LCAT) is a key enzyme of cholesterol metabolism in the blood plasma. LCAT catalyzes the esterification of cholesterol into cholesteryl esters within plasma lipoproteins, first of all within HDL.

Lipases are enzymes that help break down lipids in the body. There are several different lipases involved in the metabolism of lipoproteins, including lipoprotein lipase, hepatic lipase, endothelial lipase, PAF-AH, and LCAT.

Lipemia is presence of a high concentration of lipids in the blood.

Lipoma is a benign tumor made of lipids, primarily of triglycerides.

Lipopolysaccharides are big molecules found in the outer part of certain bacteria. They are made up of a lipid and a polysaccharide and can be toxic. They are often called endotoxins and are mainly found in certain bacteria like *E. coli* and *Salmonella*.

Lipoprotein (a) (Lp(a)) is a lipoprotein which is similar in size and composition to LDL. The main difference between them is that Lp(a) additionally contains apolipoprotein (a).

Lipoprotein lipase (LPL) is the most important enzyme that breaks down triglycerides inside triglyceride-rich lipoprotein particles like chylomicrons, VLDL and IDL. This process releases fatty acids which are taken up into the muscles and fat tissue where they are used to produce energy.

Lipoproteins are particles consisting of many molecules. They are typically shaped like spheres and made up of both hydrophilic (polar) and hydrophobic (non-polar) lipids. These lipids are kept by proteins called apolipoproteins, which have specific shapes and jobs, and also contain other proteins, sugar residues and nucleic acids.

Low-density lipoprotein (LDL) is a terminal product of the breakdown of VLDL. It can stay in the blood for a long time up to 4 days and carries a lot of cholesterol, in the form of cholesteryl ester. Average size of LDL is around 25 nm.

Low HDL-cholesterol levels are considered to be less than 40 mg/dl in men. For women, low levels are less than 50 mg/dl. These limits represent the typical range of HDL-cholesterol levels in the general population. For example, in surveys like the National Health and Nutrition Examination Survey III (NHANES-III) and the Framingham Offspring Study, the average HDL-cholesterol levels were around 44 mg/dl for men and 53 mg/dl for women.

Low LDL-cholesterol is typically defined as less than 50 mg/dl and extremely low as less than 20 mg/dl.

Nicotinamide adenine dinucleotide phosphate (NADH) is a reduced form of a coenzyme that plays a key role in the formation of ATP in all living cells. Coenzyme is a cofactor that helps enzymes catalyze biochemical reactions.

Nuclear magnetic resonance (NMR) is a physical phenomenon when the atom nuclei in a strong magnetic field are disturbed by a weak magnetic field and produce a signal with a frequency that shows the strength of the magnetic field at the nucleus. NMR spectroscopy is a technique that measures this signal.

Omega-3 and omega-6 are types of polyunsaturated fatty acids – those that contain more than one double bond. The main difference between omega-3 and omega-6 is where the first double bond happens in the molecule. In omega-3 fatty acids, the first double bond is on the third carbon atom, while in omega-6, it's on the sixth carbon atom from the molecule's end.

Oxidative stress occurs when there is an imbalance between the production and accumulation of reactive oxygen species in cells and tissues on the one hand, and the ability of the cells and tissues to inactivate them – in other words, when there are too many harmful oxygen-derived molecules in the body and not enough ways to get rid of them.

Paraoxonase (PON) is an enzyme mainly present in HDL. The enzyme breaks down different molecules like paraoxon and other toxic organophosphate chemicals, but also lactones, such as homocysteine thiolactone.

Peroxisome proliferator-activated receptors (PPARs) are a group of proteins that regulate genes related to lipid processing, lipid cell production, insulin response, and inflammation in blood vessels.

Phospholipid transfer protein (PLTP) is a plasma lipid transfer protein that exchanges phospholipid molecules across plasma lipoproteins.

Platelet-activating factor-acetyl hydrolase (PAF-AH) is an enzyme present on LDL and HDL that breaks down platelet-activating factor, a powerful bioactive lipid that mediates many cellular functions. In addition, PAF-AH cleaves other bioactive lipid molecules like oxidized phospholipids.

Primary prevention involves taking actions to prevent an illness before it starts.

Proprotein convertase subtilisin kexin type 9 (PCSK9) is an enzyme that attaches to the LDL receptor and helps break it down in the cells.

Proteoglycans are a class of high molecular weight proteins found in the extracellular space of connective tissue, which provides support to the body's structure. These heavily glycosylated proteins make up a significant portion of the extracellular matrix, filling the spaces between cells.

Reactive oxygen species are highly reactive chemicals with at least one oxygen atom. They include oxygen free radicals like superoxide anion radical or hydroxyl radical, and singlet oxygen. Reactive oxygen species are formed from oxygen, water, and hydrogen peroxide and are very reactive. They play a role in many biological functions as signals that can turn on and off different processes.

Remnant cholesterol stays for cholesterol present in remnant triglyceride-rich particles. It can be estimated as a difference between total cholesterol and HDL-cholesterol plus LDL-cholesterol. There is no test that measures cholesterol in all types of remnant triglyceride-rich particles at once.

Reverse causation is when two things are connected in a way that is unexpected. Instead of one thing causing the other, the opposite happens. For example, people might think that being overweight causes depression, but they find out that depression actually leads to being overweight. In reverse causation, the result comes before the cause, which is different from how we usually think about cause and effect.

Scavenger receptor class B type I (SR-BI) is a cellular receptor that binds HDL and many other partners like modified LDL. The receptor helps move cholesterol from HDL to cells.

Secondary prevention involves finding and treating a disease early on.

Serum amyloid A (SAA) is a group of proteins secreted during the acute phase of the inflammatory response. They are mainly transported by HDL in the blood.

Small interfering RNA (siRNA) is a type of double-stranded RNA that can stop certain genes from being expressed. It does this by breaking down the messenger RNA (mRNA) that is made after a gene is turned on.

Svedberg flotation unit is defined as a flotation rate of 100 femtometres per second measured at 26°C in NaCl solution of a density of 1.063 g/ml at a gravitational force of 1 dyne per acceleration of 1 g. The unit abbreviated as Sf is used in order to characterise plasma lipoproteins which float at different rates as a function of their physical (density and size) and chemical (lipid and protein) properties. The unit was introduced by John Gofman and colleagues in 1951.

Very low-density lipoprotein (VLDL) is a large and light lipoprotein of human plasma. Its size is from 35 to 80 nm. VLDL is rich in triglyceride.

Xanthoma is a benign tumor primarily made of cholesterol. It is a discrete, raised yellow deposit underneath the skin, which occurs when circulating levels of cholesterol are elevated. Both lipoma and xanthoma are deposits of lipids.

Index

Page numbers referring to figures are *italics* and those referring to tables in **bold**.

FIELD, *see* Fenofibrate Intervention and Event
   Lowering in Diabetes trial
Fielding, Christopher   188, 206, **303**
Fielding, Phoebe   206
Finland   89, 90, 147, **292**
First International Congress of Biochemistry   62
Fish oil   5, 8, 238, 306, 311, 313
Fish roe   237
Fish   42, 132, 133, 217, 236, 237, 304, 306
Fish, Pierre   **271**
Fish-eye disease   163, 223
Fizzy beverages   41
Flaxseeds   132
Fleming, Alexander   138
Flexibility   94, 95
Flexible parts   95, 190
Flotation   76, 77, 80, 274
Fluid balance   19
Fluidity   5, 99, 183, 186, 212, 216, 282
Fluorescence spectroscopy   192
Flushing   312
Fluvastatin   135
Foam cells   25, 43, 112, 126, 127, 210, 230, 301,
   302, 315
Follicular fluid   59
Food(s)   4, 5, 10, 16, 20, 41, 42, 46, 55, 113, 115,
   132–134, 153, 238, 268, 271, 284, 301, 306,
   312, 322
Forte, Trudy 79
FPLC, *see* Fast protein liquid chromatography
Frameshift mutations   117
Framingham Heart Study   34, 75, 79, 80, **86**, 87,
   88, 91, 159, 213, 234
Framingham Offspring Study   160, 234
Framingham Risk Prediction Tool   159
Framingham Risk Score   88, *91*
Framingham   **86**
France   **12**, 47, *60*, 61, 62, 90, 130, 167, *173*, 309,
   310
Franceschini, Guido   **164**
Fredrickson, Donald   116, **120**, 121, 122, *123*,
   124, 128, 133, 212, 276
Free radical(s)   24, 214–216, 230
French Academy of Sciences   13
French National Institute of Health and Medical
   Research (INSERM)   187
Friberg, Sven 128

Frictional force   27
Friedewald's formula   83
Friedewald, William   83
Fructose   238
Fruits   41, 42, 46, 50, 132–134, 139, 236, 306
Fungi   138, 151

g

Gage, Simon   **271**
Gain-of-function mutations   118
Gallbladder   13, 19, 20
Gallstones   *13*, 20
Gangliosides   99, 183
Gargnano–Riva road   164
Garlic   239
GDP, *see* Gross domestic product
Gel filtration   101, 182
Gemfibrozil   242, 308
Gender   35, 37, 74, 88, 101, 160, 161, 168
Gene cluster   300
Gene therapy   239
Gene transcription   99
Genetic deficiency   21
Genetics   36, 81, 87, 115, 128, 130, 134, 162, 165,
   166, 252, 262, 264, 300
Genetic variant(s)   81, 118, 162–164, 210, 222,
   223, 225, 252, 264, 297, 313
Genotype   122
Geography of milestones in the field of
   lipoprotein research   xi
Germany   **28**, 43, 123, 207
Giblet   5
Giovanelli, Rosa   164
Gjone, Egil   201
Gland   4, 10, 19
Globulin(s)   61, 62, 76
Glomset, John   28, *201*, 202, 206, 314
Glucose   9, 24, 40, 133, 142, 158, 219, 220, 227,
   229, 238, 243, 250, 267, 284, 289
Glutathione peroxidases   179
Glutathione selenoperoxidase   178
Glycation   24, 229, 230
Glycemic index   238
Glycerol   8, 15, 20, 288, 289
Glycogen   9
Glycolipid(s)   *10*, 11, 100, 183
Glycolysis   9